Richard Hutten

WORKS IN USE

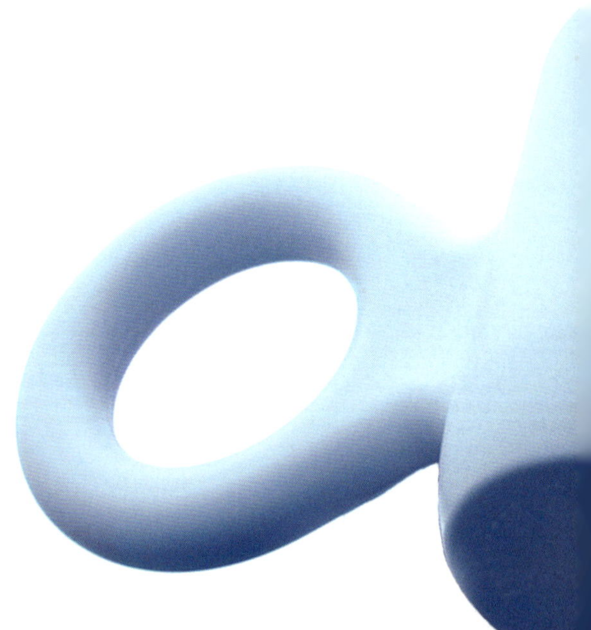

Richard Hutten

FOR YOU

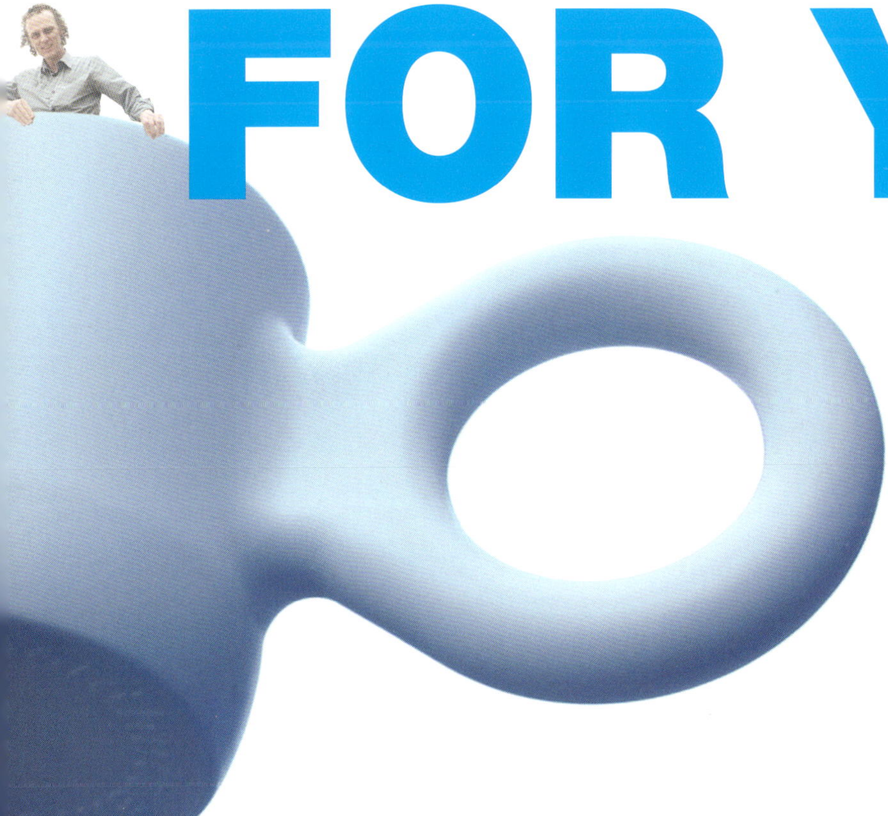

Texts by
Brigitte Fitoussi
Aaron Betsky

stichting kunstboek

Book concept
Richard Hutten advised by Roelof Mulder

Box design
Richard Hutten

One of the interns called Richard Hutten Studio a Rock&Roll band. I'm the singer songwriter but I need the rest of the band to make music, so thank you: Jeroen, Mischa, Brenno, Claudio, Raymond, Charlotte, Paultje, Rodney, Jos, Catheleine, Ilse, Saskia, Silvia, Olivier, Marije, Kim, Gerard, Breg, Huib, Noriko, Masano, Kotoho, Thijs, Gijs, Vincent, Heiko, Billy, Joris, Mijndert, Joch, Sandra, Catherien, Ivar, Jeroen, Stephan, Thomas, Martijn, Dominique, Joost, Bouke, Femke, Nadine, Adrian, Michiel, Sjoerd, Jurre, Christian, Anke, Pieter, Esmee, Laurentiu, David...

Texts
Brigitte Fitoussi
Aaron Betsky

Photography
Rene Koster, Hans van der Mars, Boris Braakhuis, Richard Hutten, Yasumasa Yonehara, Willem de Roon, Humberto Campana, Michael Young, Christophe Marchand, Ed van Hinte, Jacob van Rijs, Pierre Fantys and the users.

Final editing
Mieke Dumon
Karel Puype

English translation
Paul Belle (text B. Fitoussi)

French translation
Dominique Mailleux (text A. Betsky)

Layout
Graphic Group Van Damme, Oostkamp (B)

Printed by
Graphic Group Van Damme, Oostkamp (B)

Published by
Stichting Kunstboek
Legeweg 165 - B-8020 Oostkamp
T. **32 50 46 19 10
F. ** 32 50 46 19 18
www.stichtingkunstboek.com

This book was published with the support of
Grand-Hornu Images
www.grand-hornu.be

TPG Post, Nederland

D/2006/6407/19
NUR: 656
ISBN10: 90-5856-176-3
ISBN13: 978-90-5856-176-3

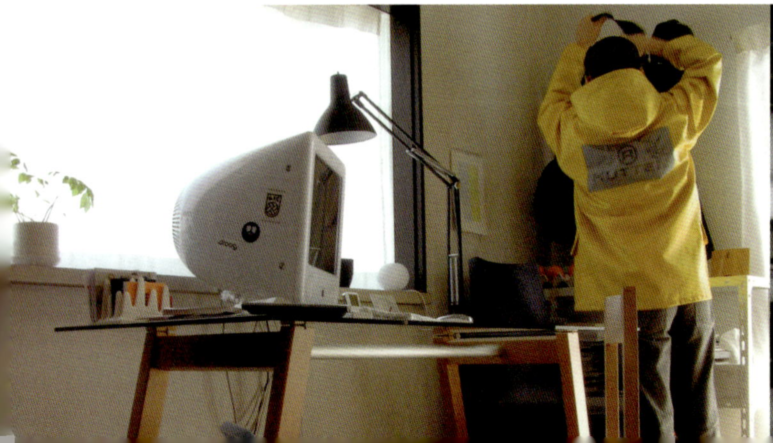

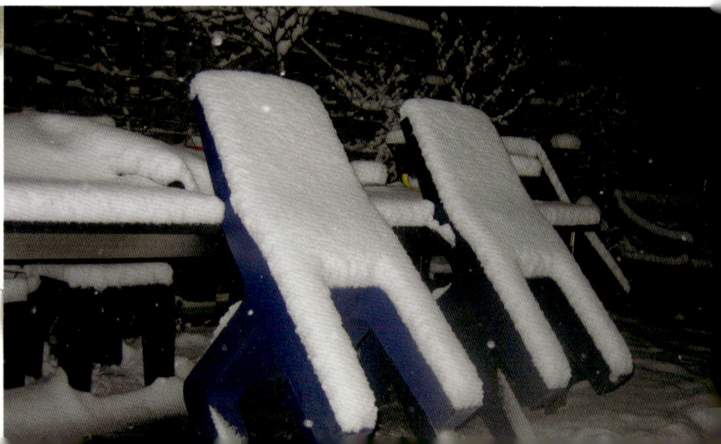

From Primitive Hut(ten) to Rotterdam Romance: Why I Love and Live with Richard Hutten ('s Furniture) **6**
Aaron Betsky

Works in use **8**

In use. A new look at users **94**
Brigitte Fitoussi

Users about Hutten **122**

Interview with Richard Hutten **146**
Brigitte Fitoussi

Catalogue **196**

Selection of books and catalogues **226**

Others **227**

Selection of exhibitions **228**

Selection of works in public collections **229**

In use. Nouveau regard sur l'usage **232**
Brigitte Fitoussi

Entretien avec Richard Hutten **234**
Brigitte Fitoussi

De Hutte(n) primitive à une romance de Rotterdam: pourquoi j'adore et vis avec (du mobilier) Richard Hutten **239**
Aaron Betsky

Special thanks **240**

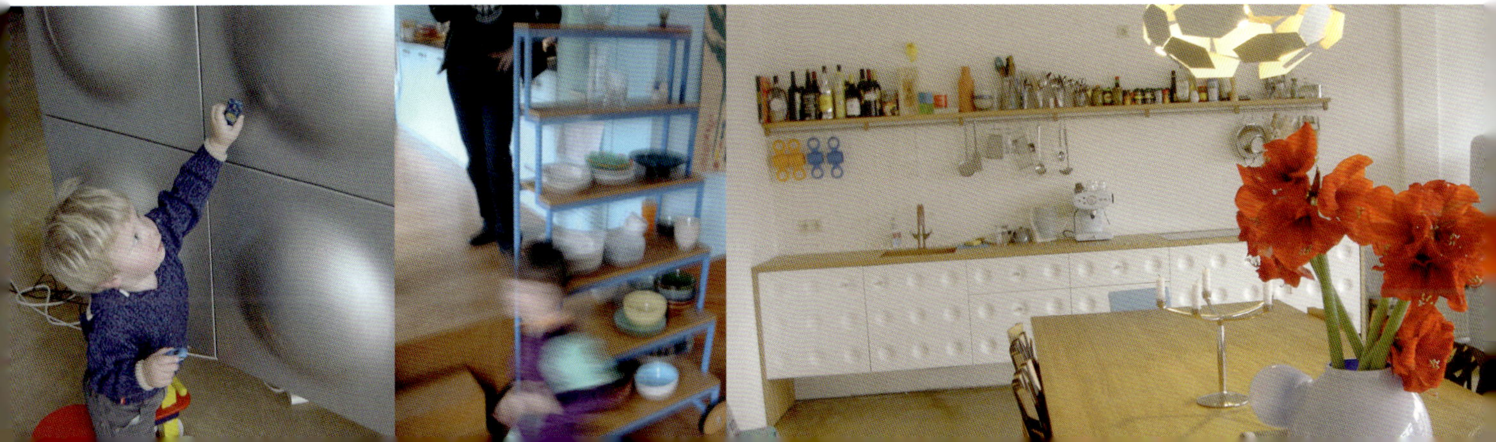

Aaron Betsky

From Primitive Hut(ten) to Rotterdam Romance:
Why I Love and Live with Richard Hutten ('s Furniture)

I live in Rotterdam, along with Richard Hutten and with a fair amount of Richard Hutten-designed furniture and objects. That seems only fitting to me. Rotterdam thinks of itself as 'Manhattan on the Maas', but to my mind its only true skyscraper is *Skyline Rotterdam*, Richard Hutten's 1994 hybrid armoire. Tall, modern, gridded and yet falling apart into the complex constellations of functional spaces that serve every-day life, it embodies the very excitement born out of rationality for which New York has come a symbol. It is a very Rotterdam version of that urbanity, however, in which a pared-down simplicity and anonymous appearance gives an object devoted to storage (this is, after all, a town born out of transshipment) a sense of no-nonsense elegance.

It is also typical of Richard Hutten's work. For the last fifteen years, he has been reassembling the basic building blocks out of which furniture is made into forms that open up a new kind of space. Sometimes, those spaces can be as small as slight dip in the surface of table that allows it to a fruit bowl at the same time (*Table-Bowl*, 1995), and sometimes they imply a completely new and almost religious sense of gathering in the public arena, as in the *Cross Bench* of 1994. Always they have the purity of an orthogonal geometry and a complexity born from the designer's desire to alter simple social rituals through a rearrangement of the spaces in which they take place.

This seems to me to be the core of Richard Hutten's work. It reaches its most elemental perfection in the series of highchairs he designed around 1996 for his young children, in which

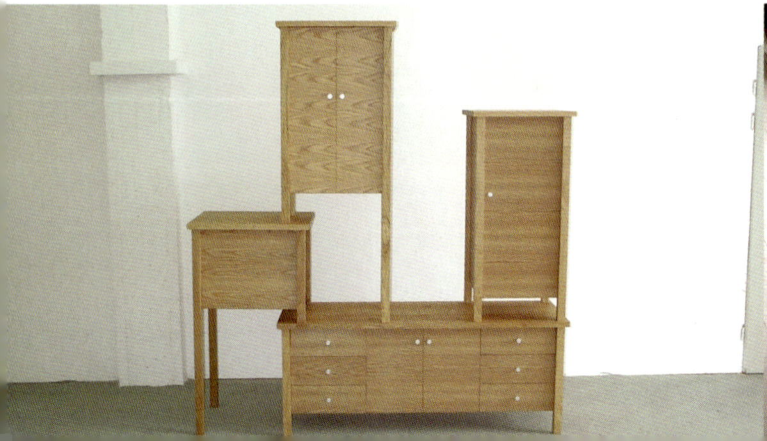

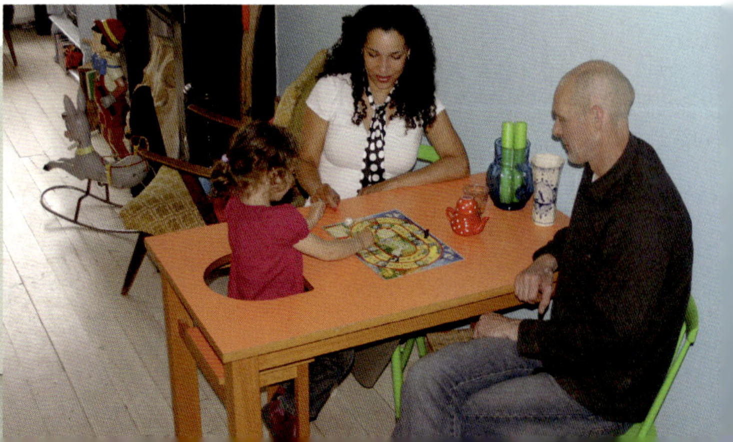

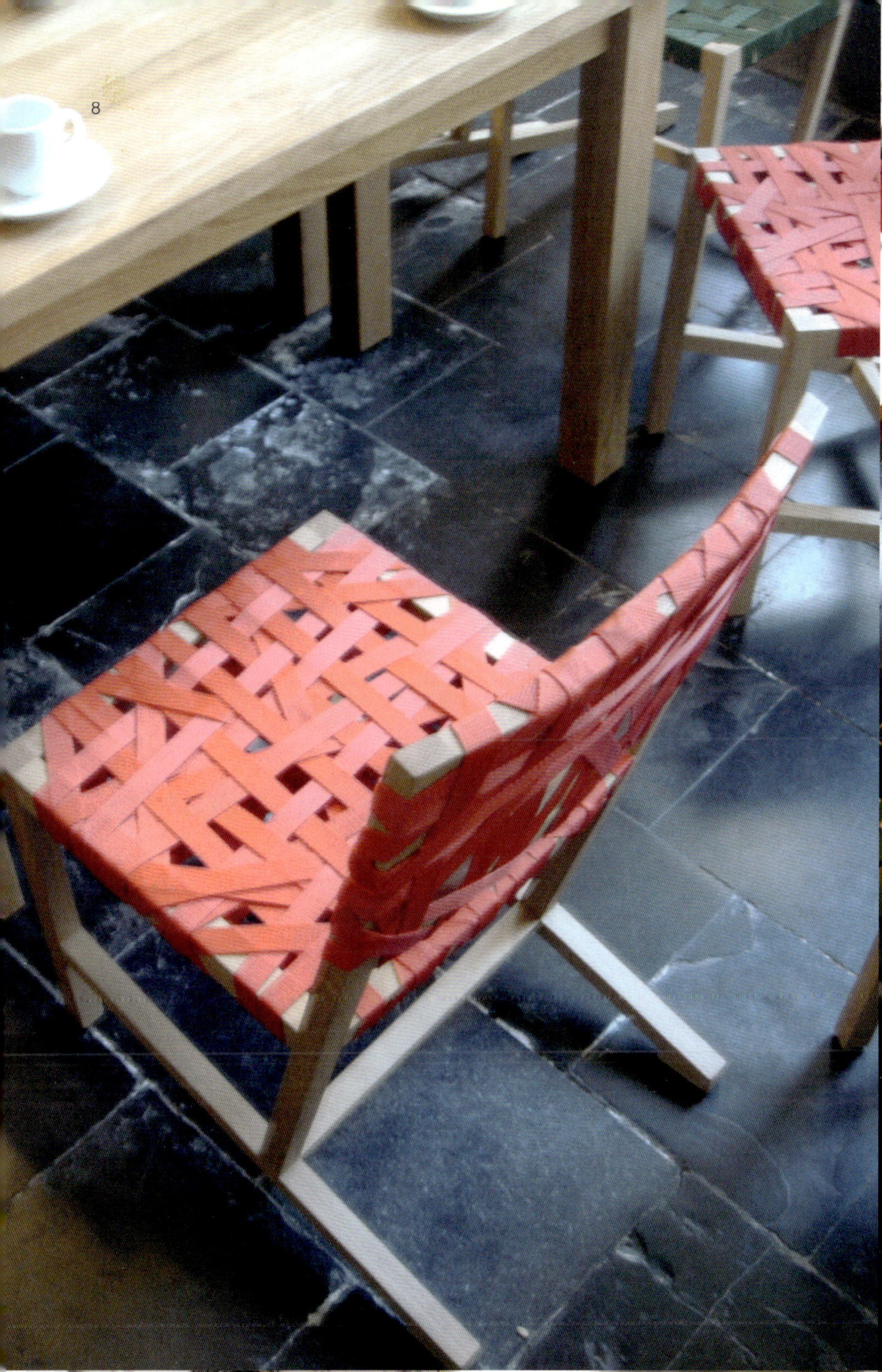

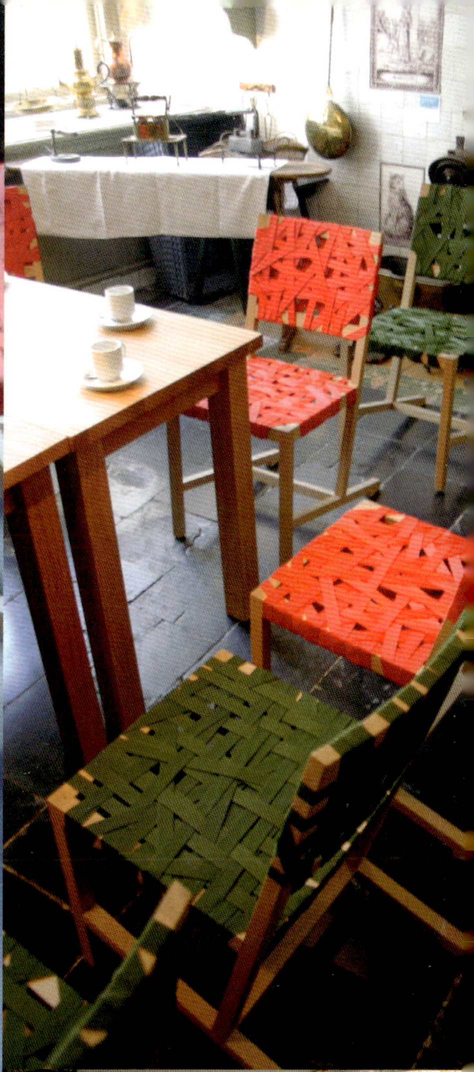

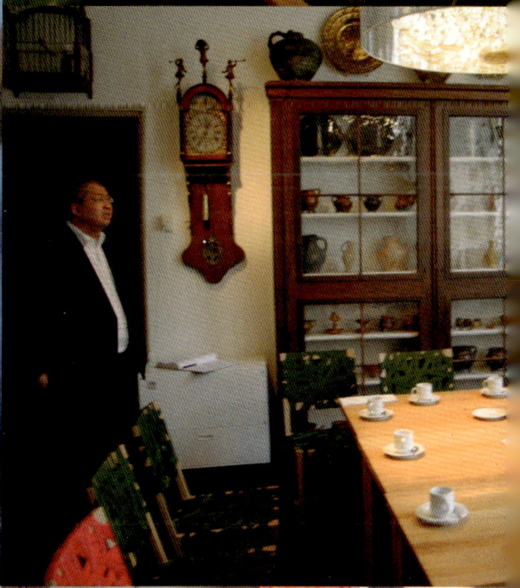

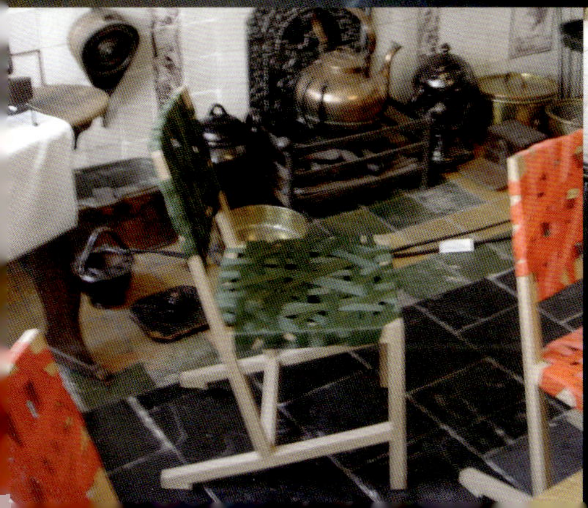

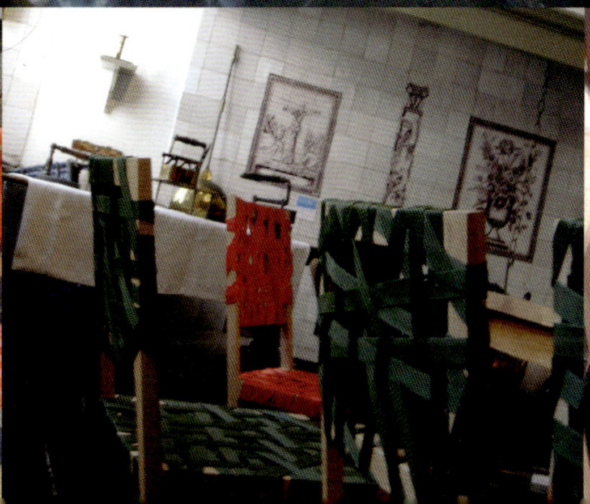

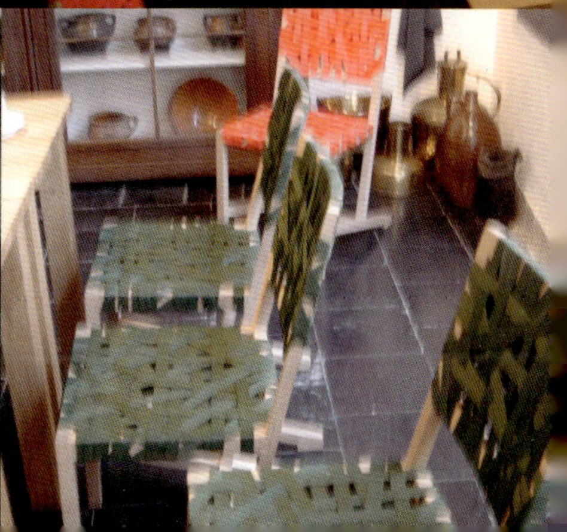

the recombination of familiar forms to create a dignified place in which a small child can partake in family meals has an almost lyrical innocence. Here furniture goes back to its roots, just as it serves the roots of at least one form of social relation at the core of our society.

It would be wrong, of course, to identify Hutten only with his earliest and most widely known work. There is another aspect to his designs (at least), one that concerns itself with the possibilities of new materials, particularly with plastics. Whether that leads to the undulating forms of his Bronto chairs of 1997 and onward, or the *Fruit Tray* (1995), you can still see the drive to a simplicity and generic quality, though now carried out in forms that have less of the romantic recall to older elements in our domestic environment that feed the lyricism of the wood furniture. The two come together in Hutten's reinvention of the 'Skippy Ball' as an outdoor seating system (*Zzzidt Ball*, 2000) – at least for those of us who grew up with those toys as part of our childhood.

In Hutten's more recent work, he seems to be moving away from a reliance on echoes of familiar forms and thus either social (historical) or personal memories. Yet there are always echoes of something, even in the abstracted chandelier of circular cut-outs orbiting around a central light that recall their crystal cousins; and even in the chair he has just produced for Lensvelt and which consciously reduces the experiments by Charles and Ray Eames and Arne Jacobsen to a foundation for the making of a pared-down and almost generic element of seating. Richard Hutten cannot escape from the past, whether his own or our society's, and doesn't want to. Instead, his particular achievement consists of continually querying those memories to get at their useful, playful, delightful, and above all else open essence, so that he can make things that are deeply rooted in what we know and yet open to new interpretations.

Aaron Betsky

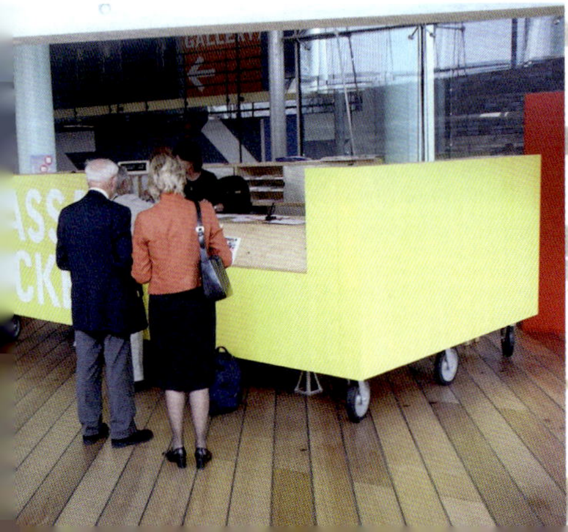

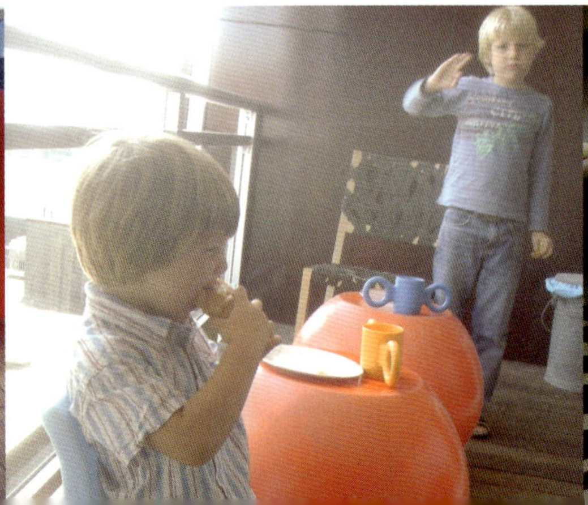

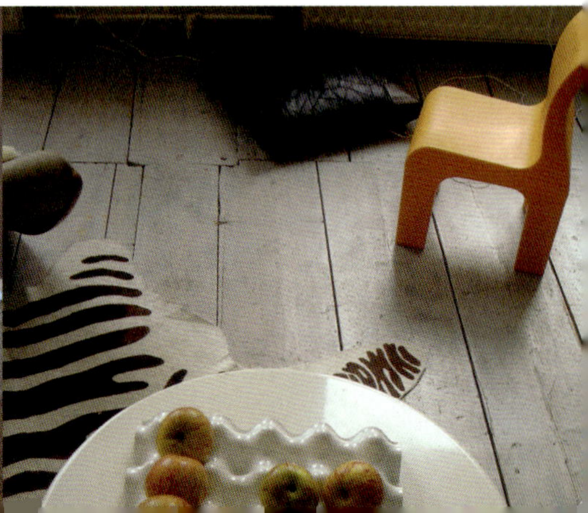

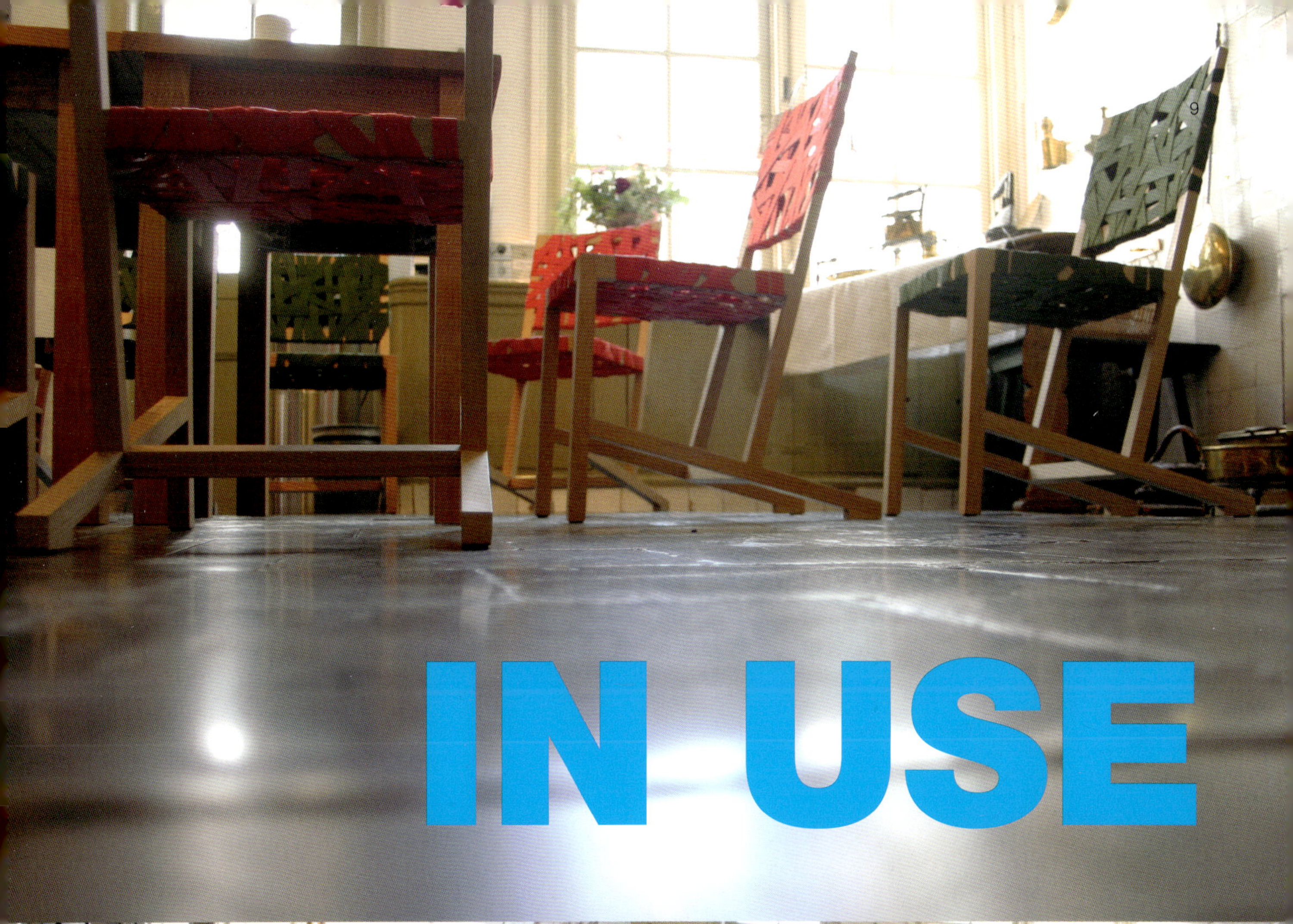

IN USE

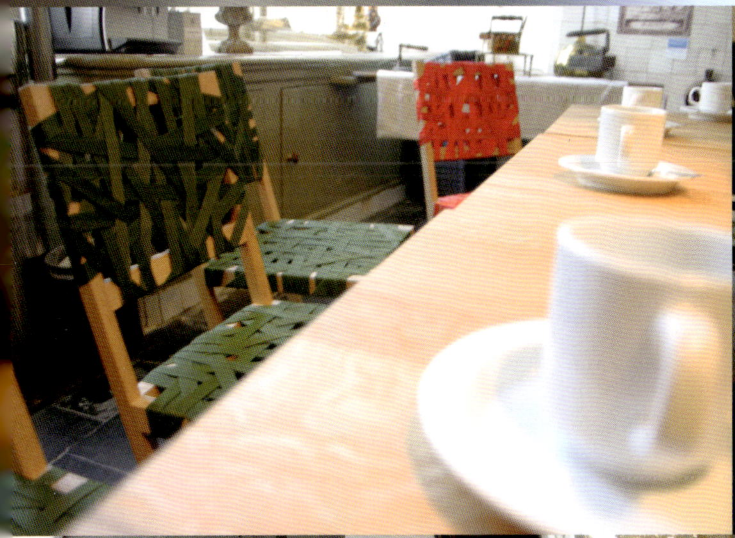

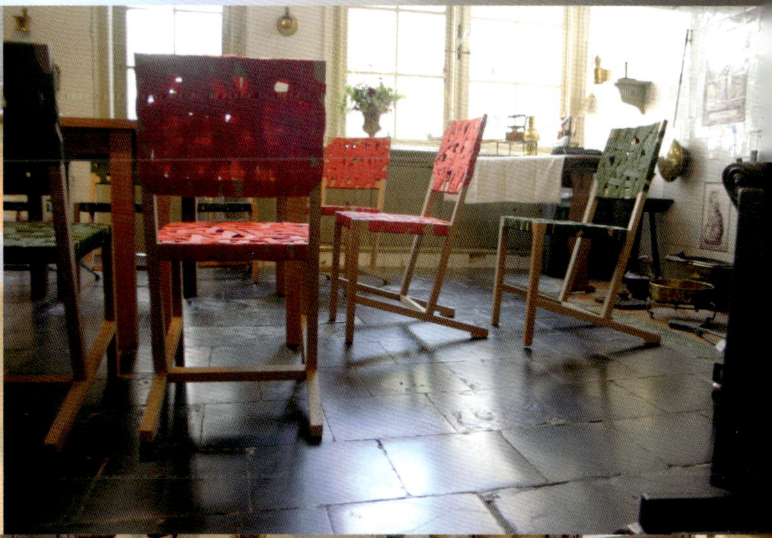

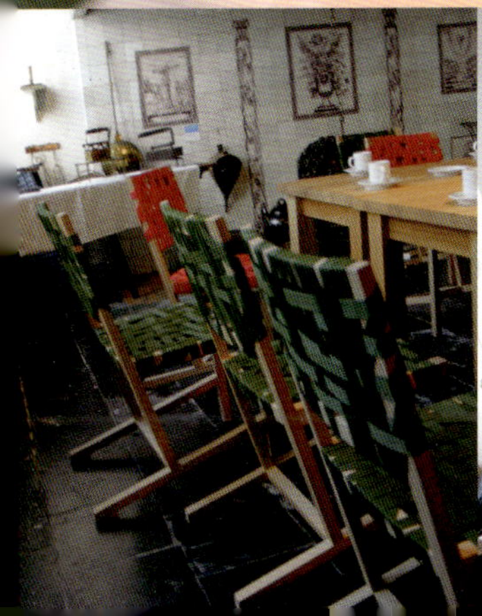

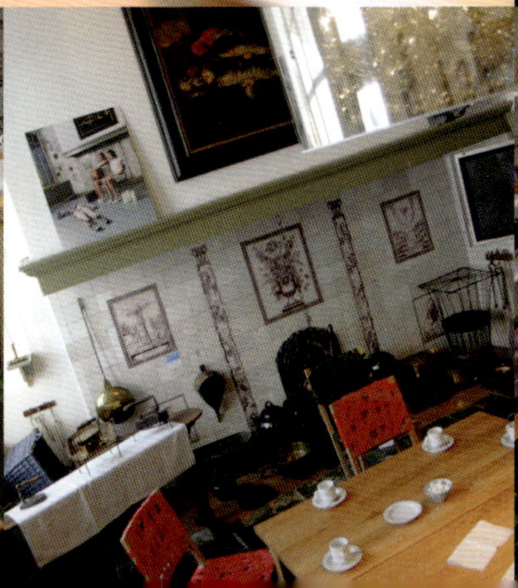

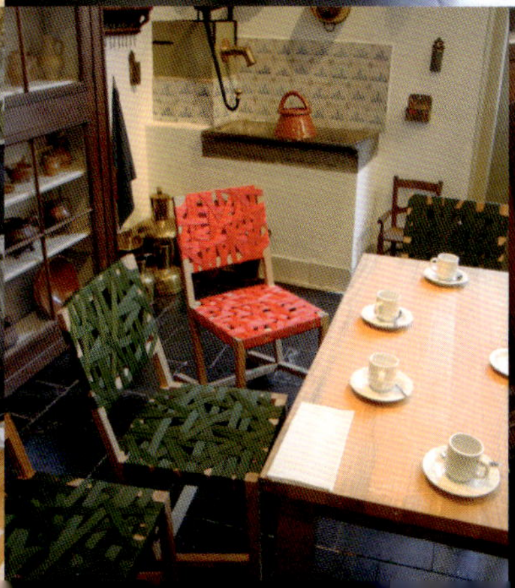

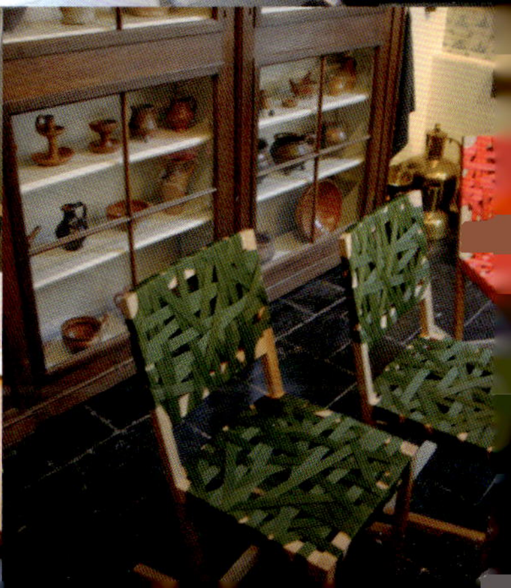

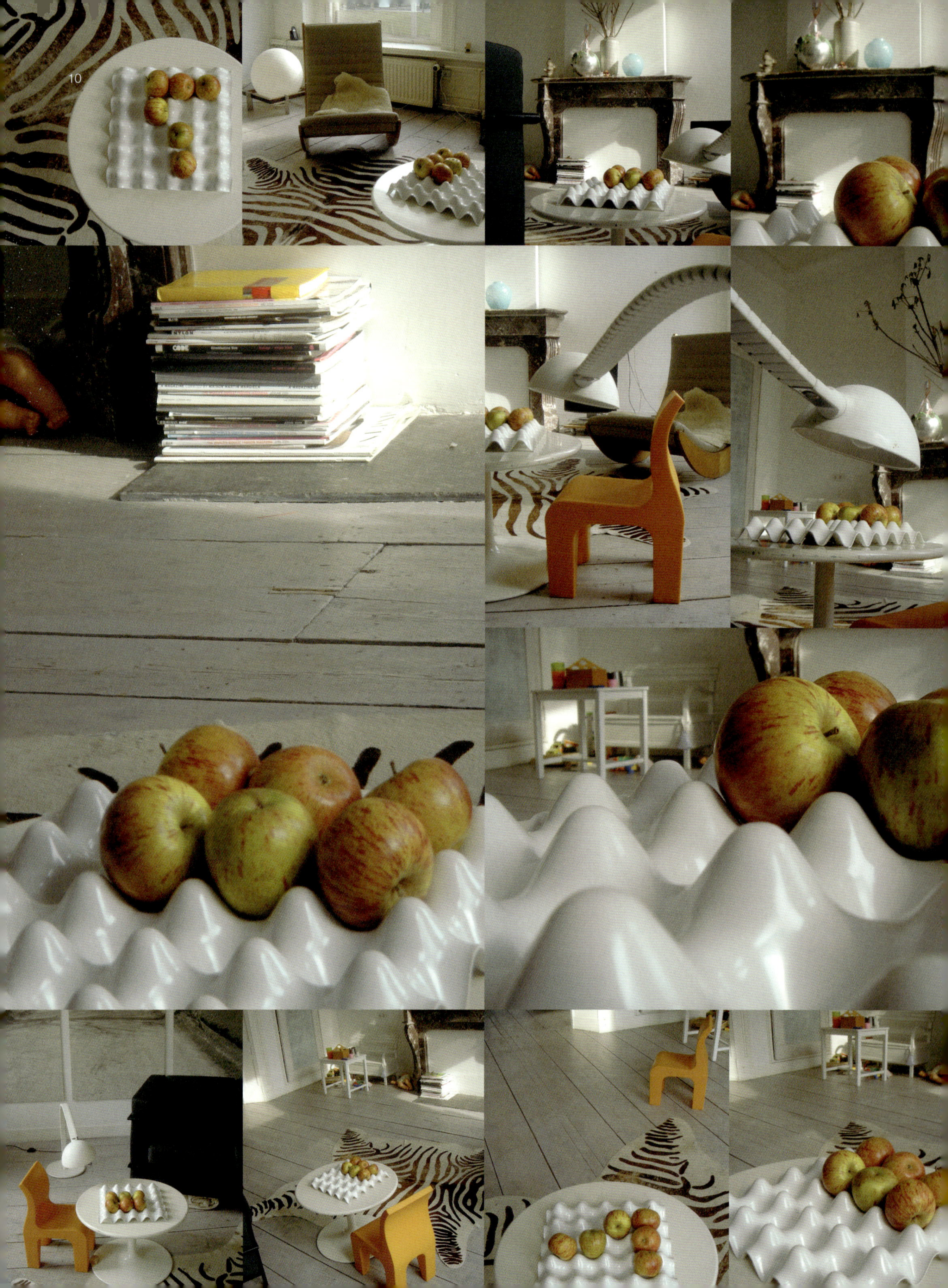

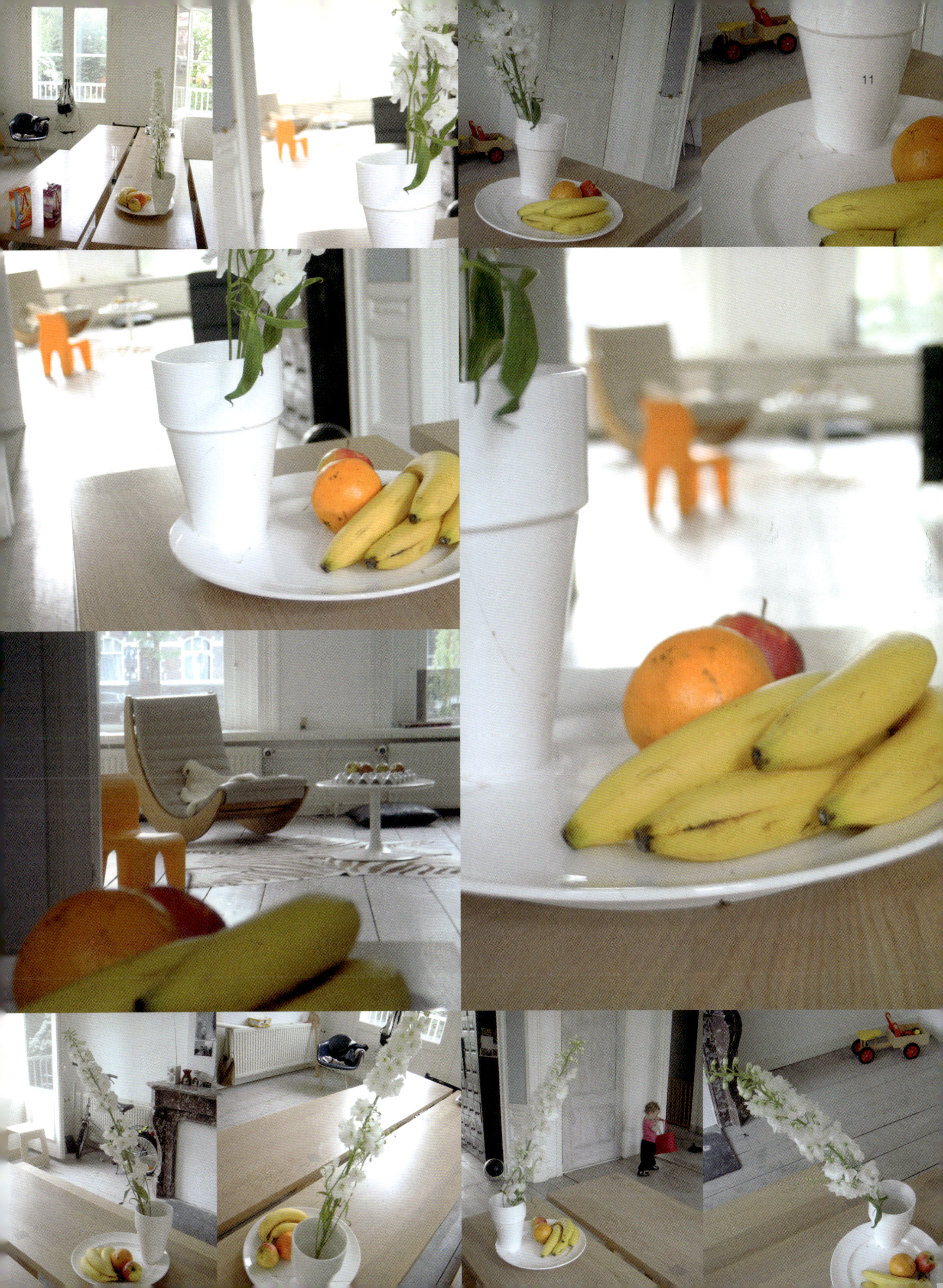

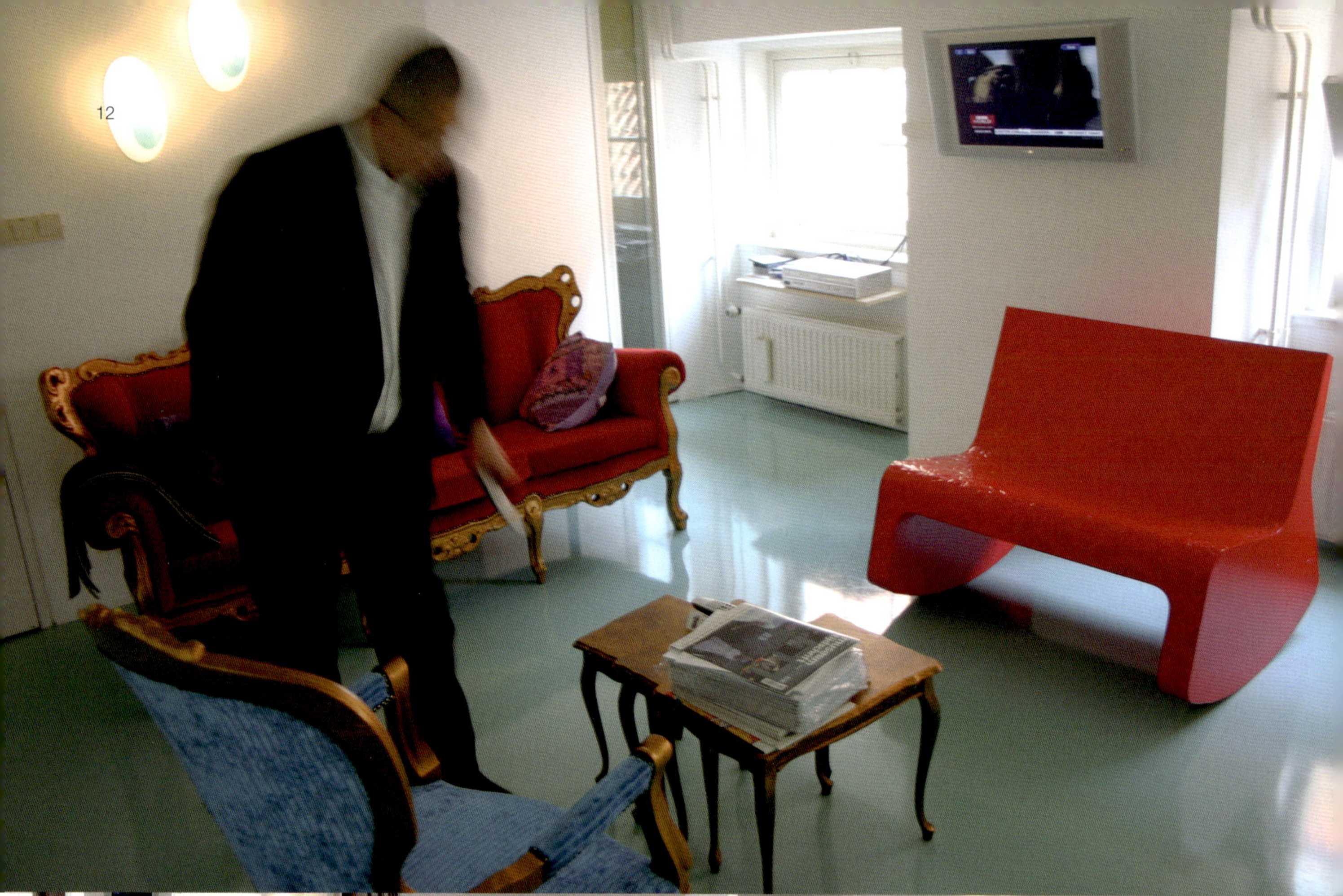

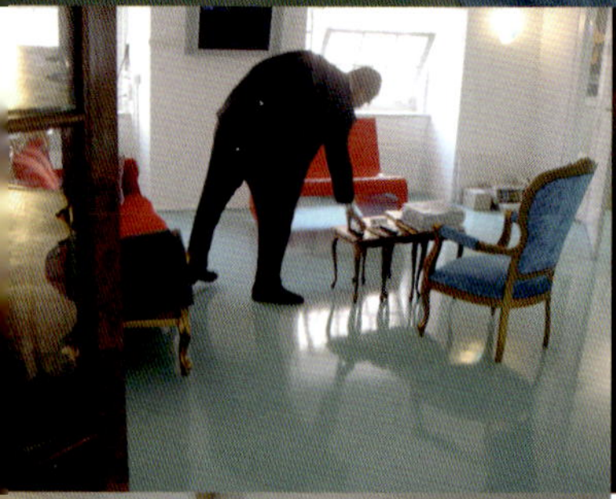

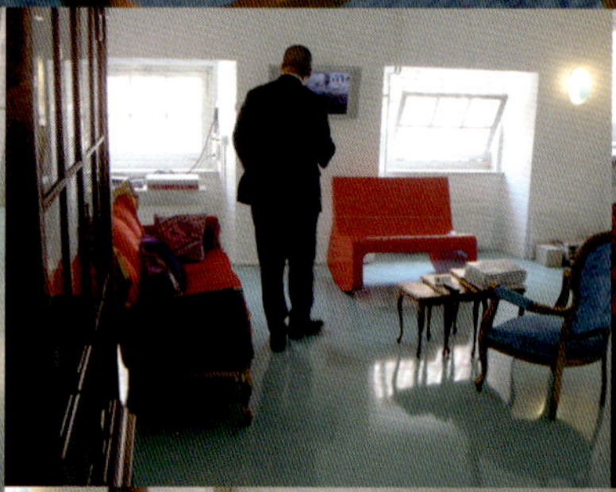

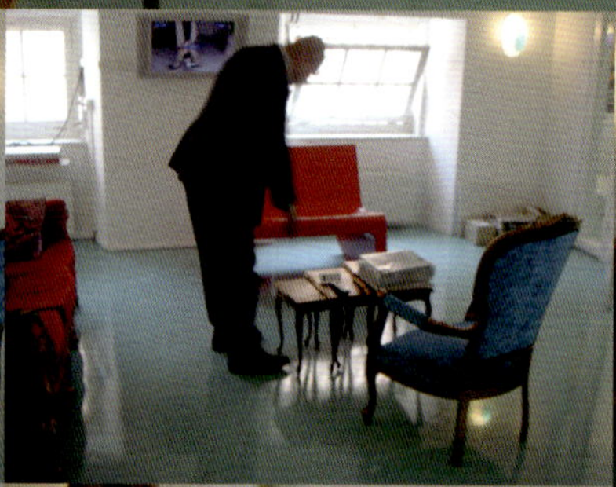

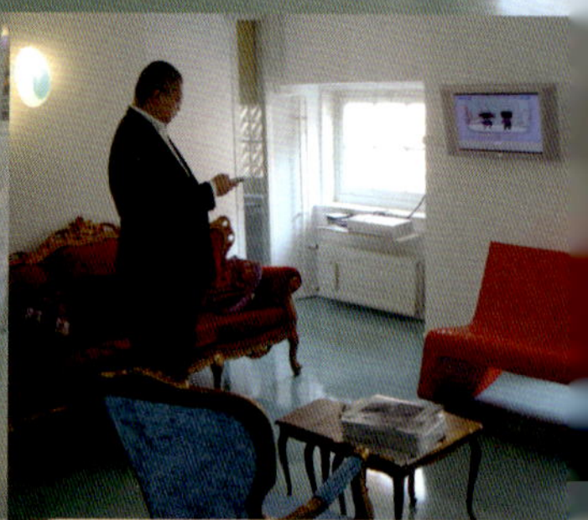

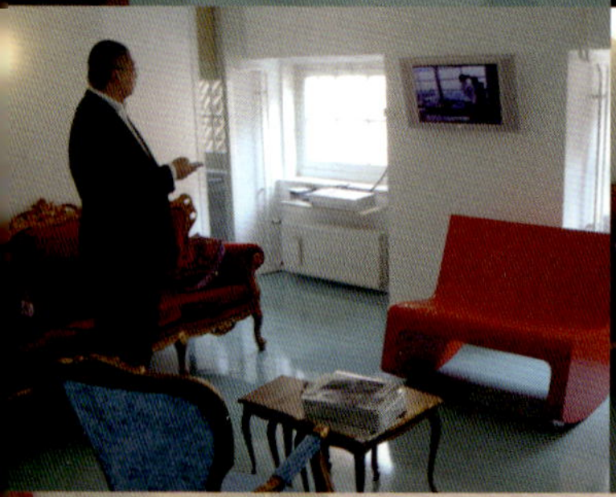

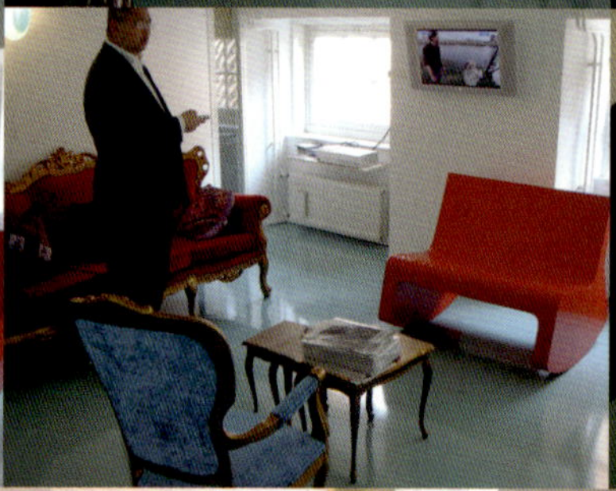

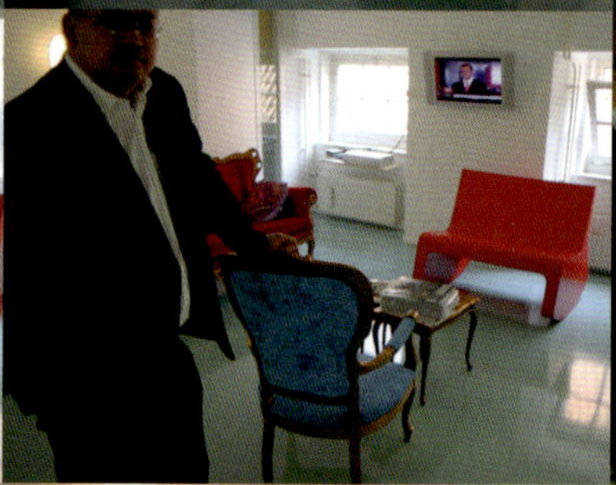

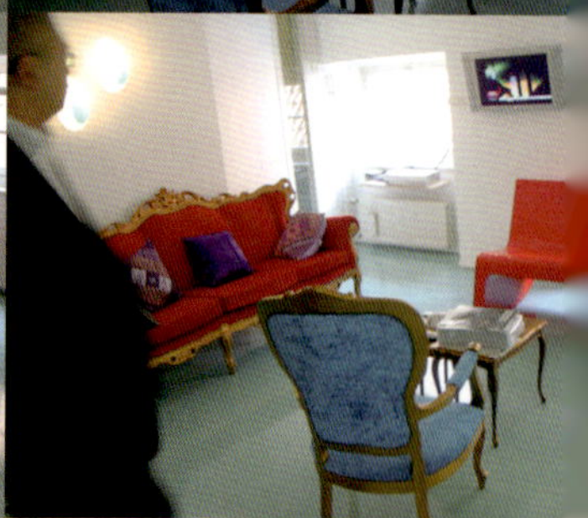

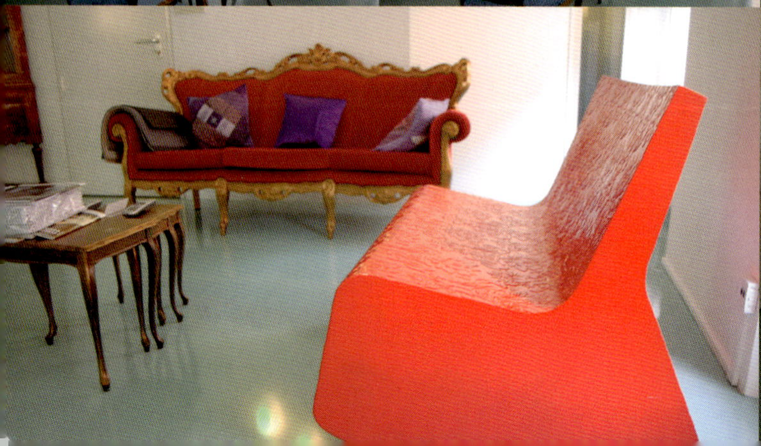

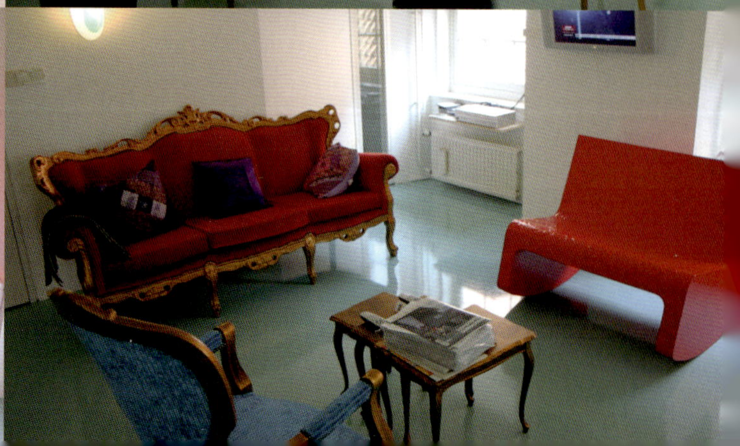

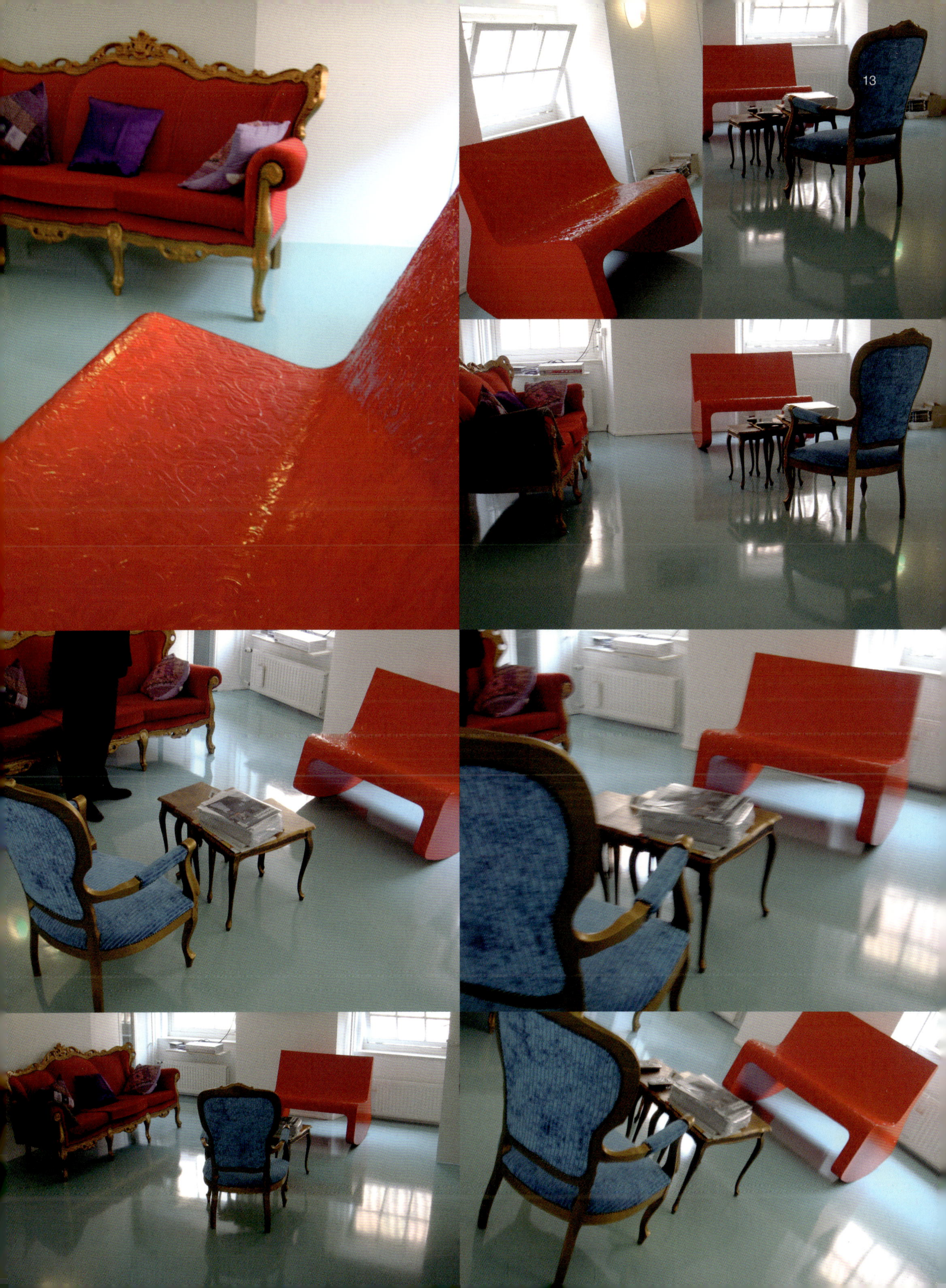

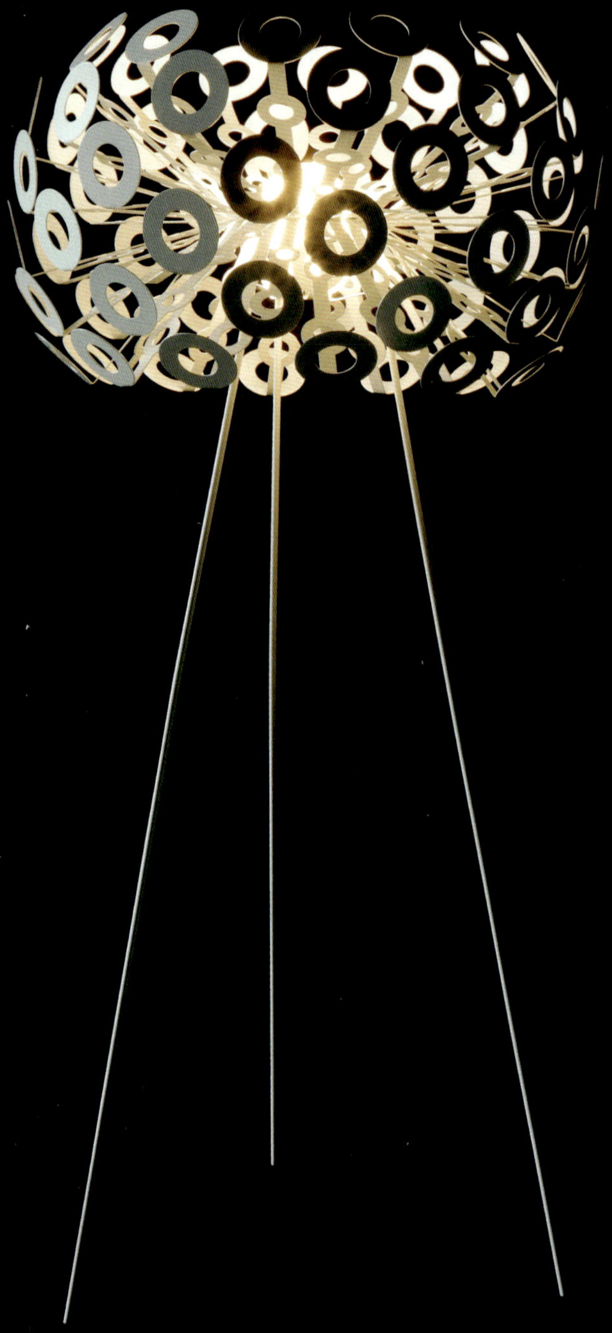

Dandelion — 2004

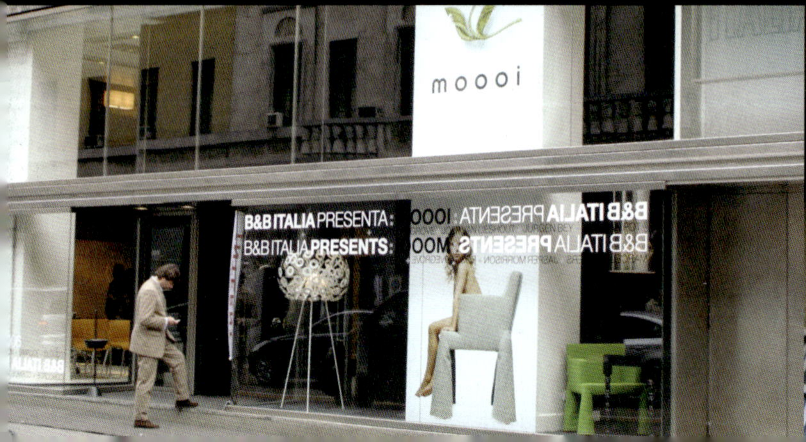

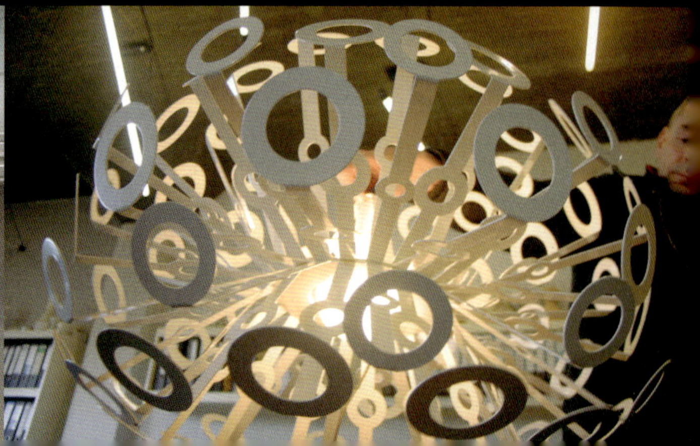

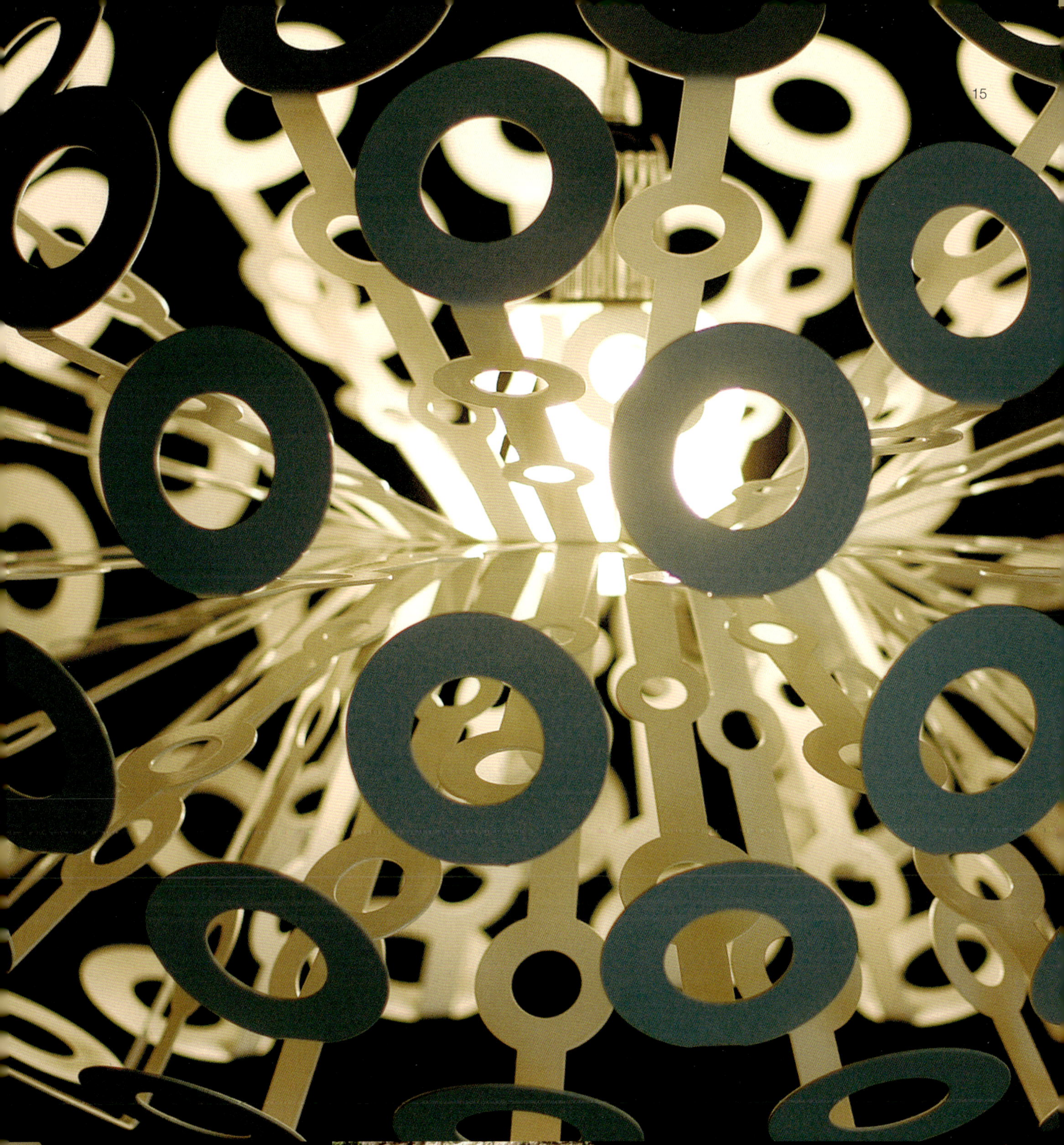

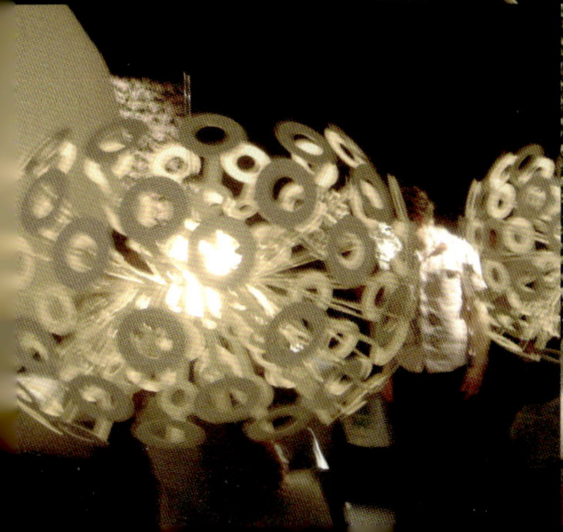

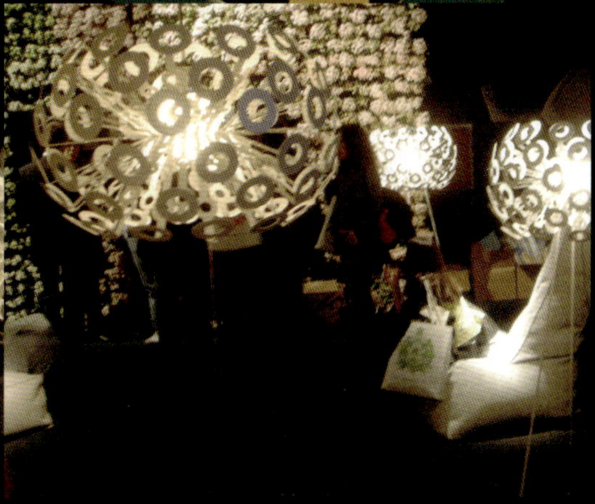

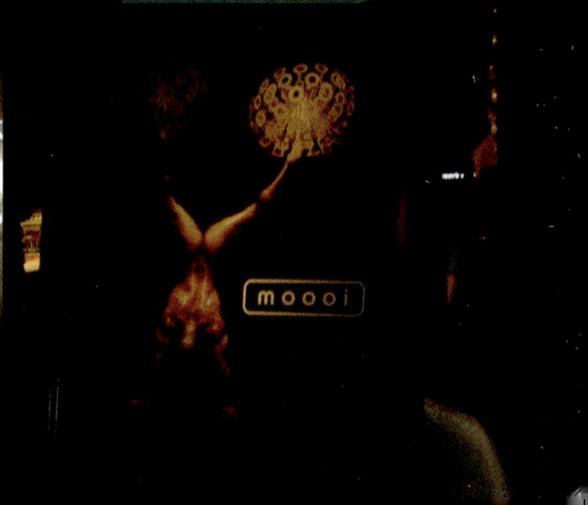
moooi

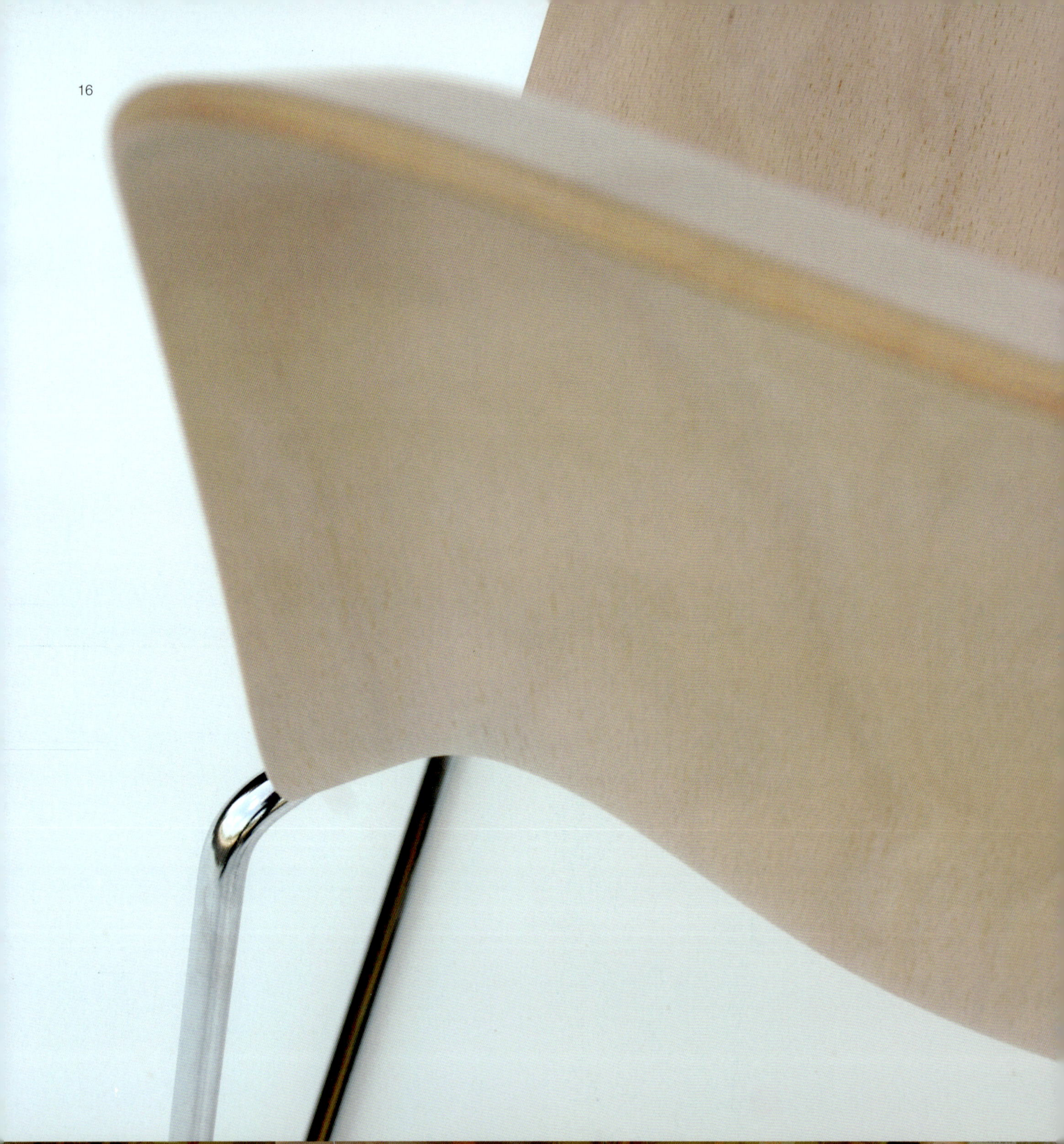

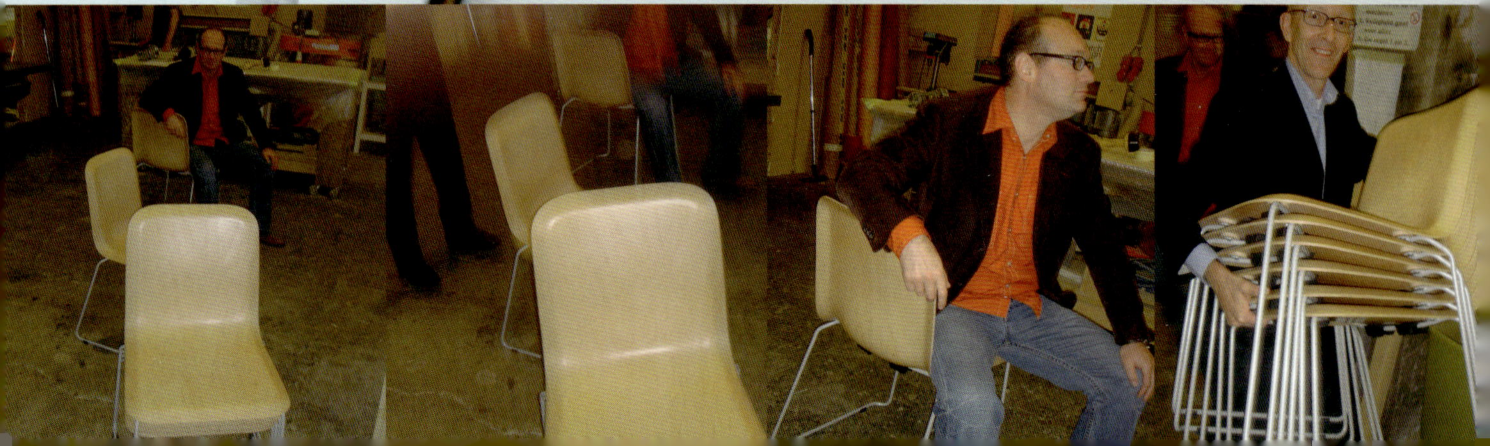

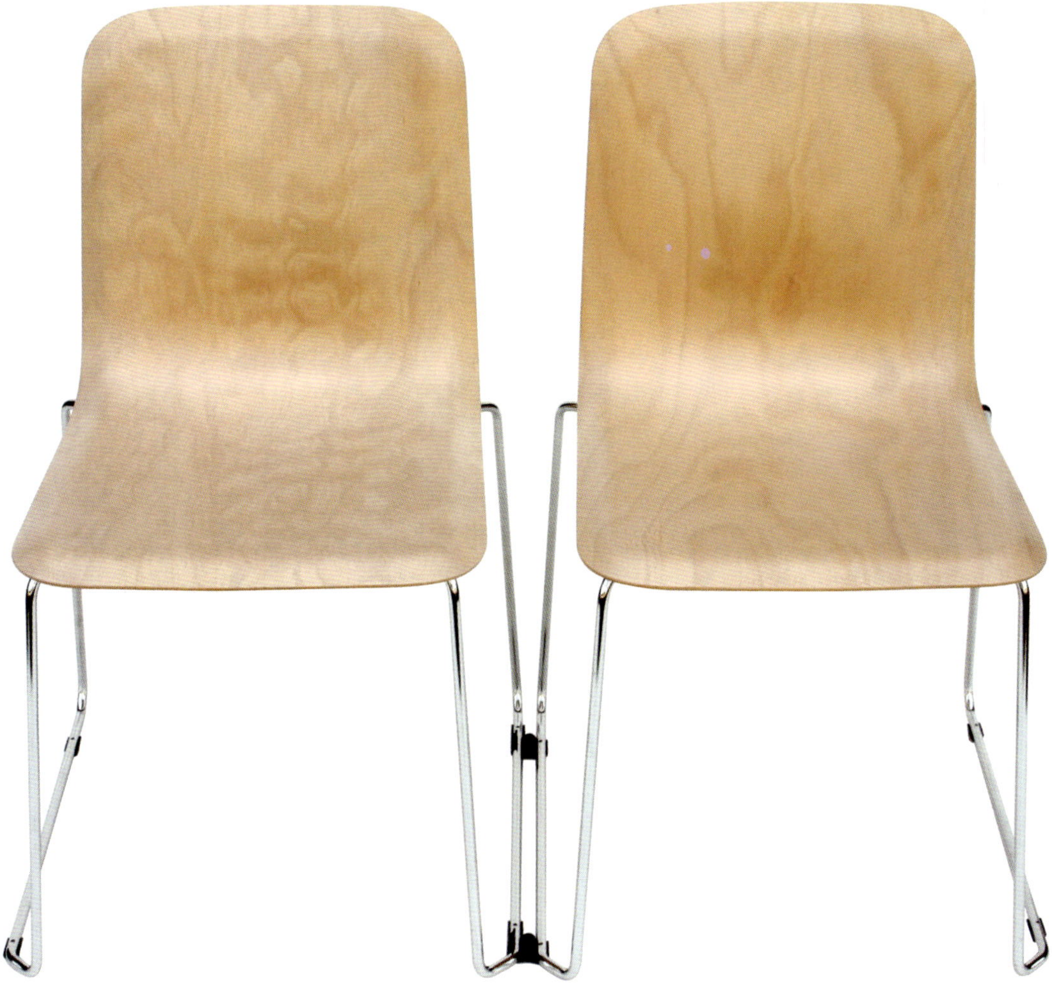

This chair — 2004

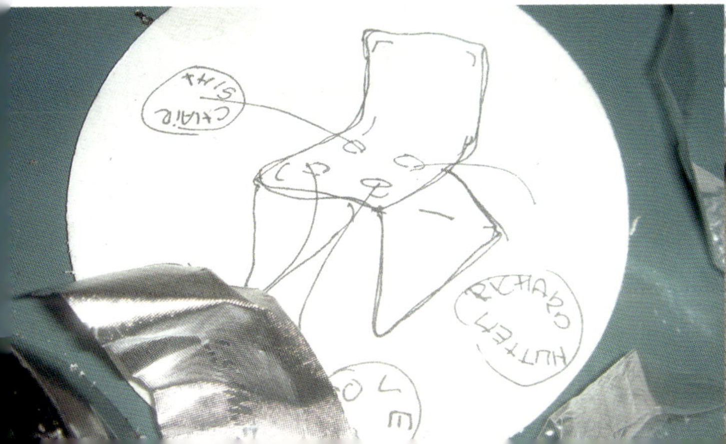

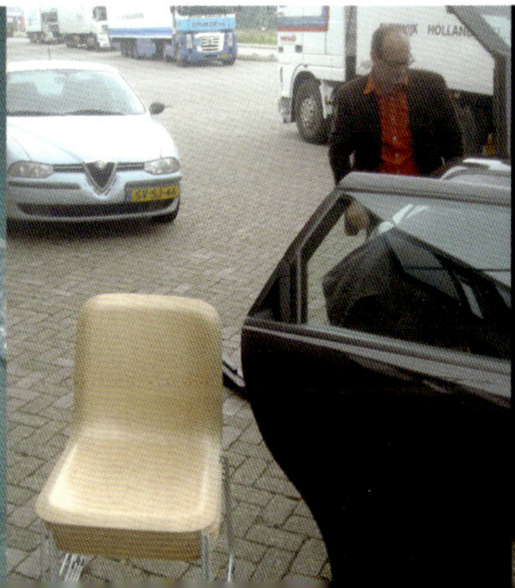

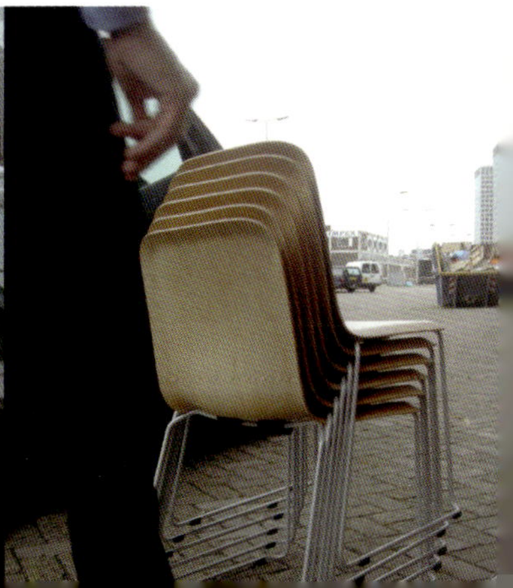

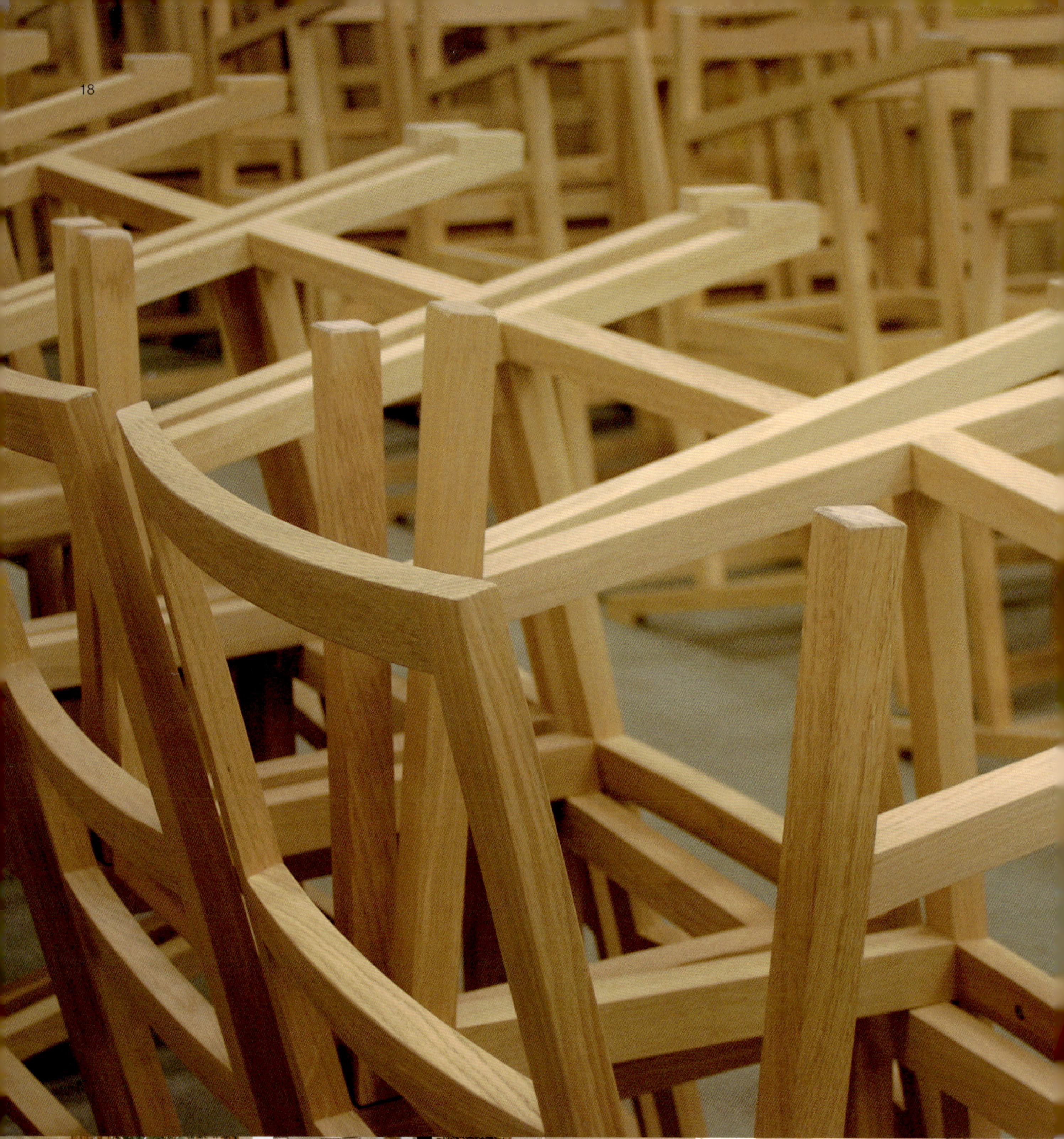

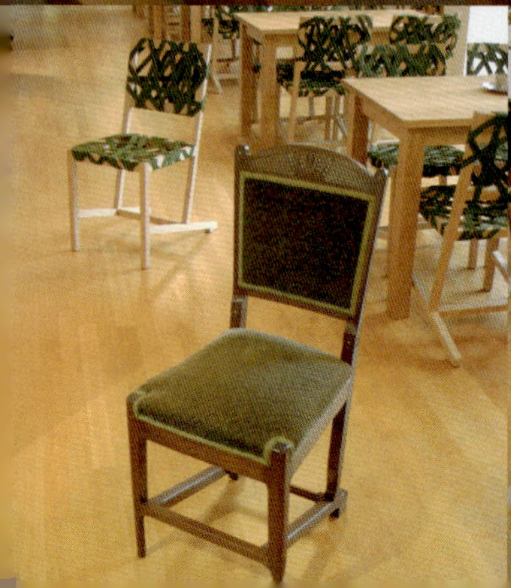

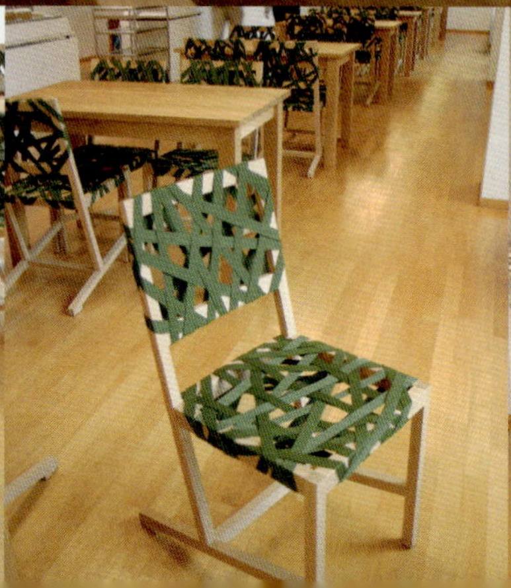

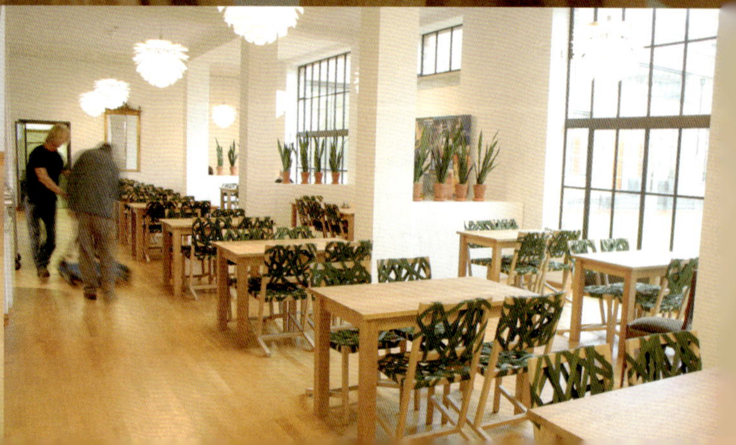

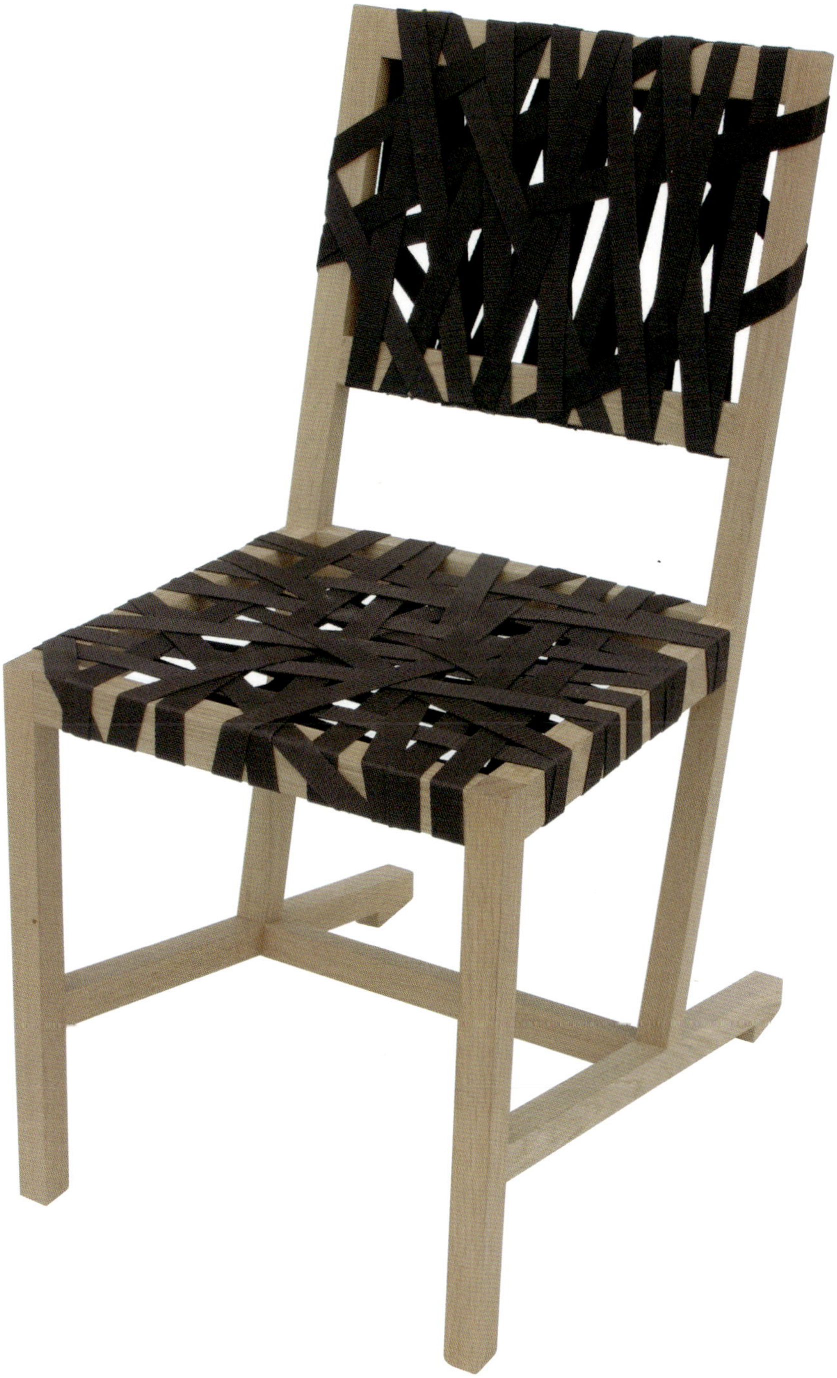

Berlage chair — 2004

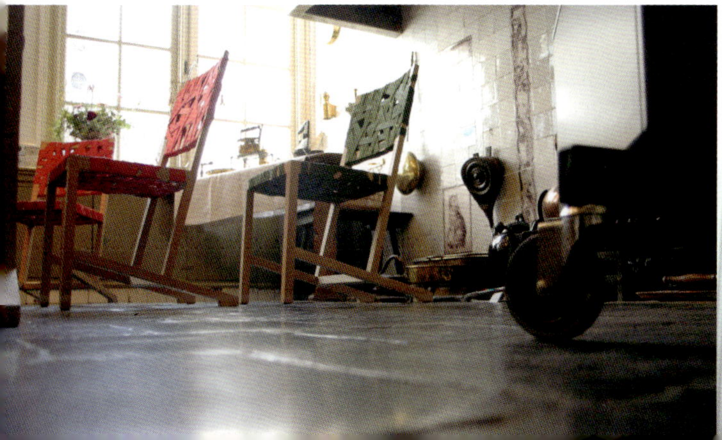

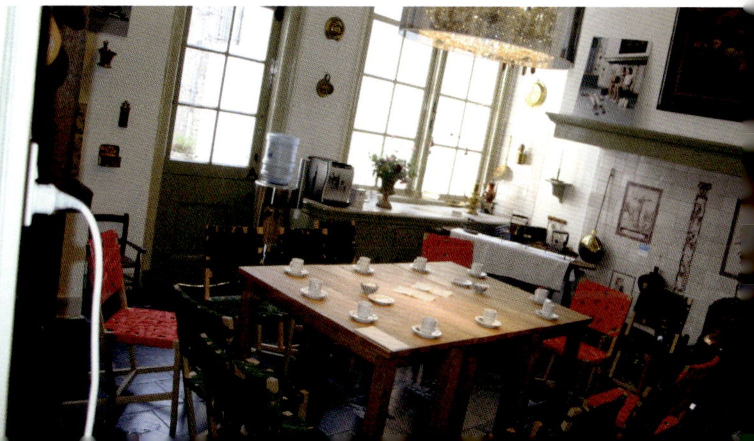

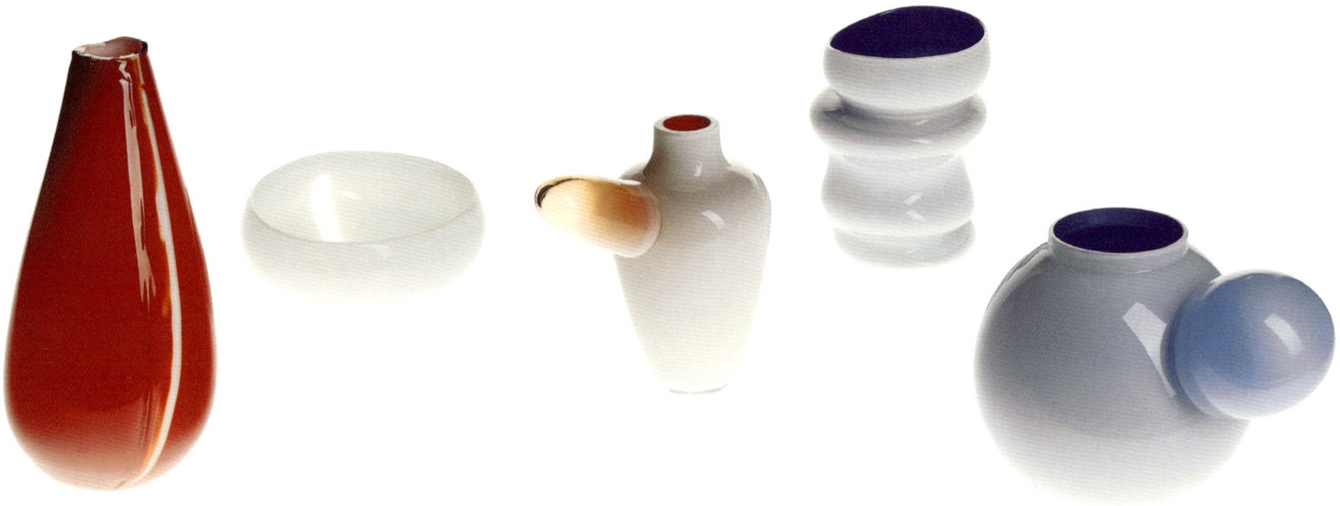

What happened? Series of vases and a bowl for Covo — 2001

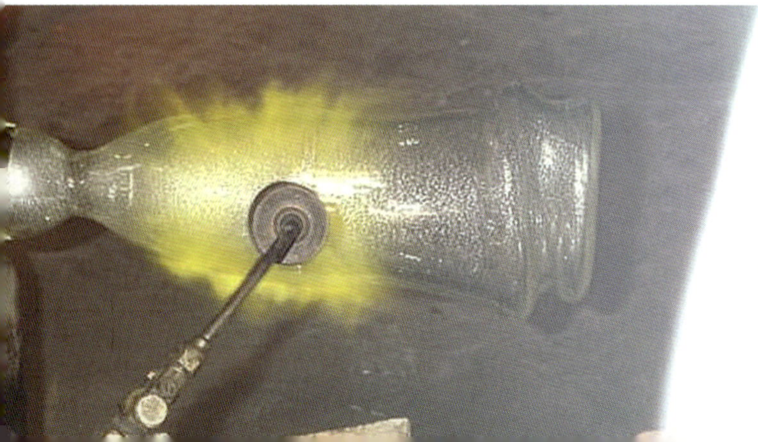

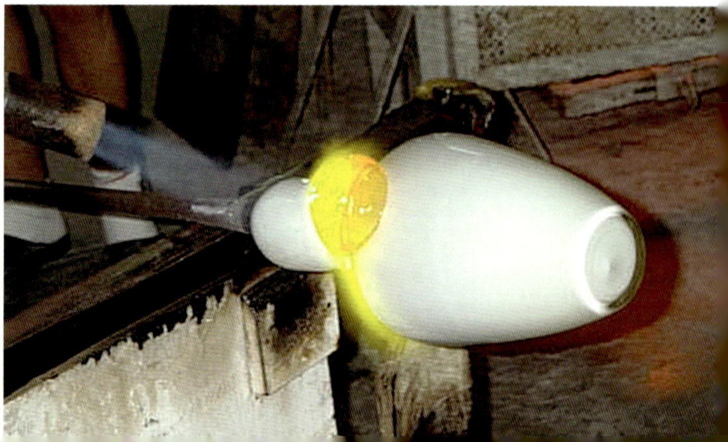

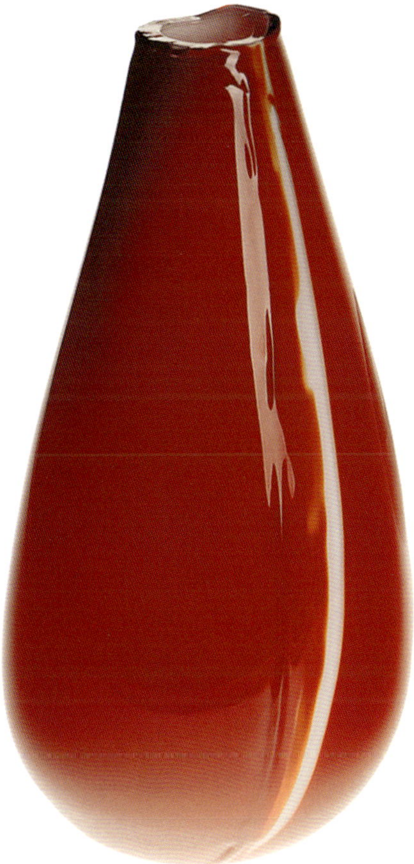

What happened? - Scarvase — 2001

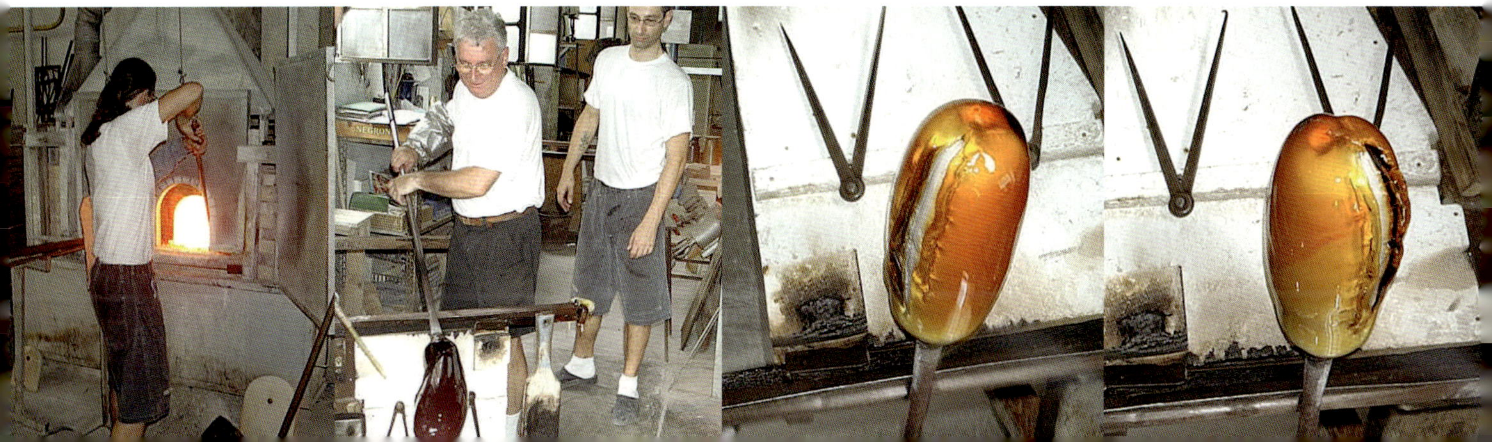

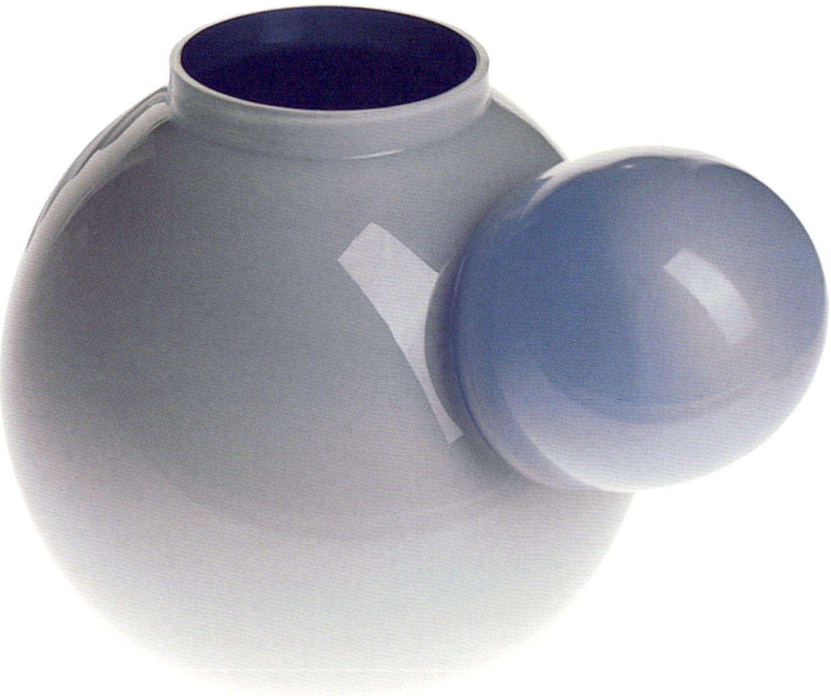

What happened? - Upwelling Round — 2001

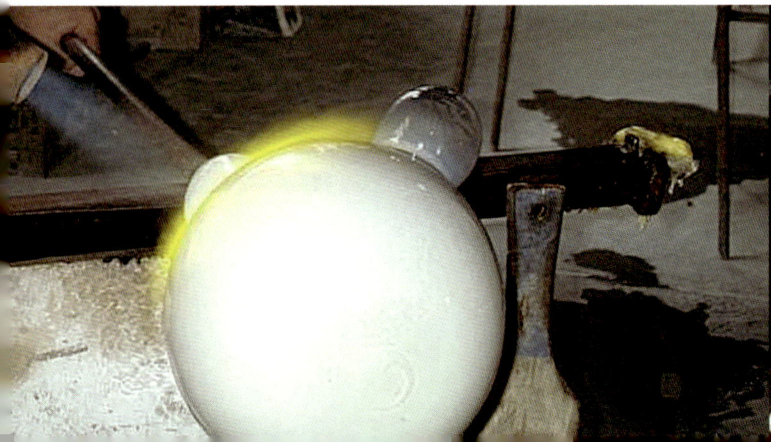

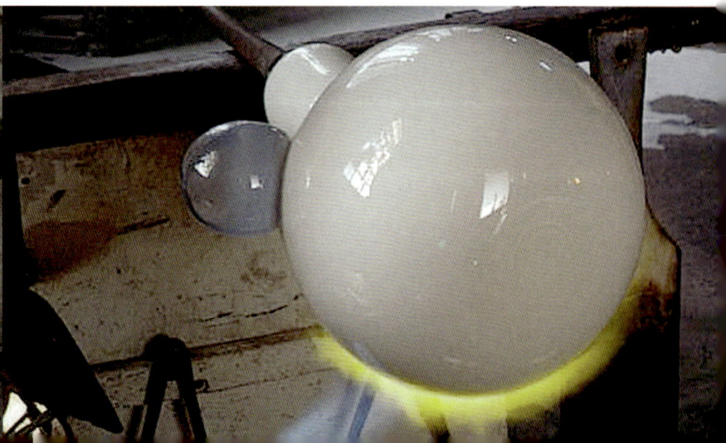

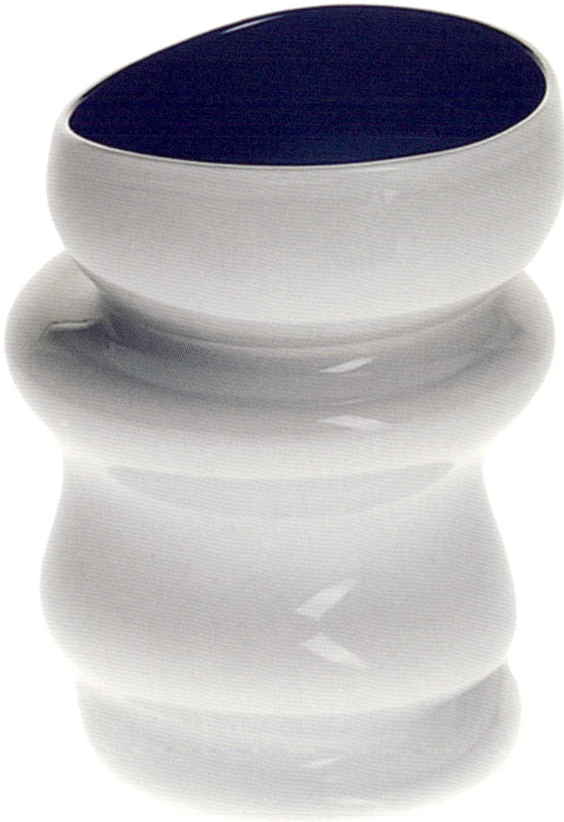

What happened? - Collapse vase — 2001

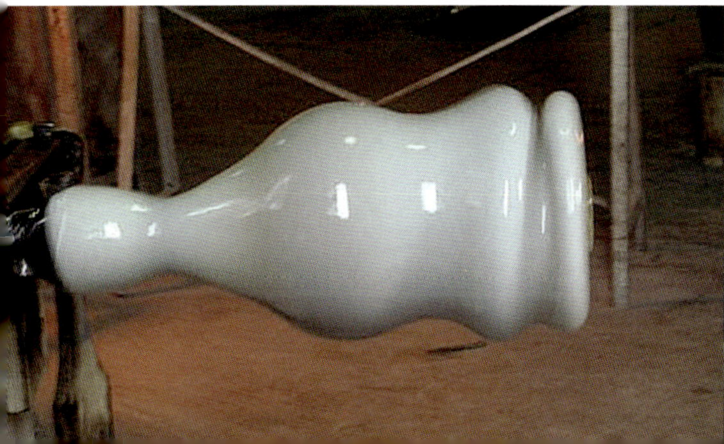

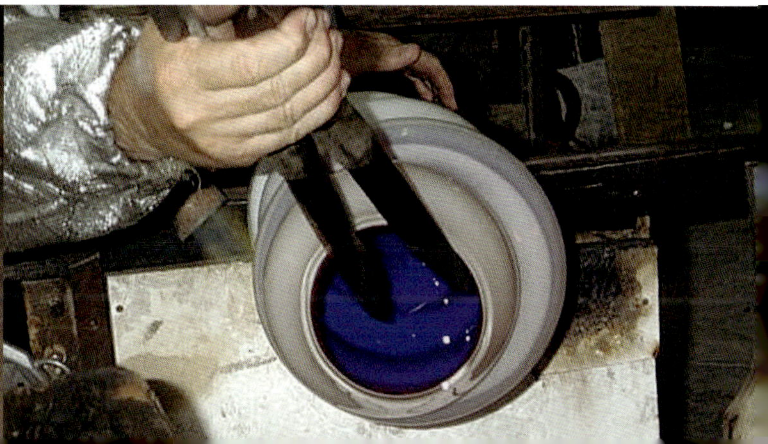

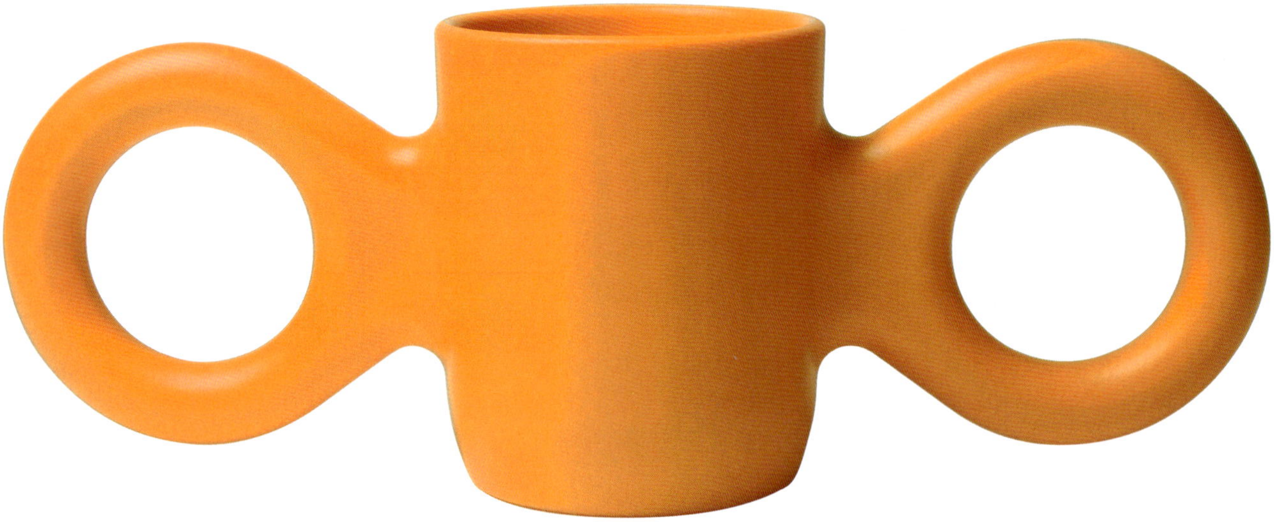

Domoor - Dombo — 2001

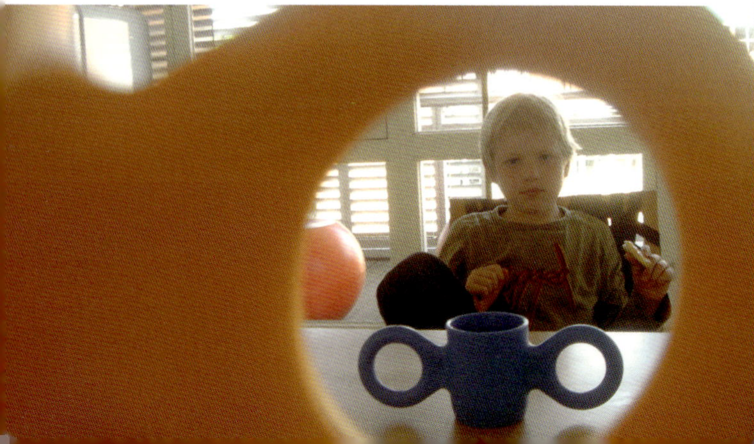

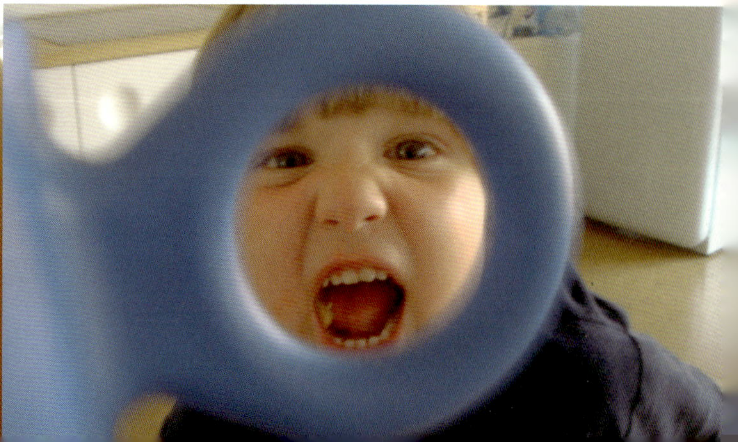

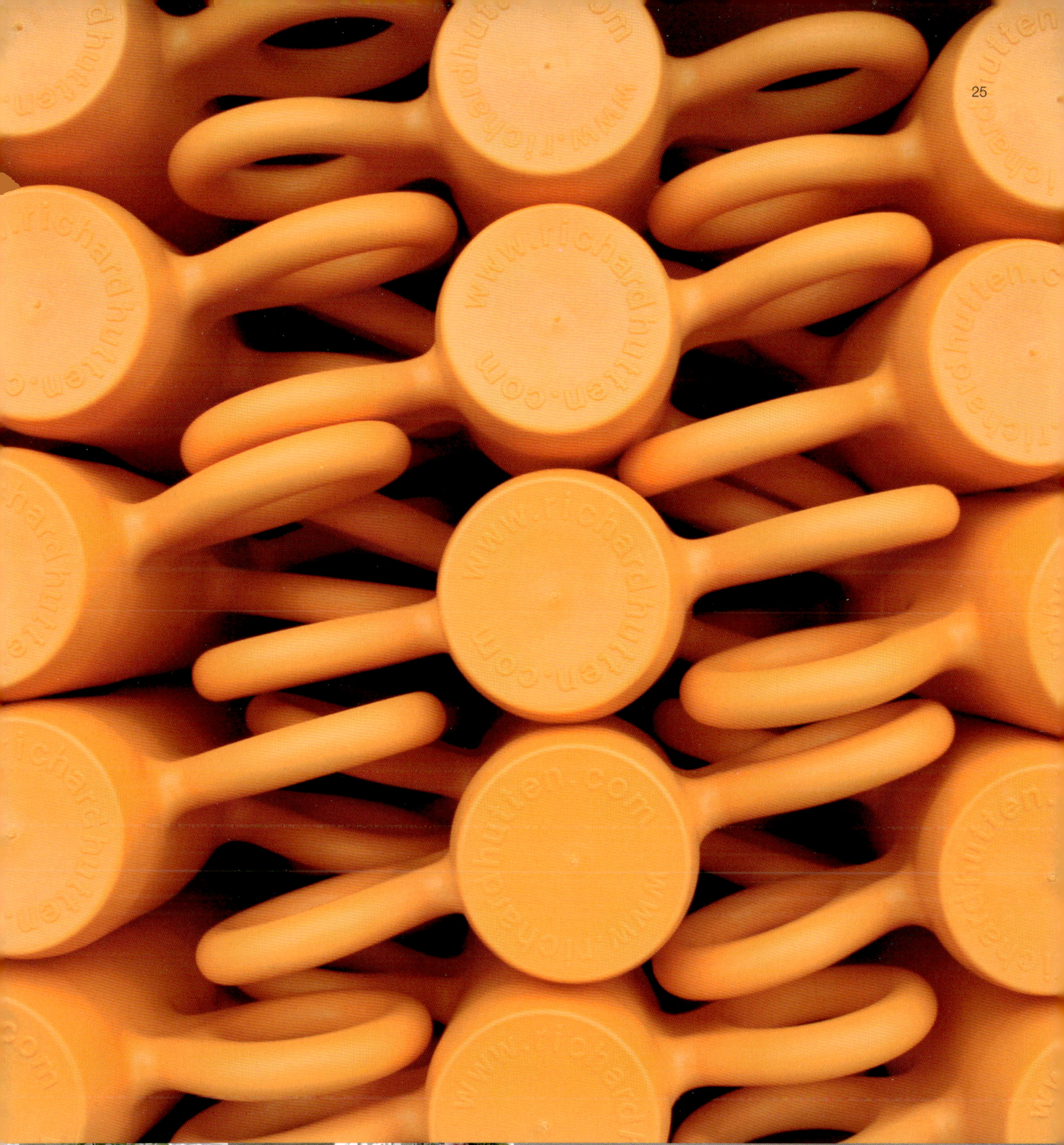
www.richardhutten.com

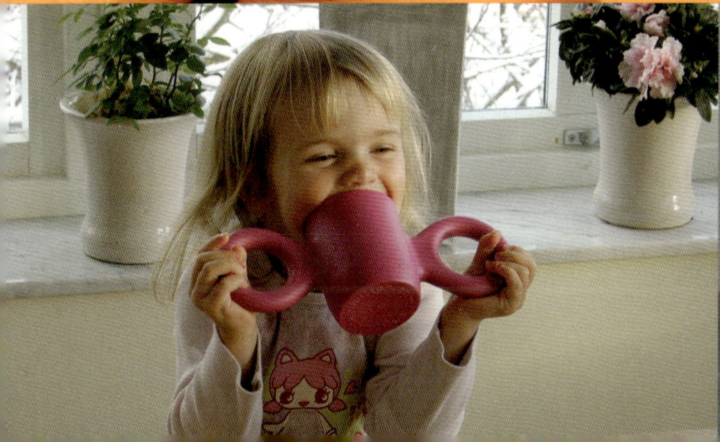

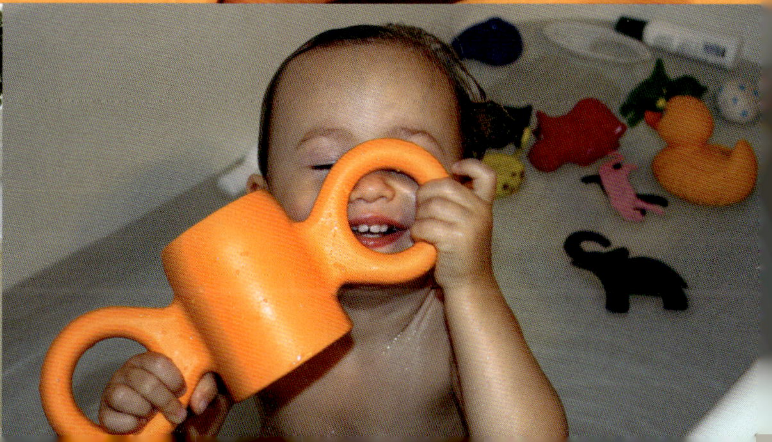

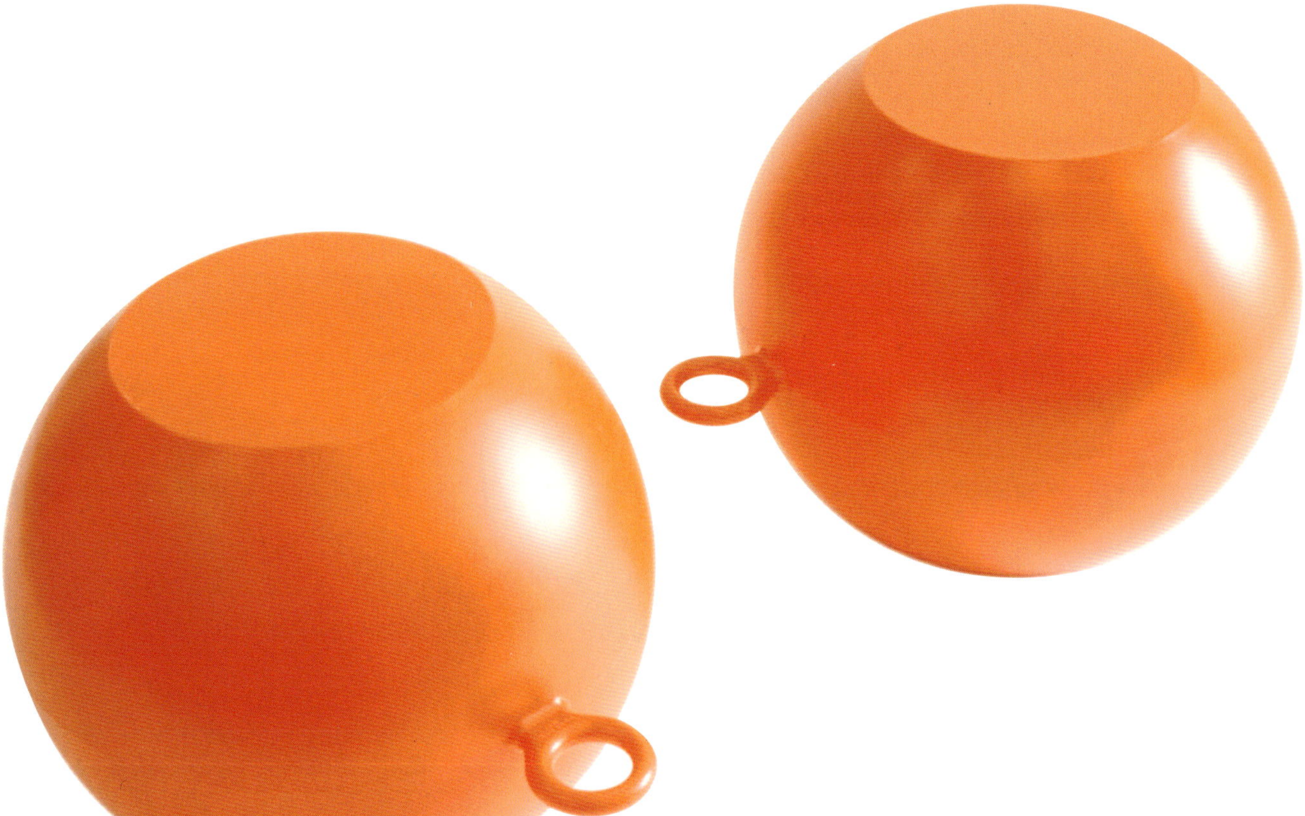

Zzzidt-object (Skippy) — 2000

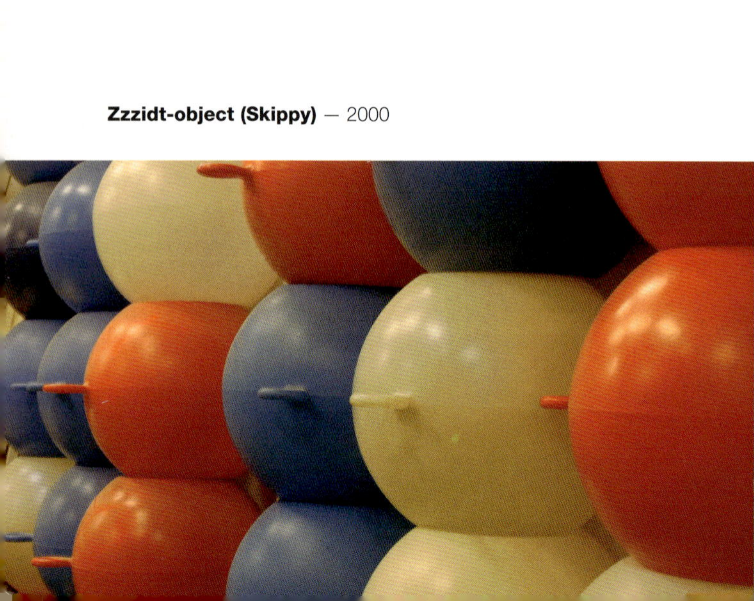

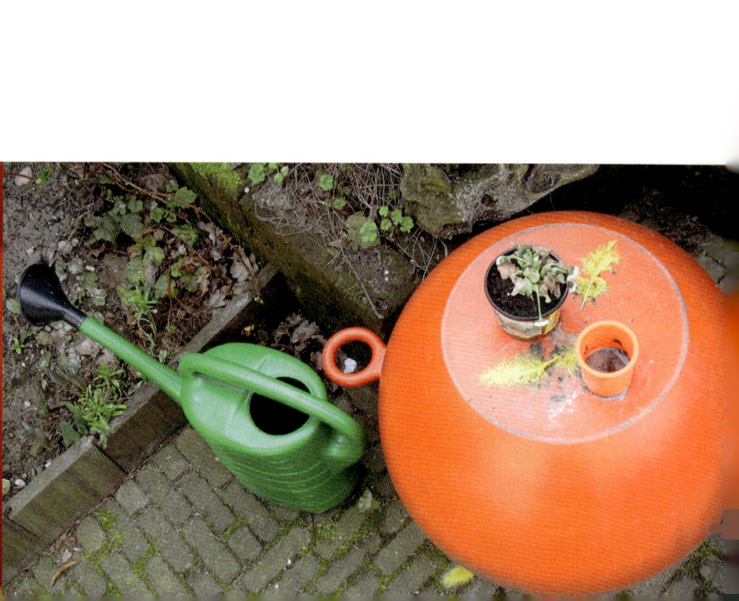

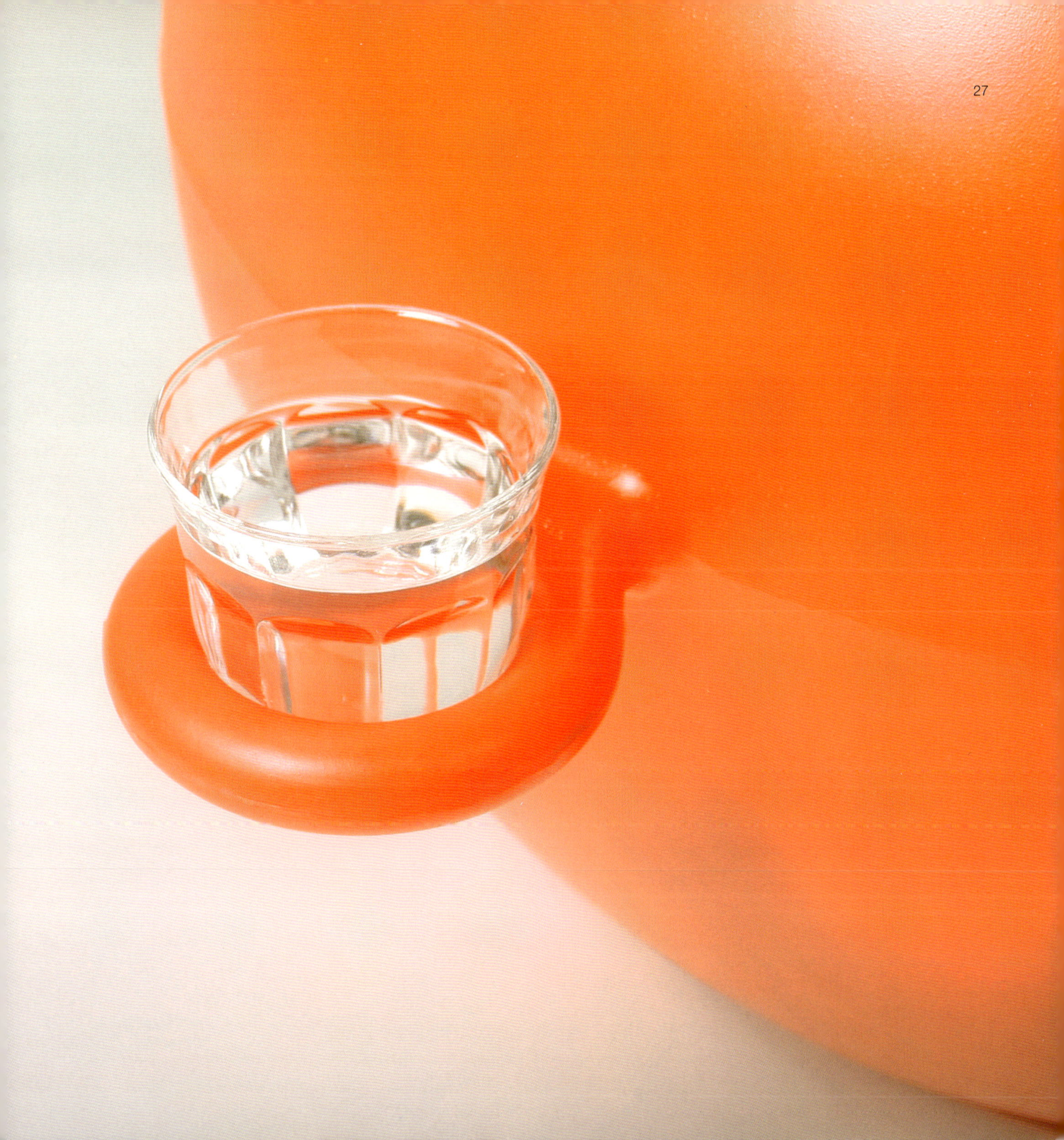

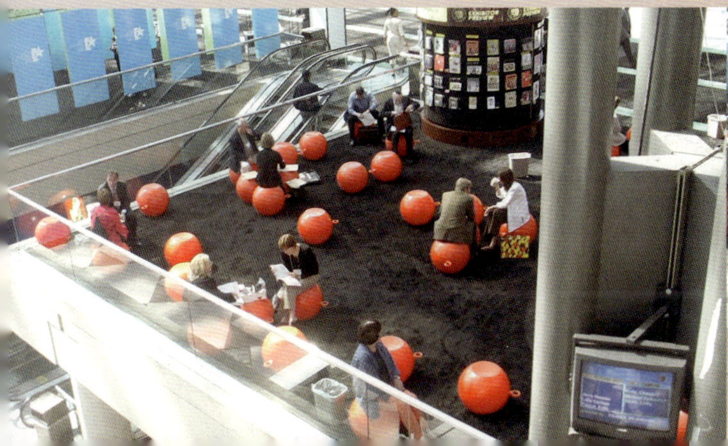

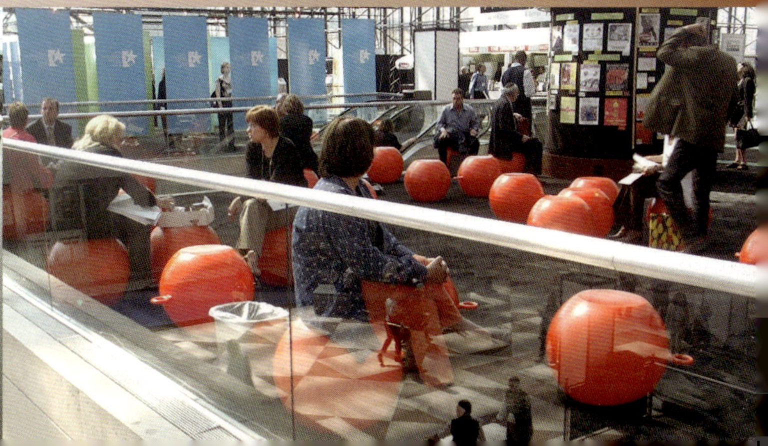

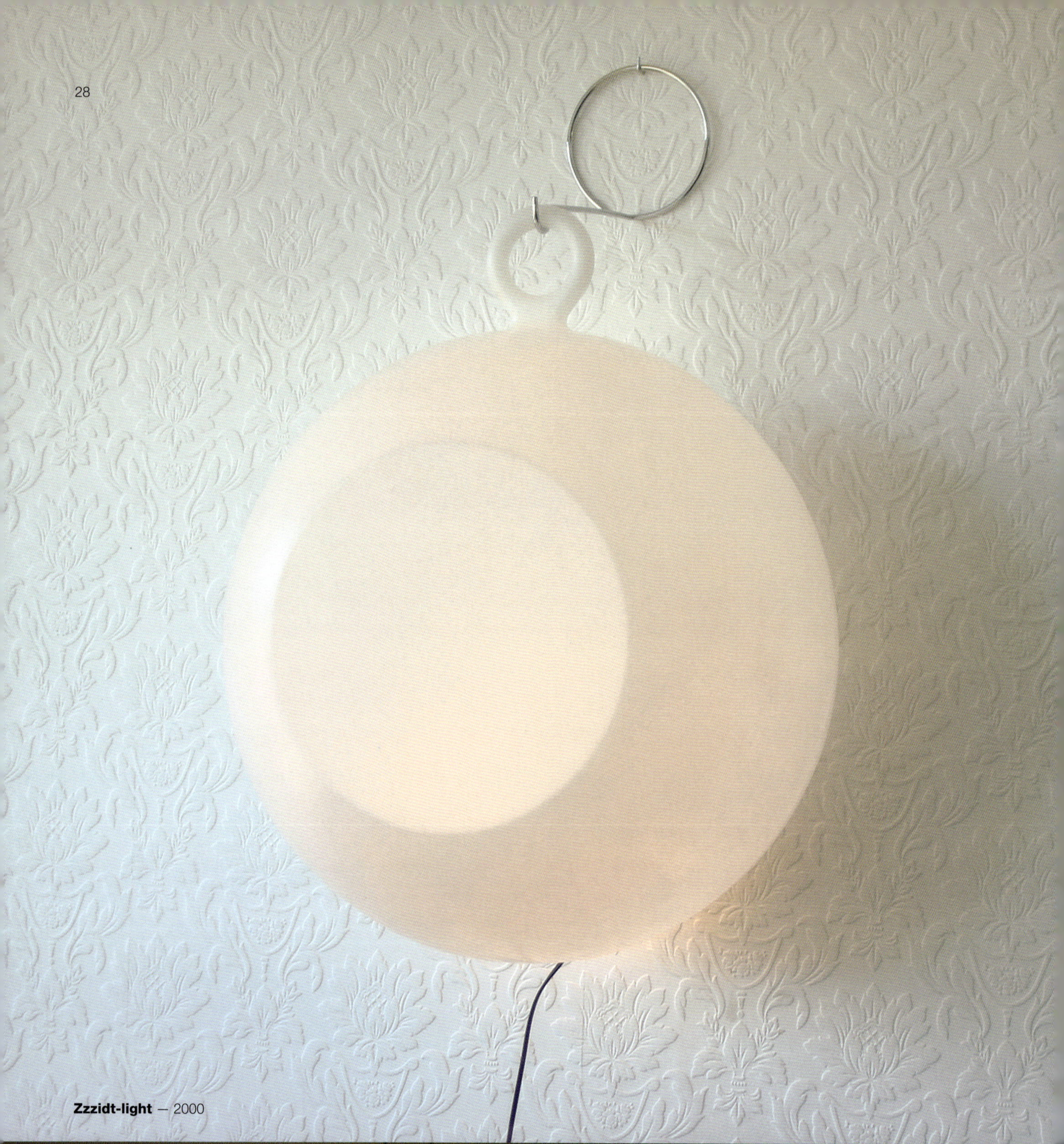

Zzzidt-light — 2000

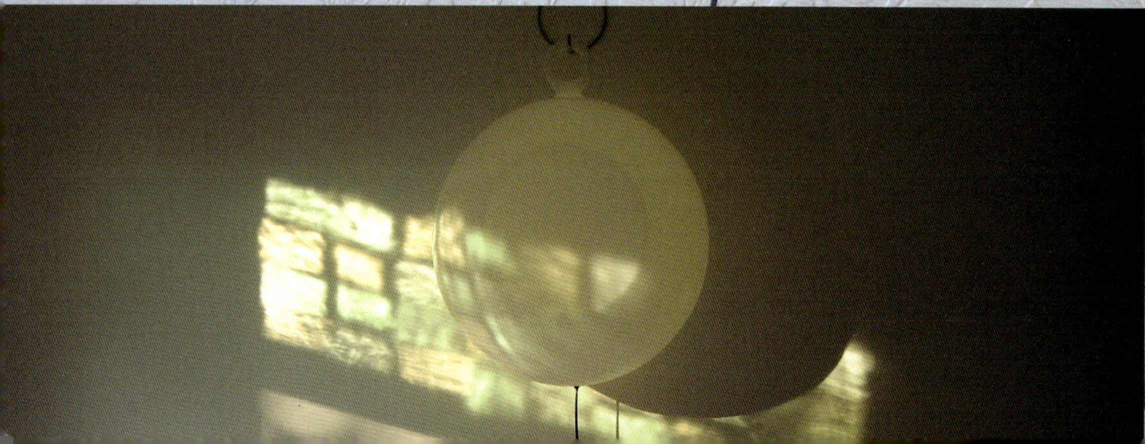

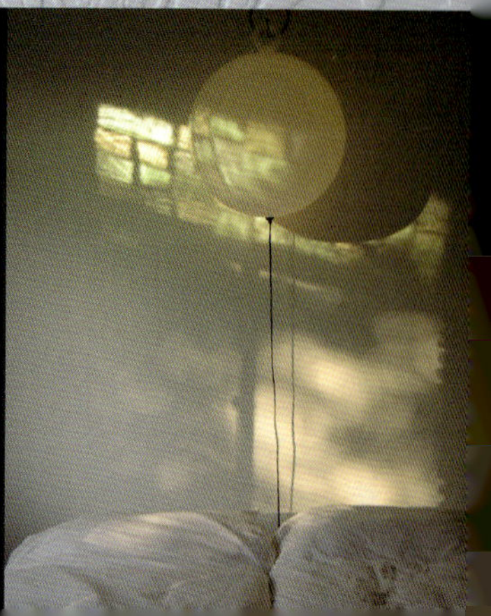

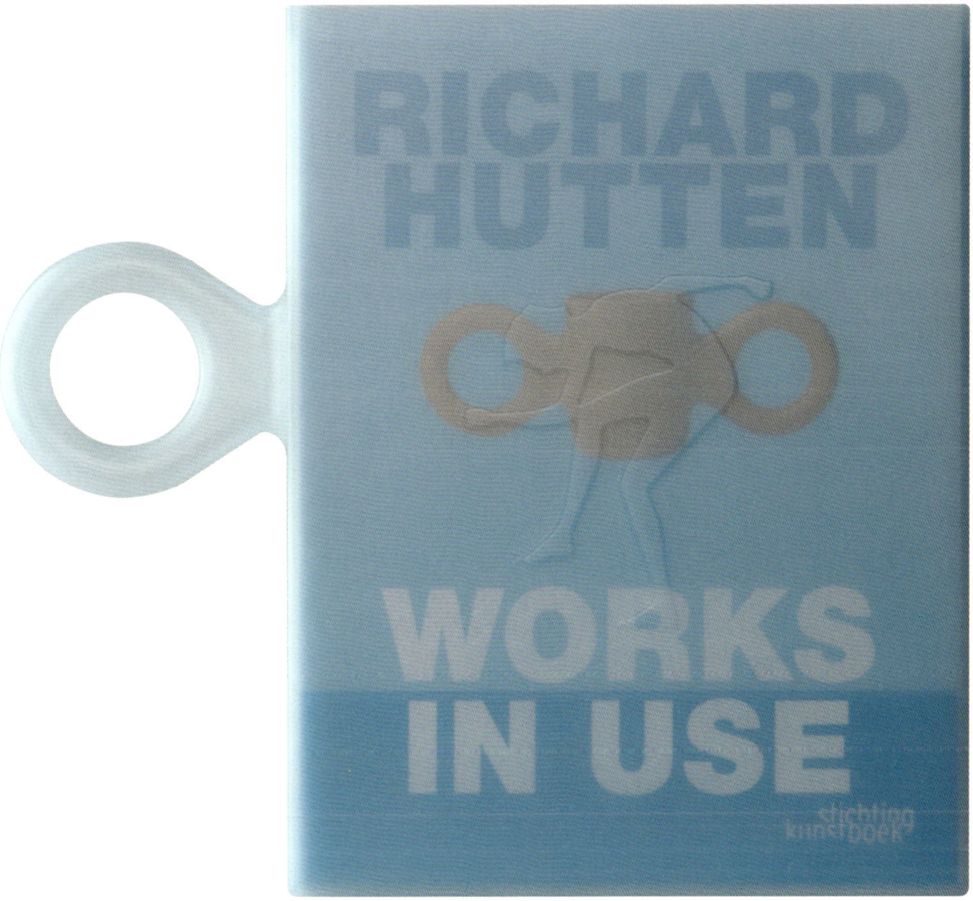

Book **'Works in use'** — 2006

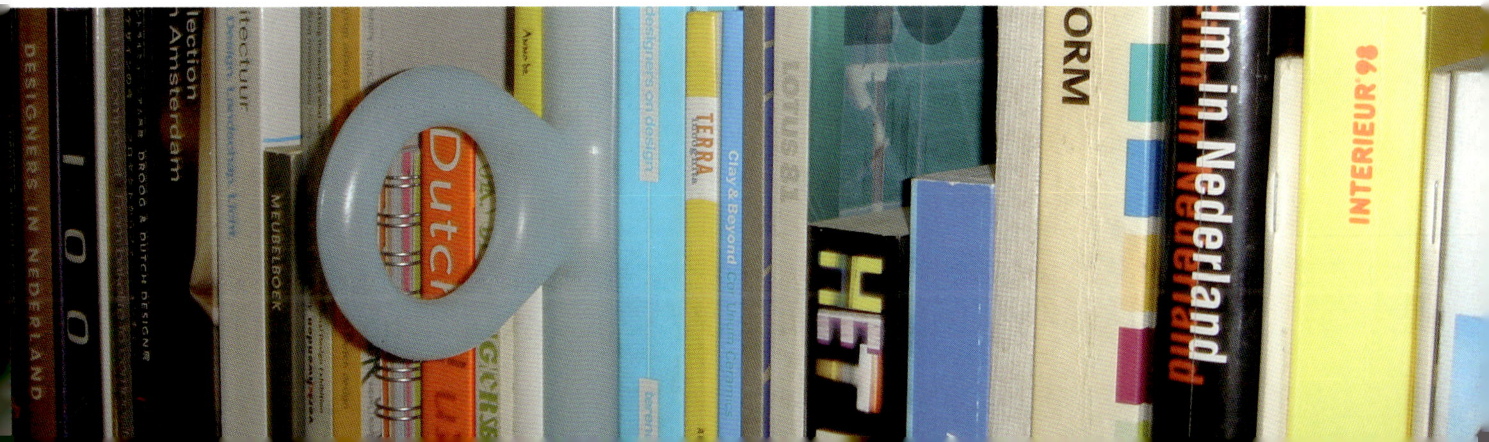

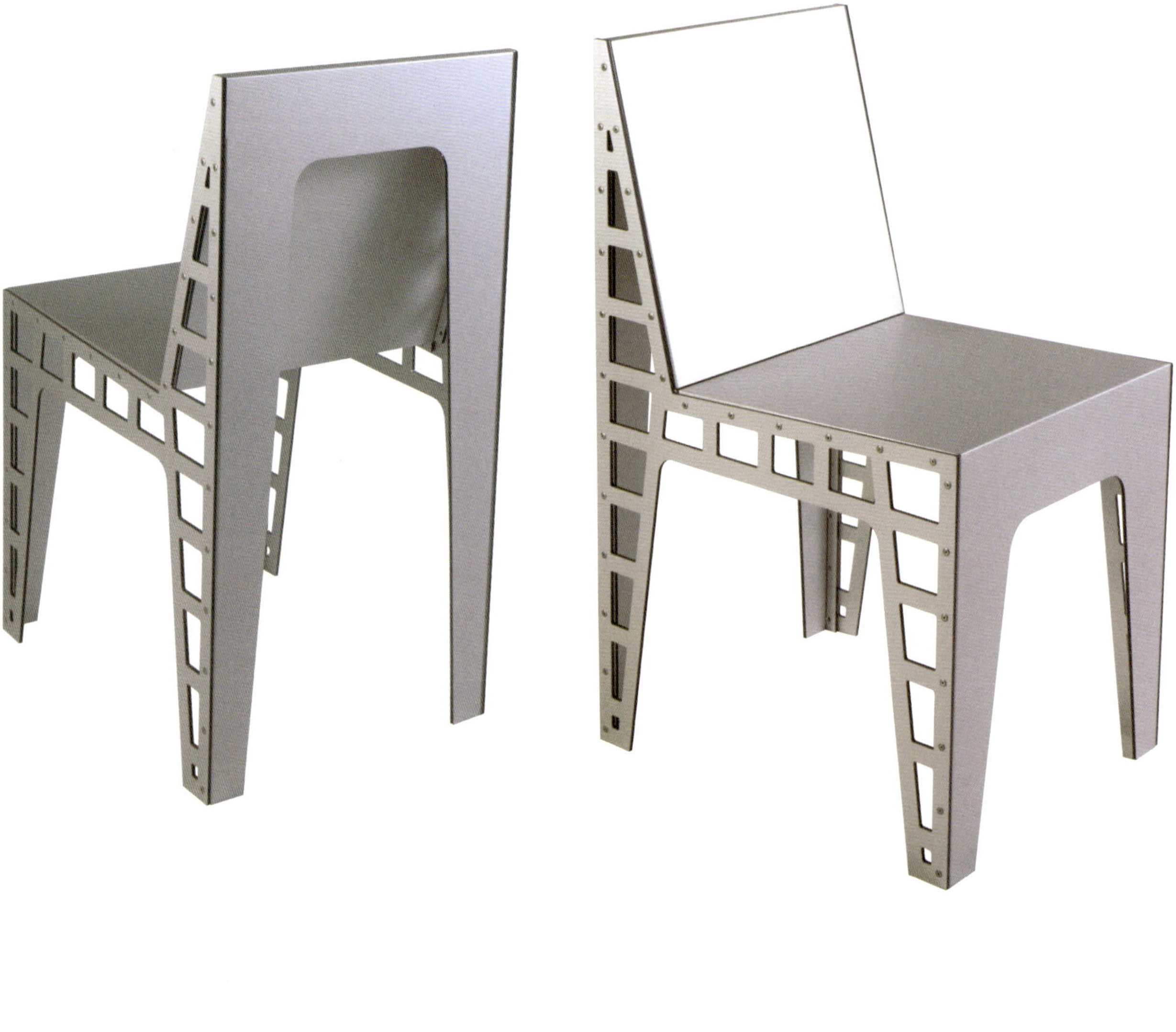

One of a kind dining chair — 1999

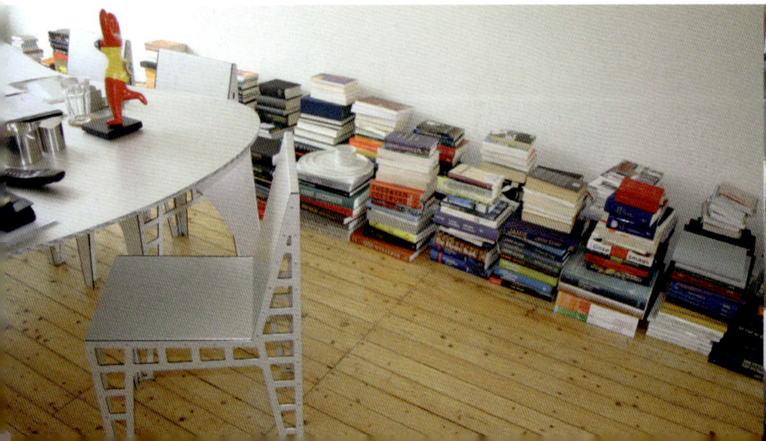

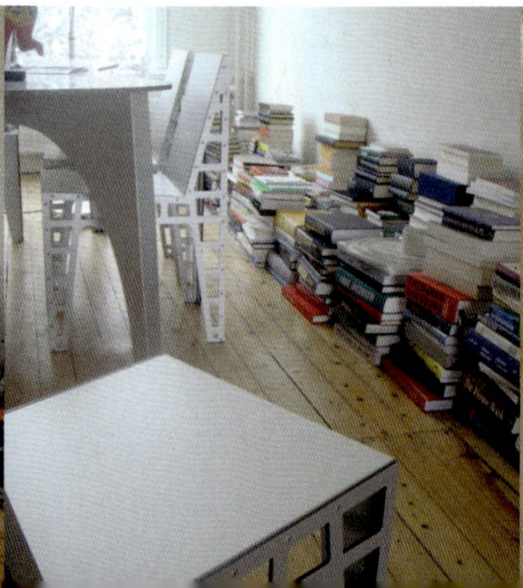

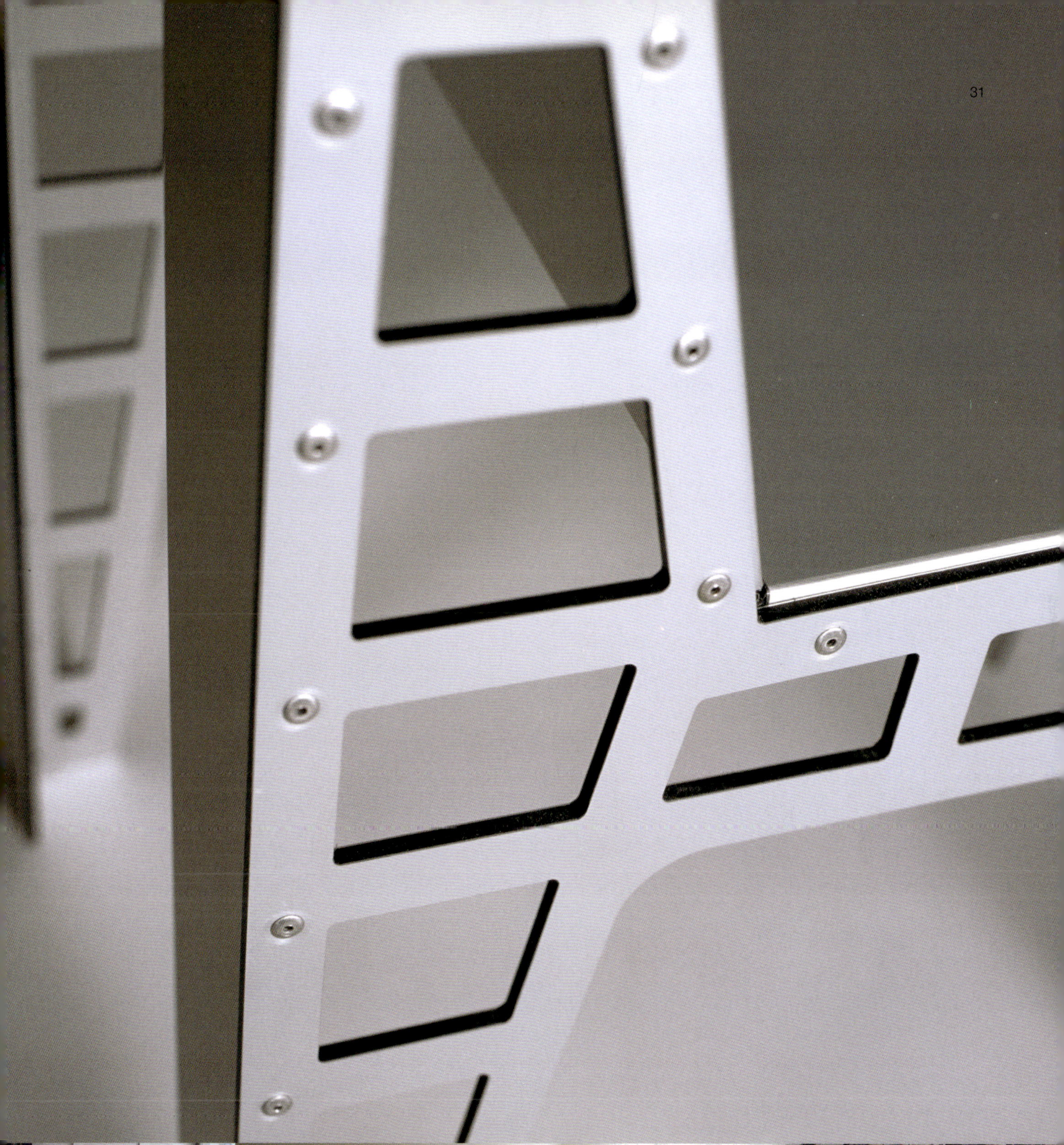

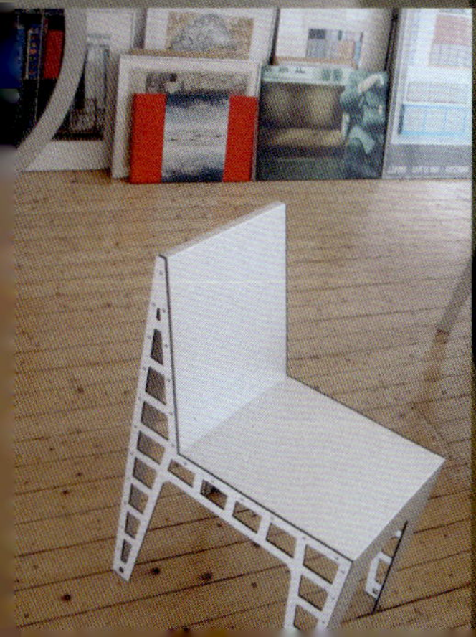

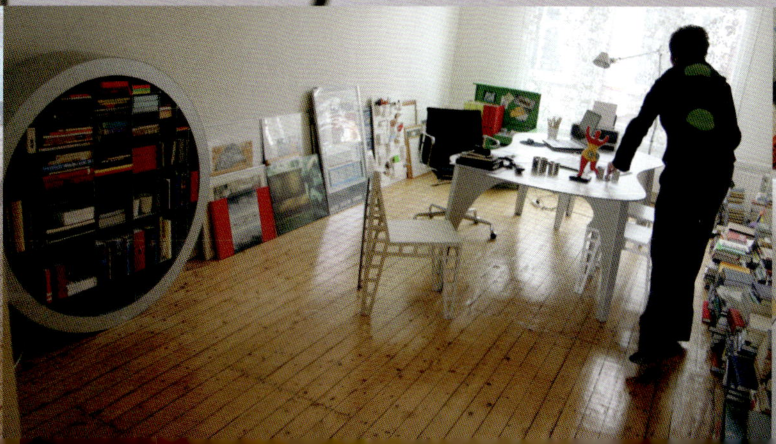

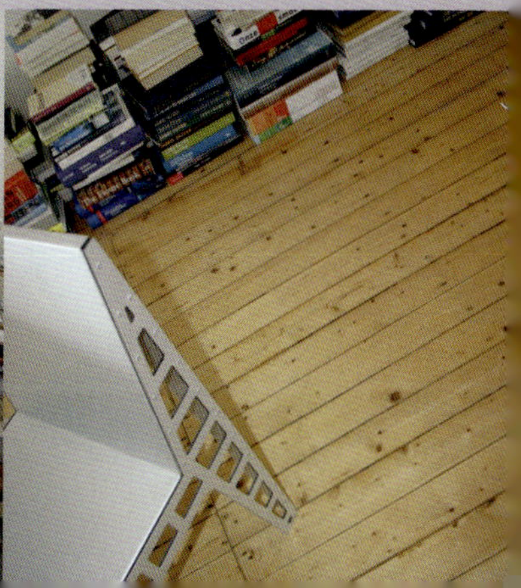

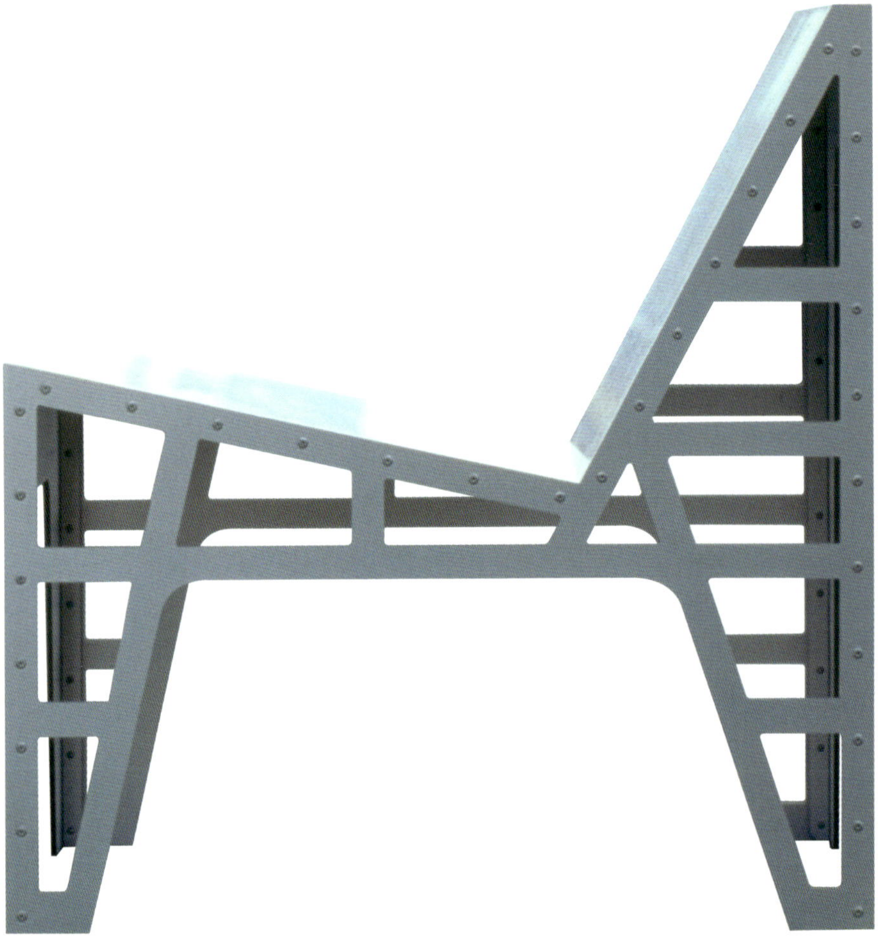

One of a kind lounge chair — 2000

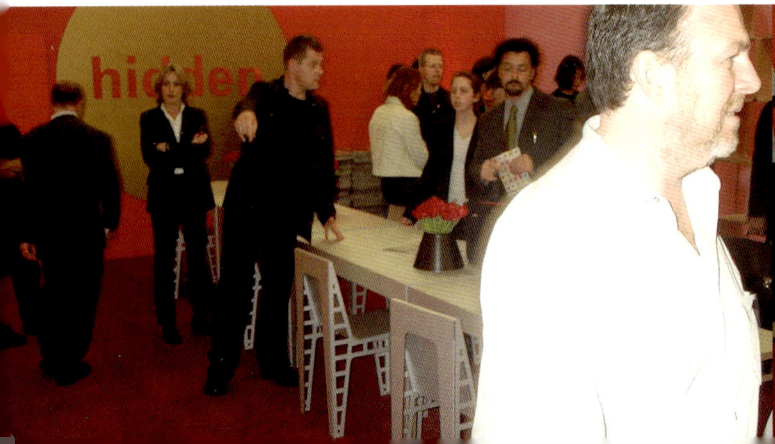

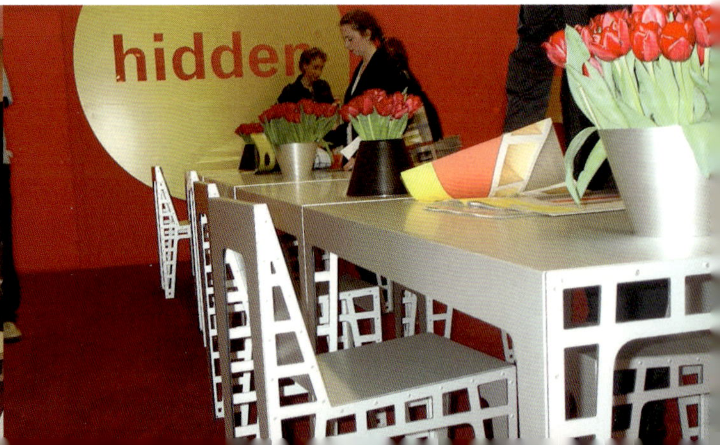

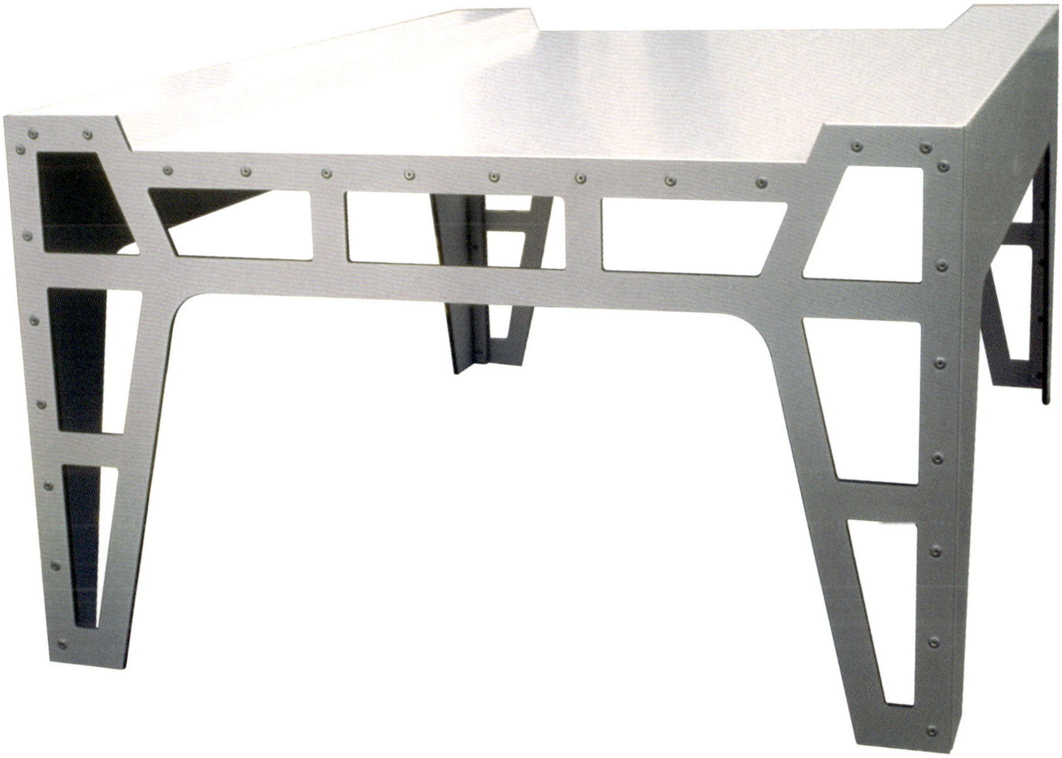

One of a kind coffee table — 2000

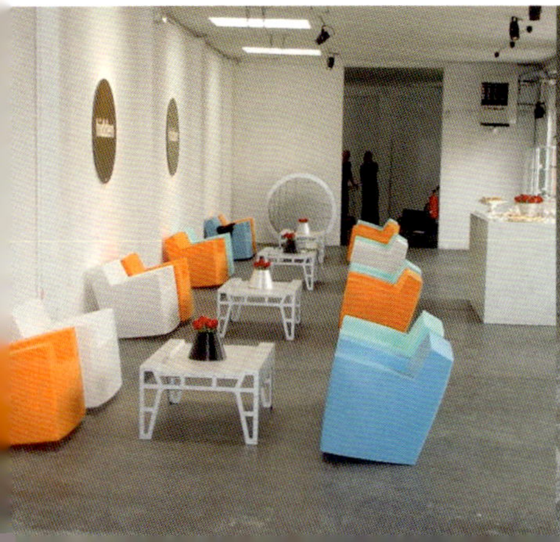

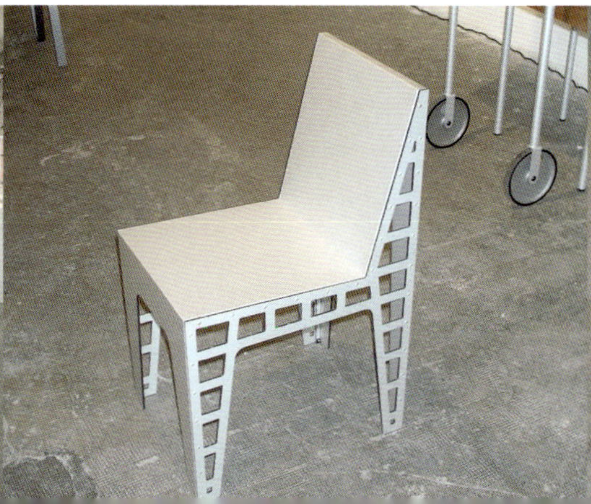

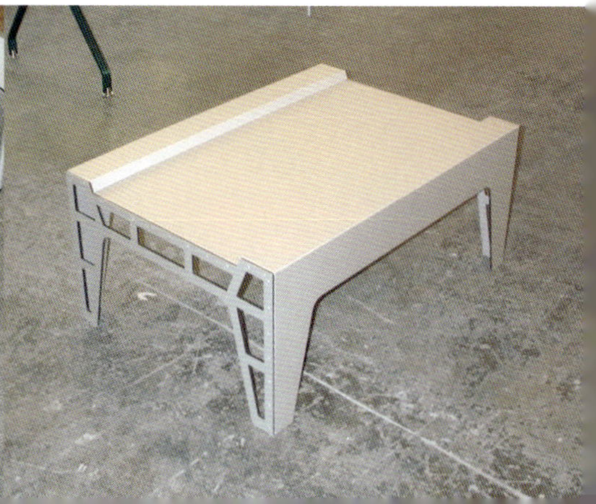

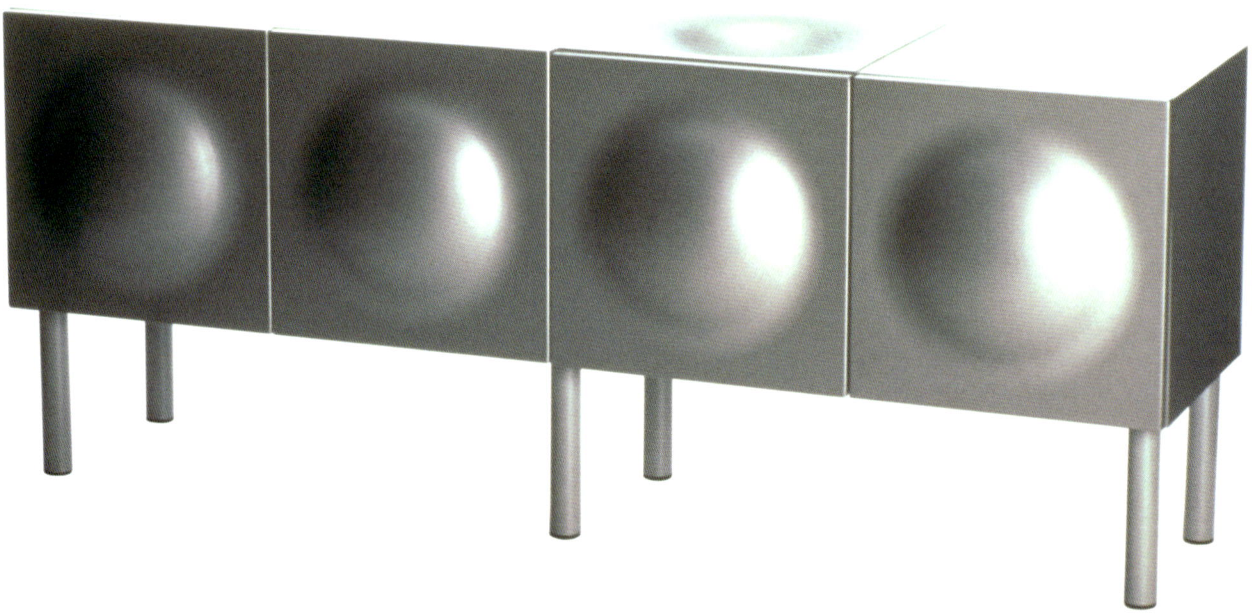

Stouwkast — 1999

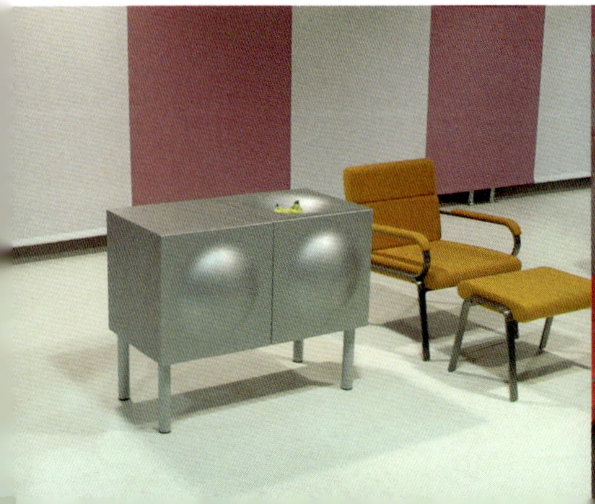

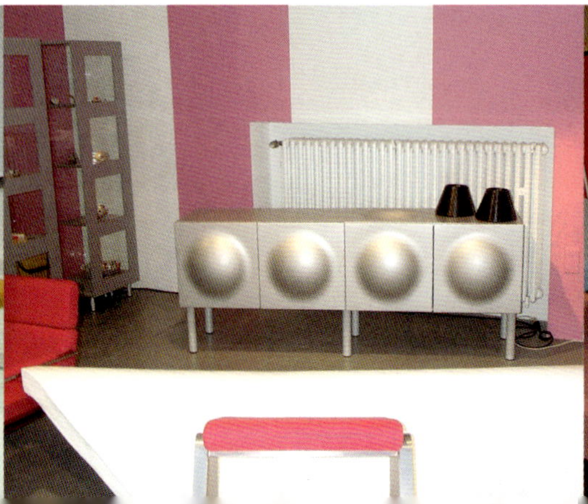

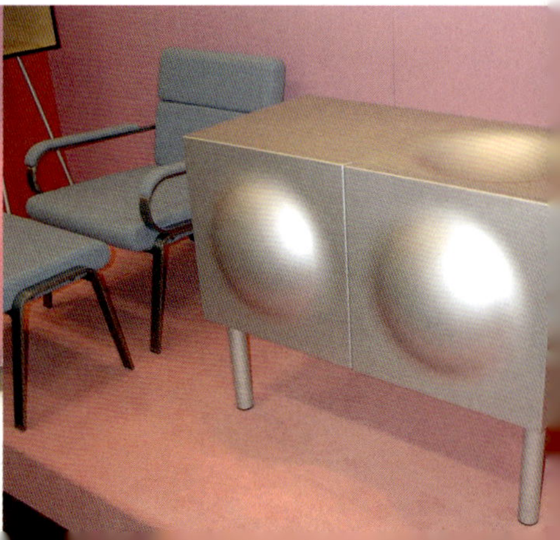

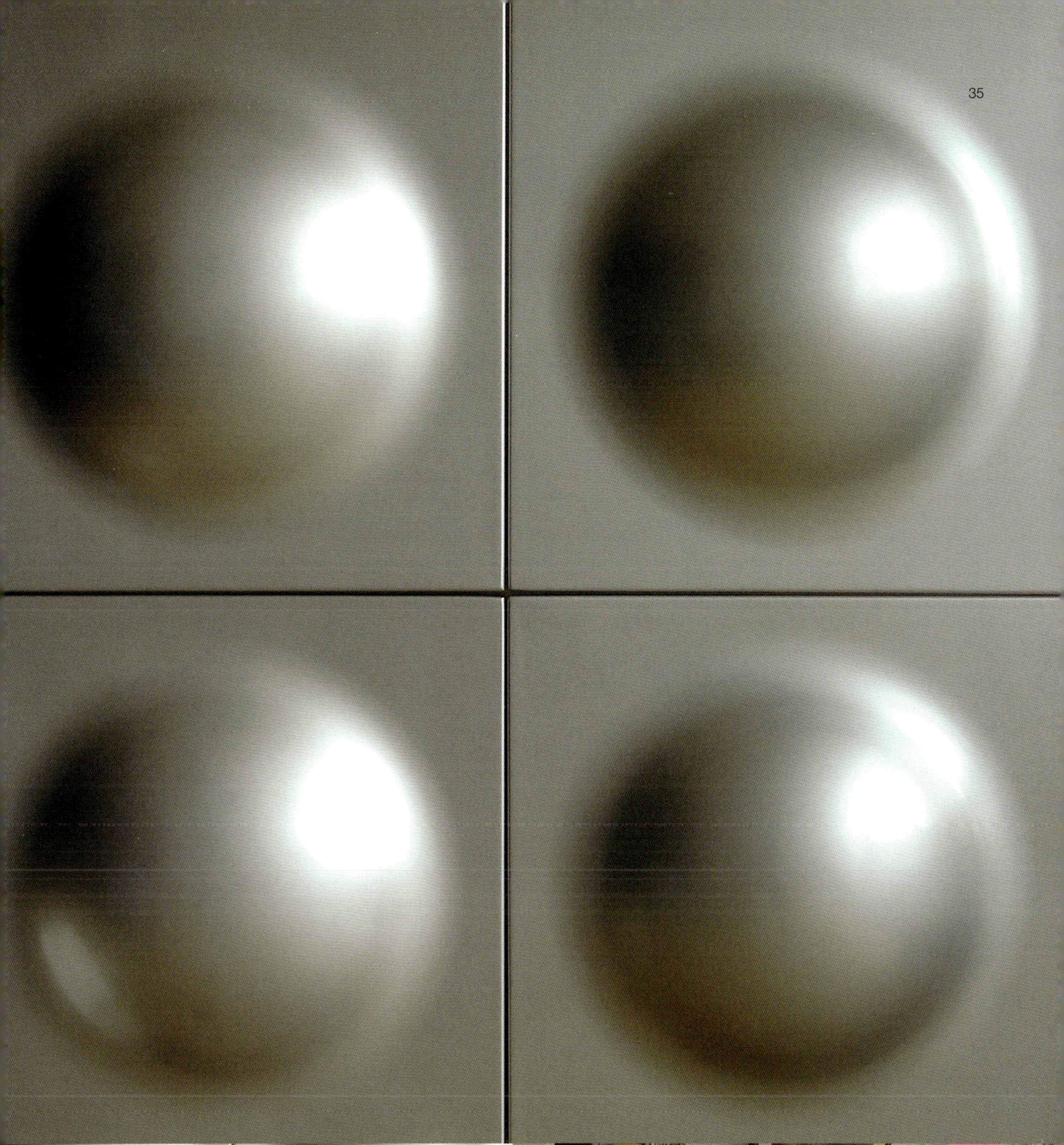

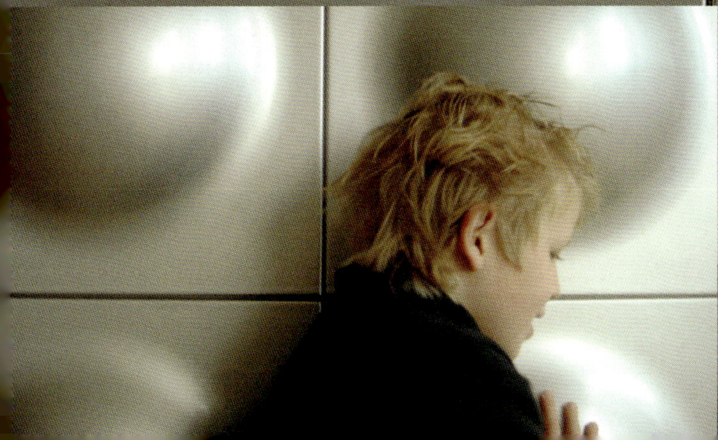

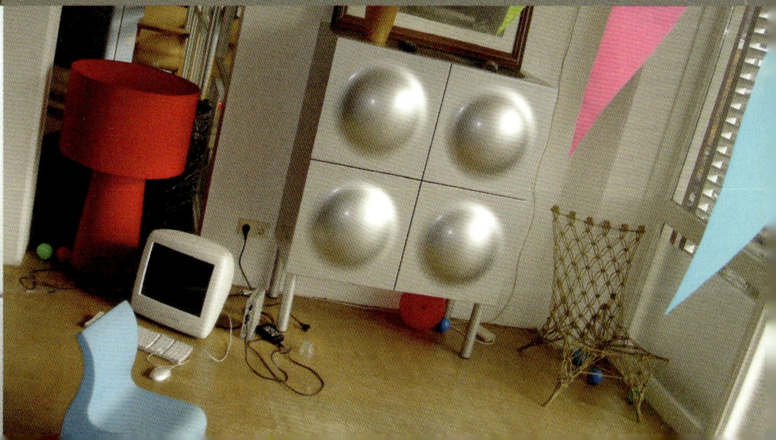

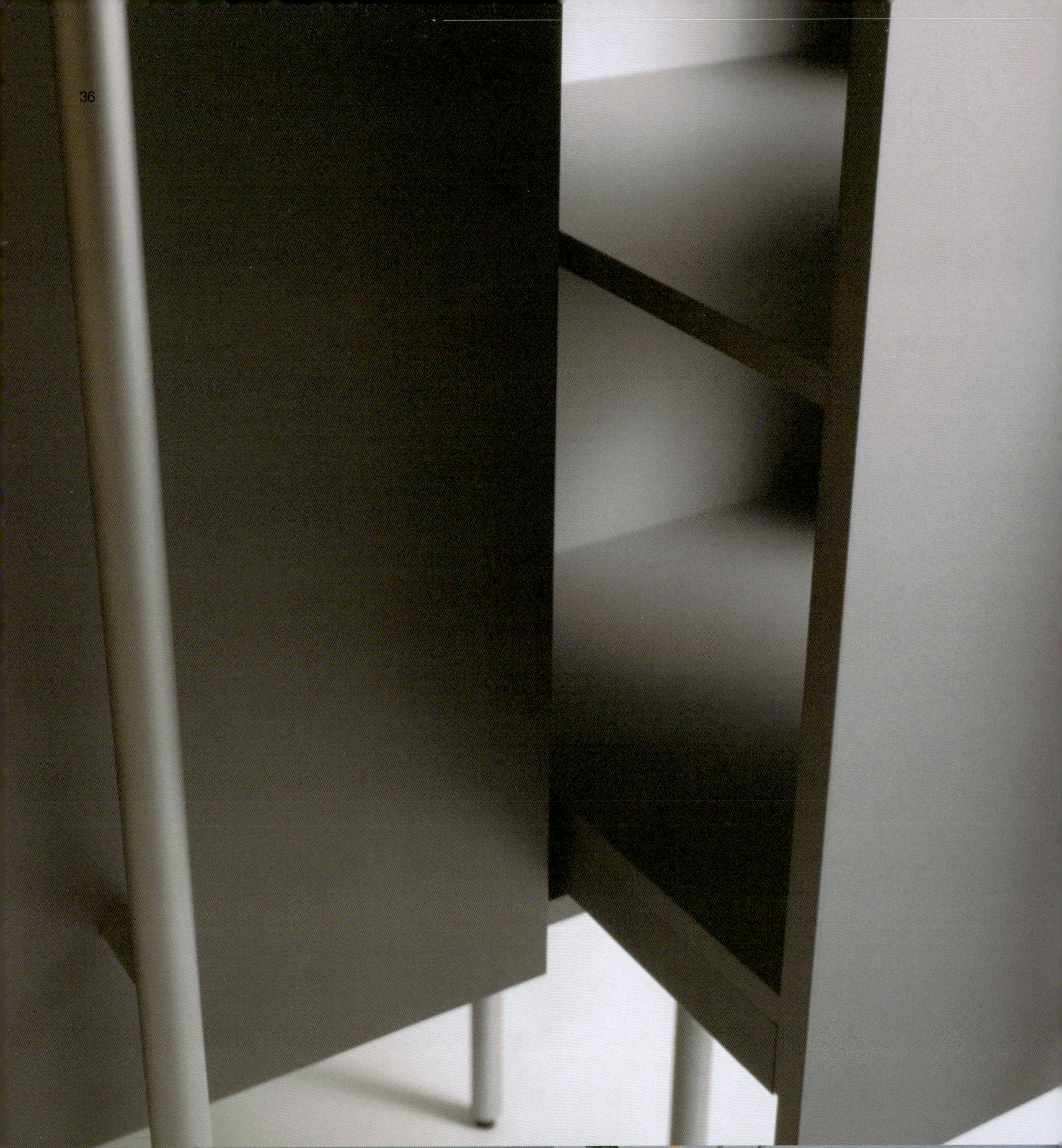

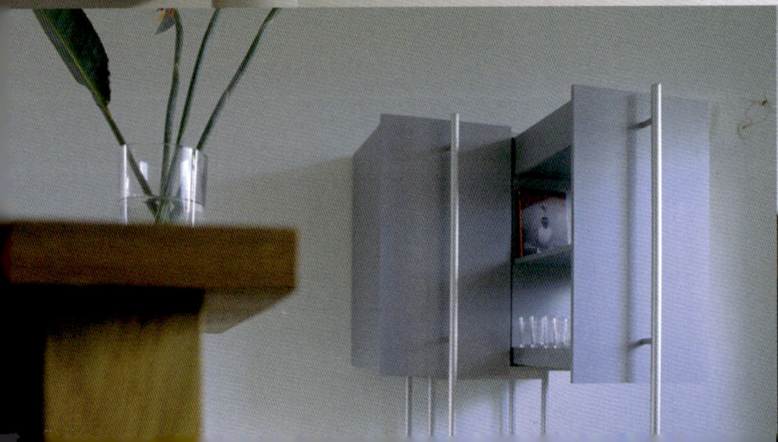

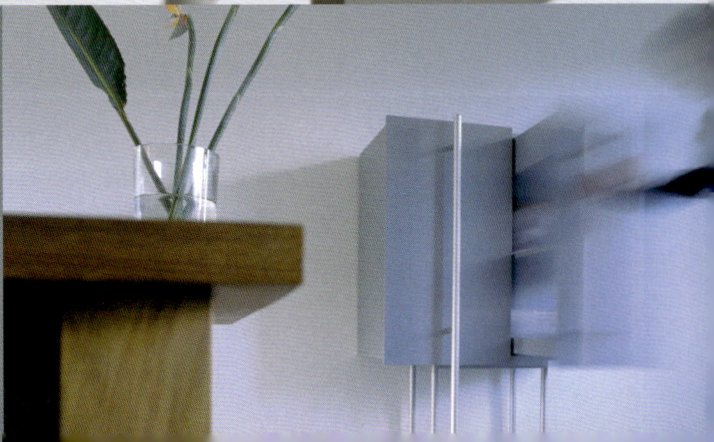

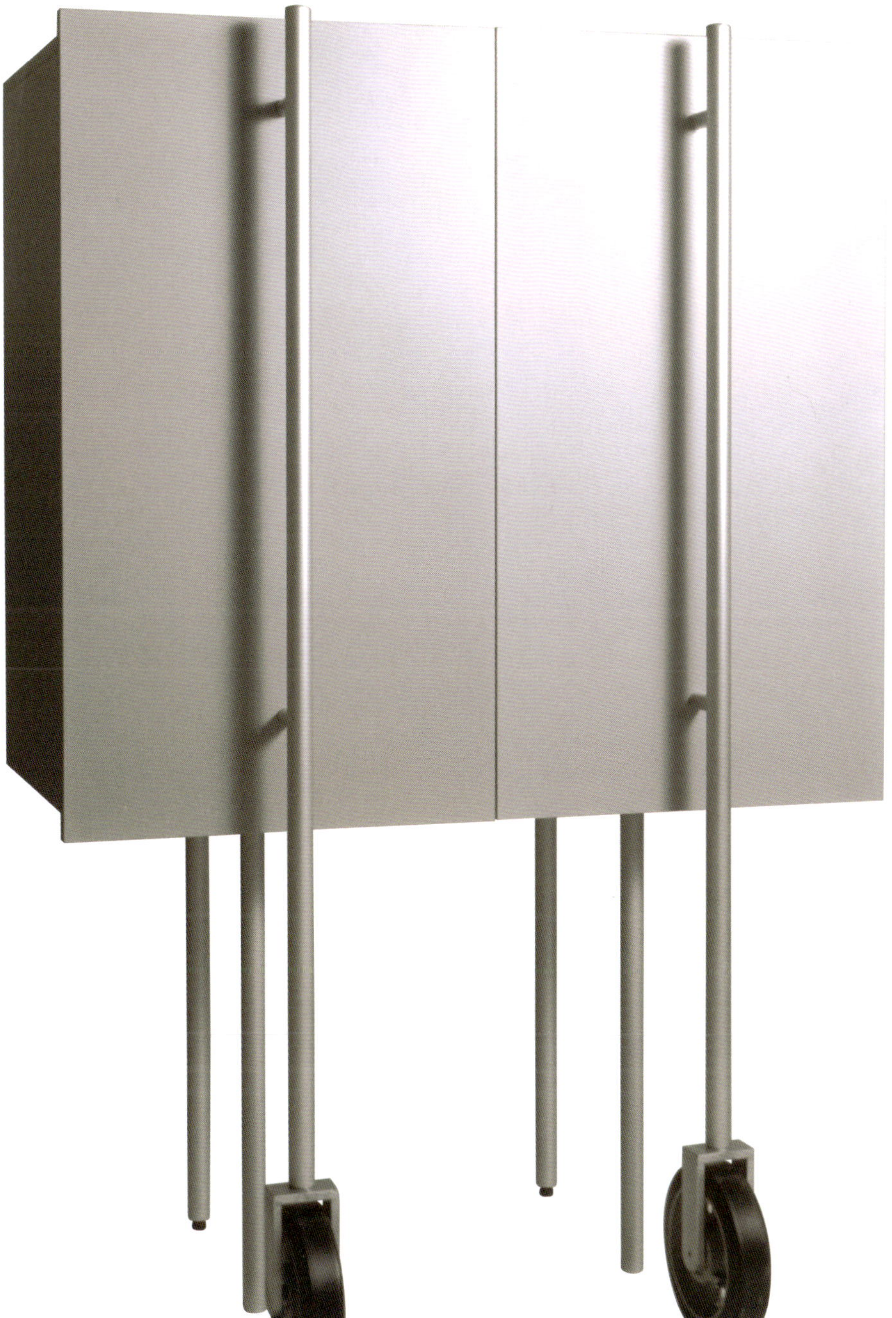

Wheels cabinet — 2000

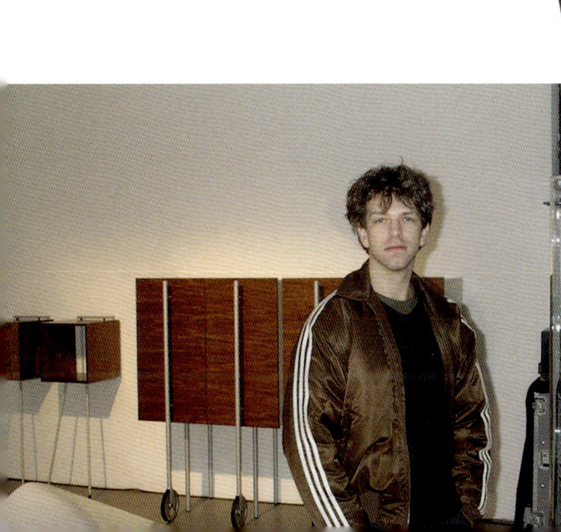

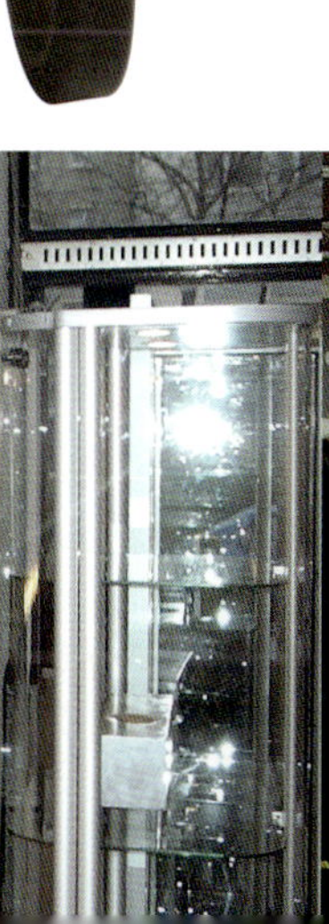

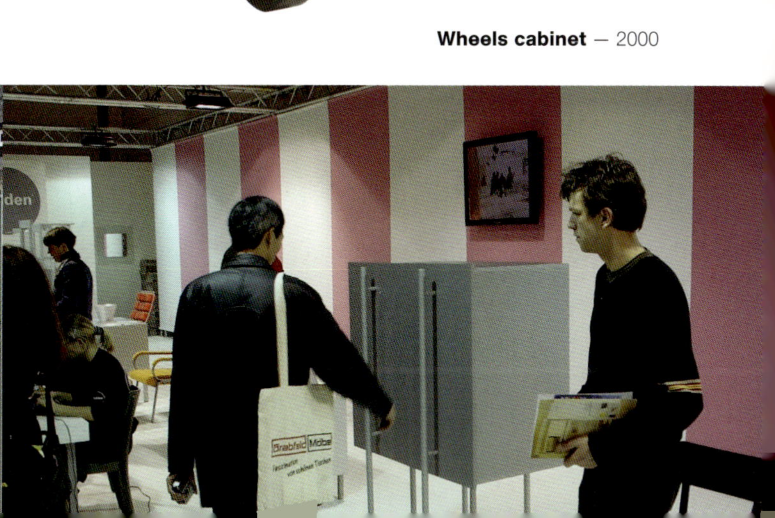

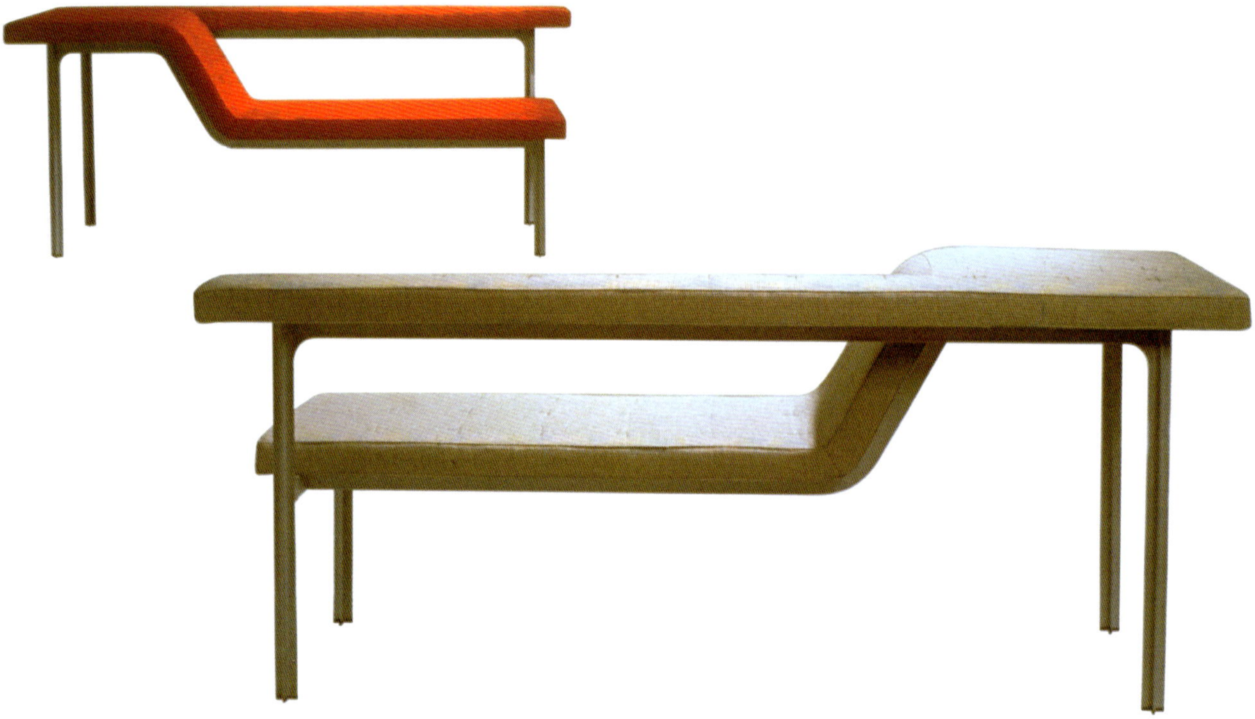

Table-bench — 2000

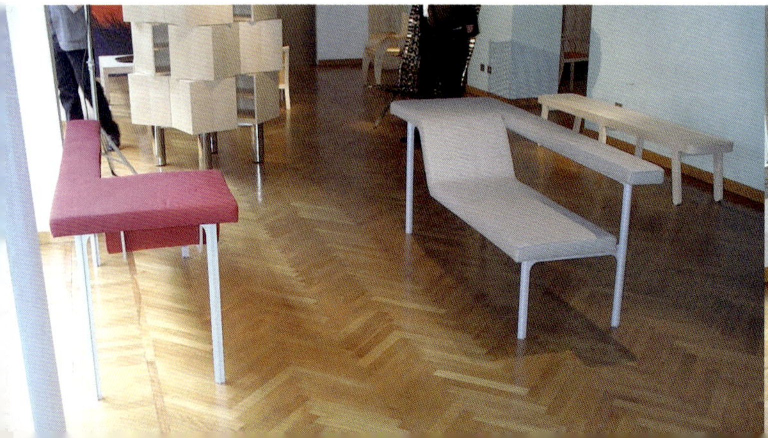

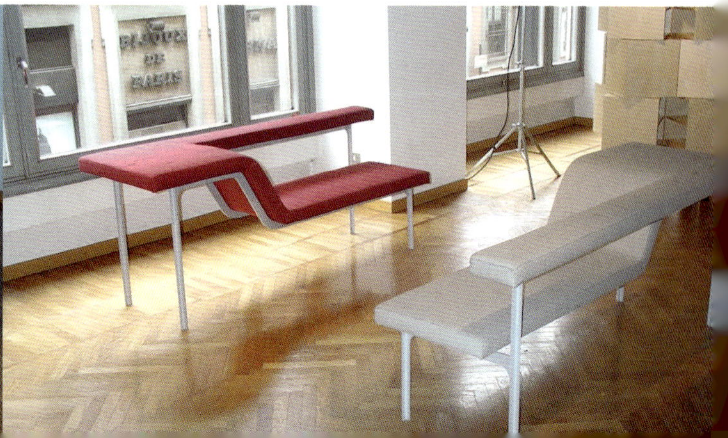

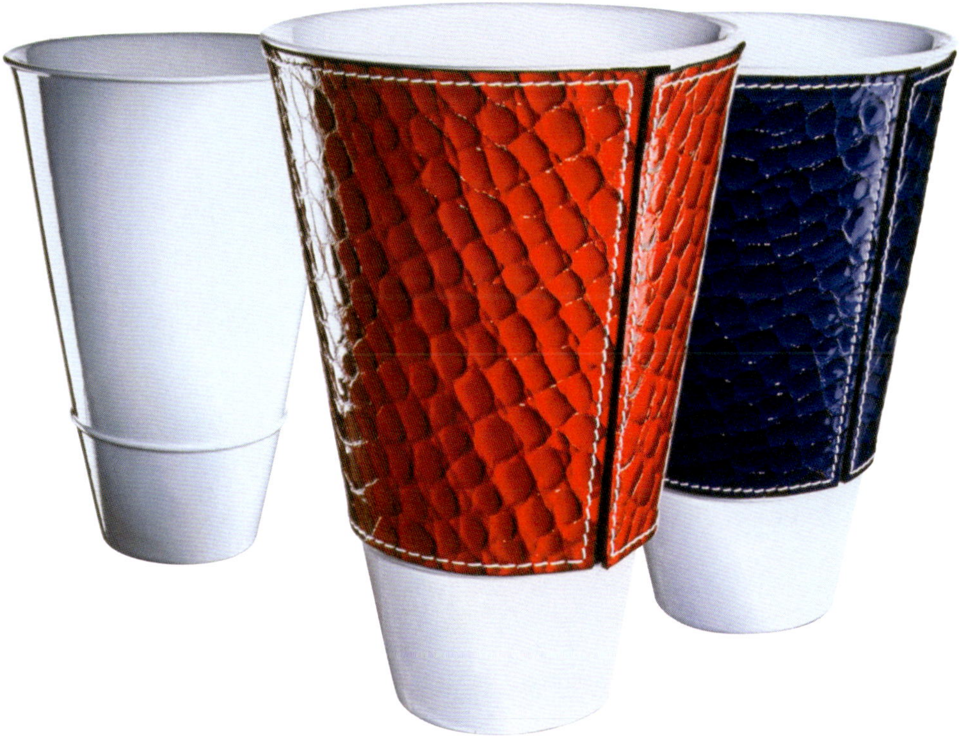

Dutch vase — 2001

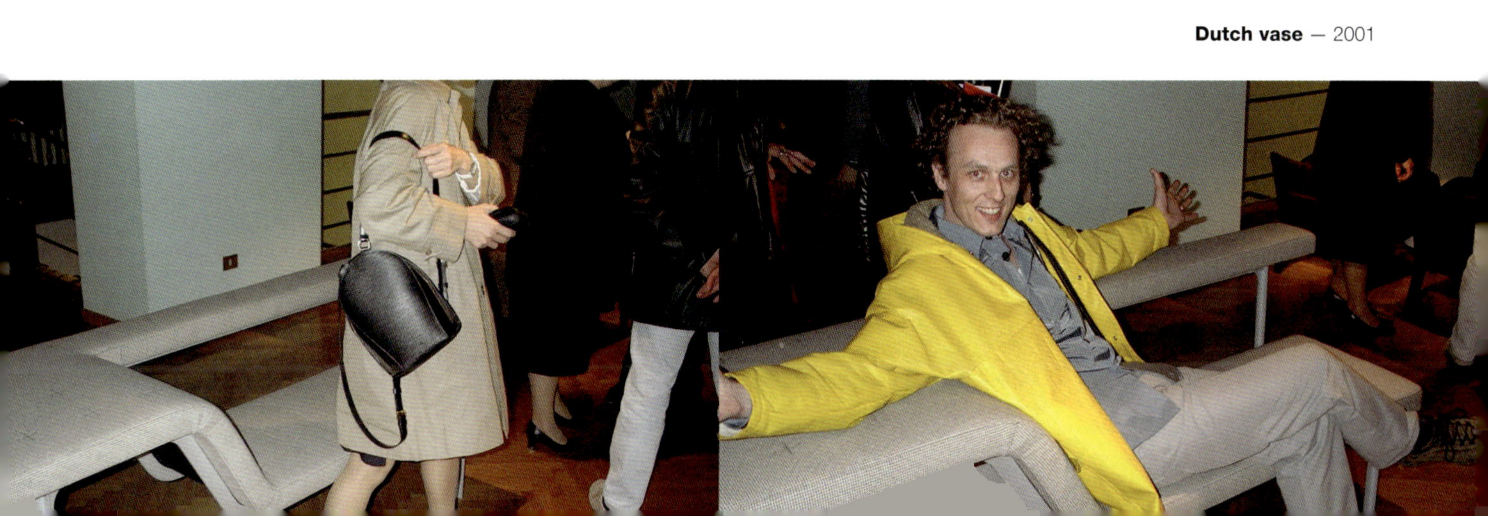

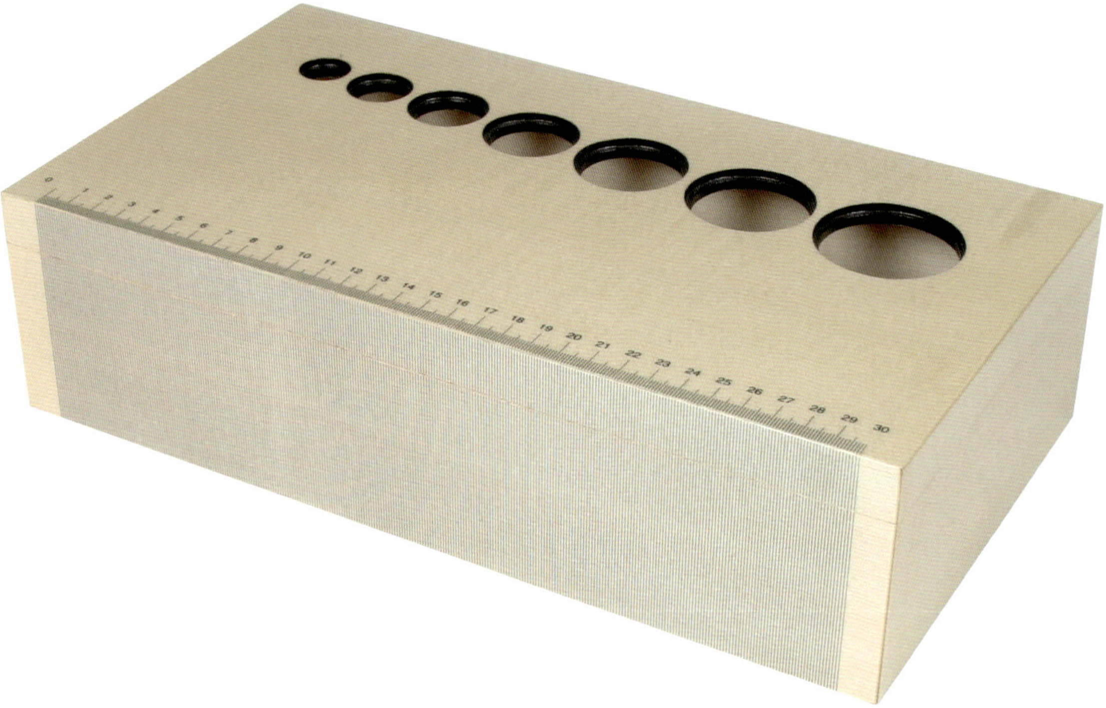

Drawing box — 2001

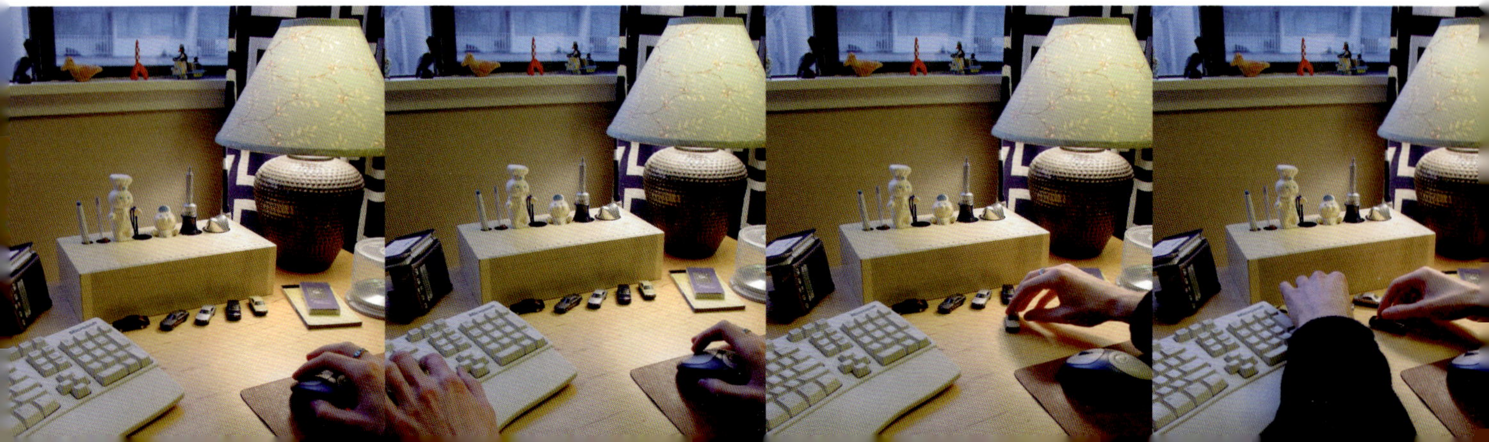

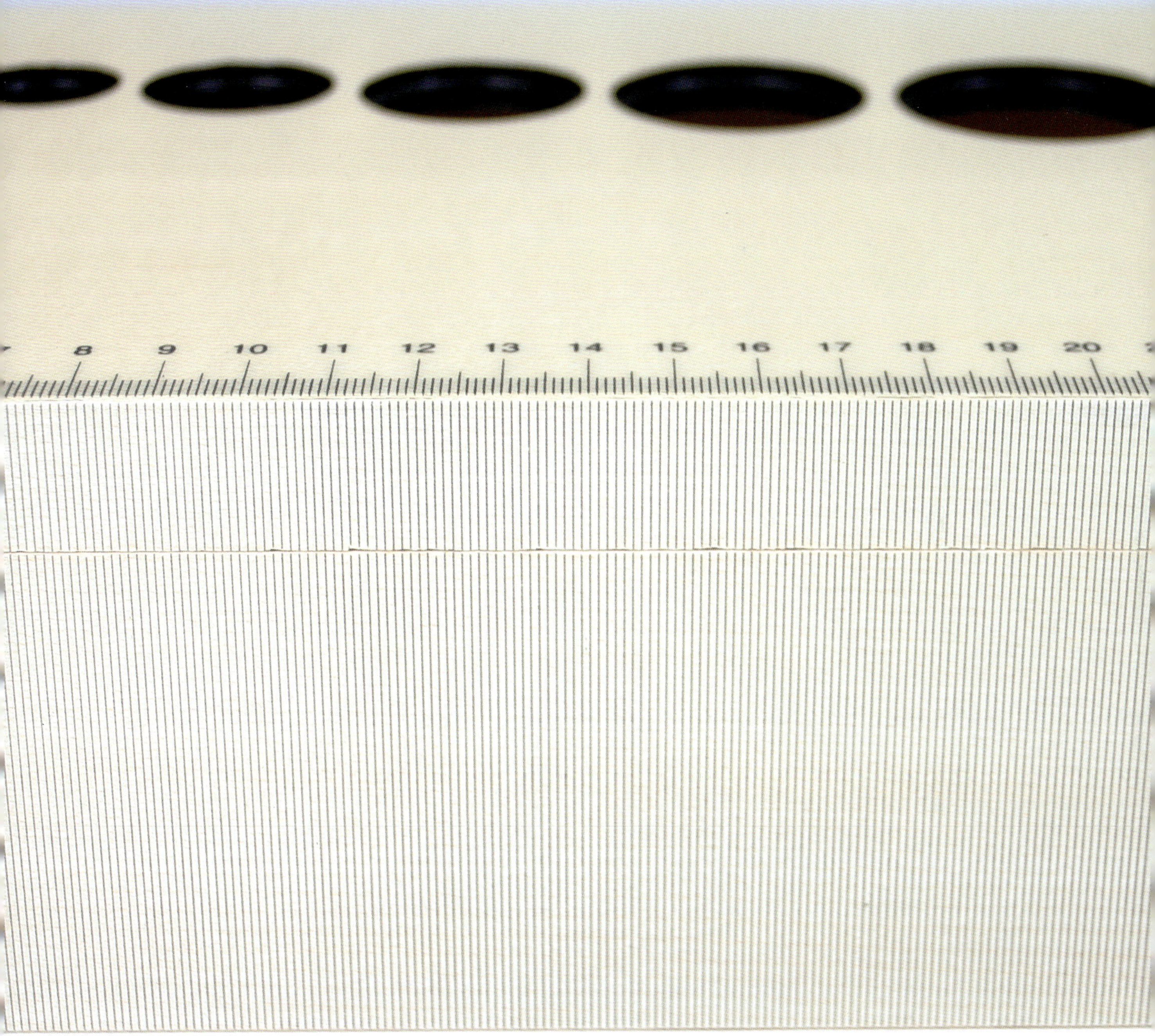
8 9 10 11 12 13 14 15 16 17 18 19 20

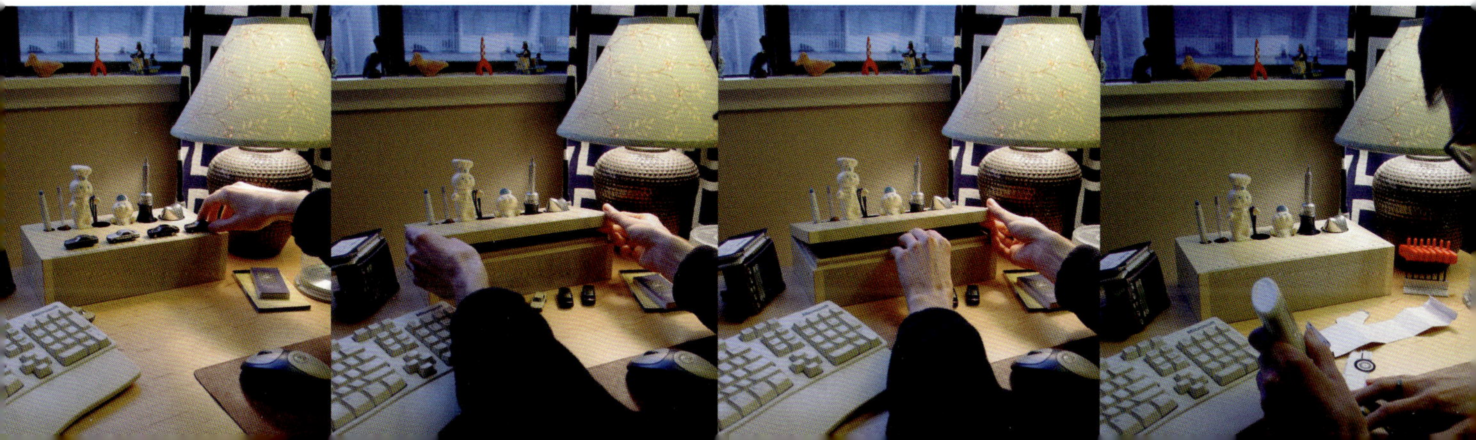

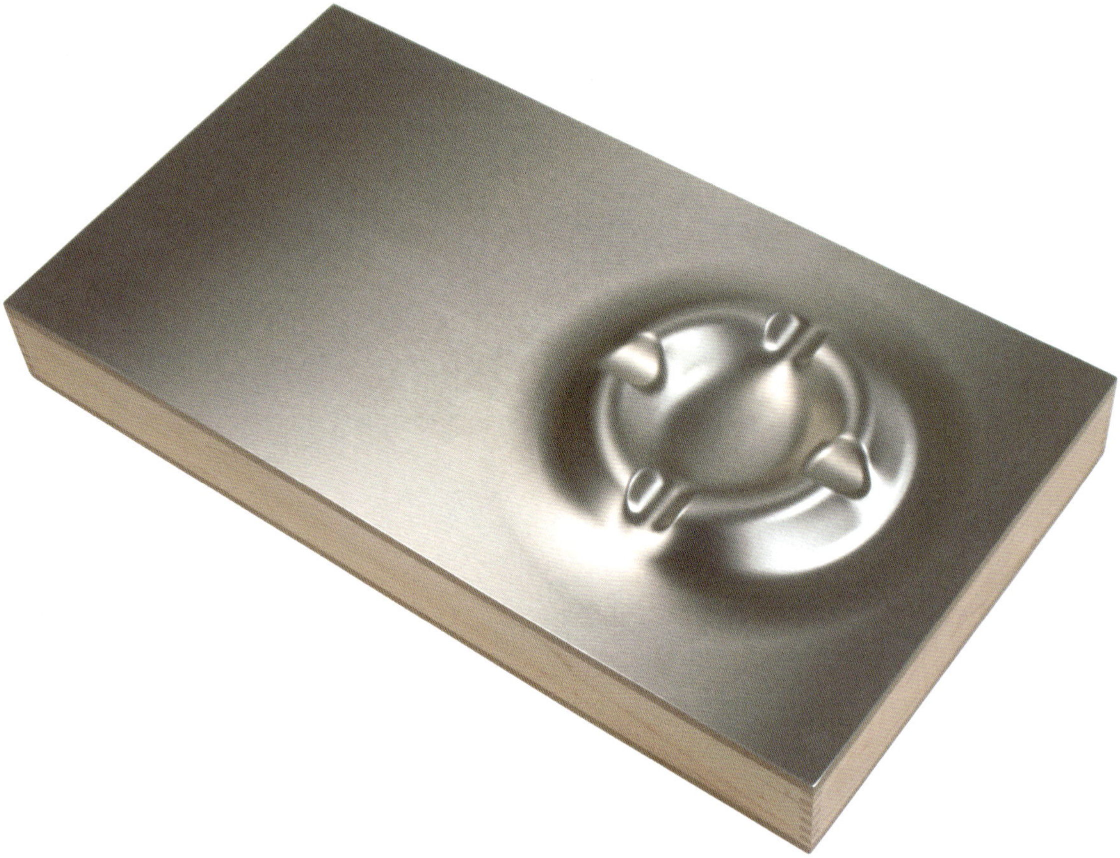

Cigar box — 2001

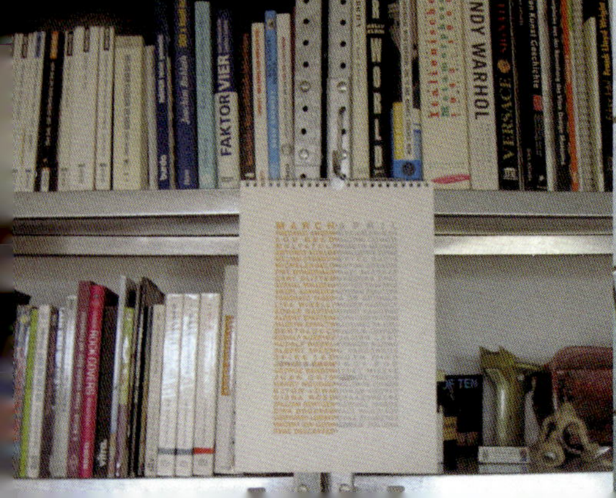

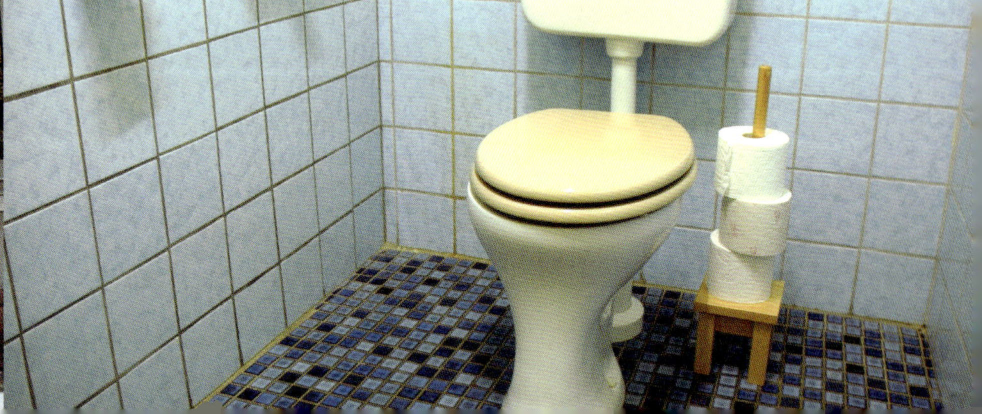

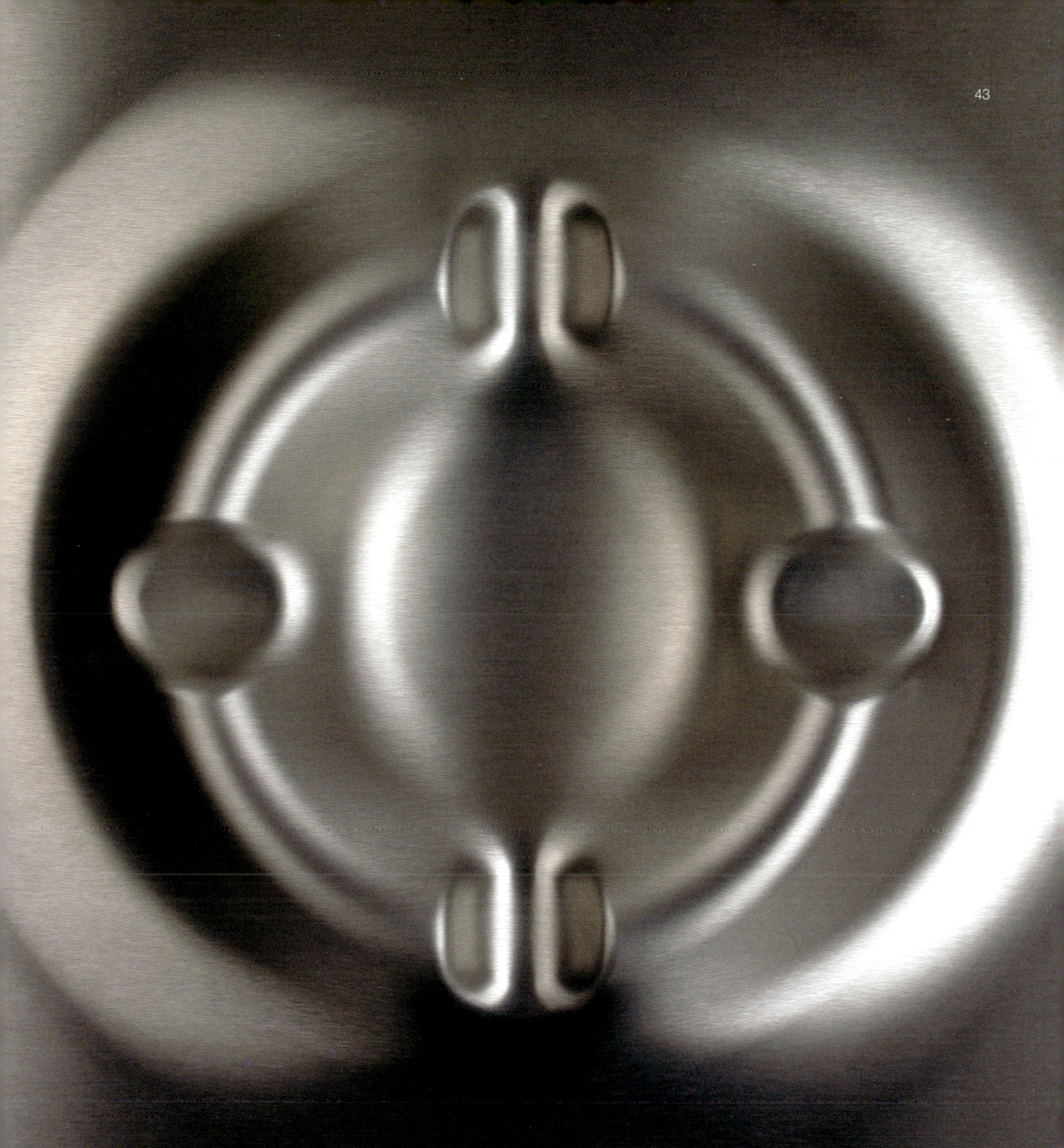

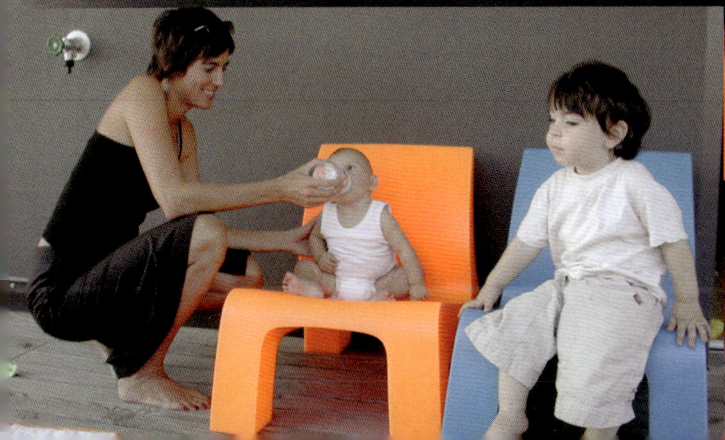

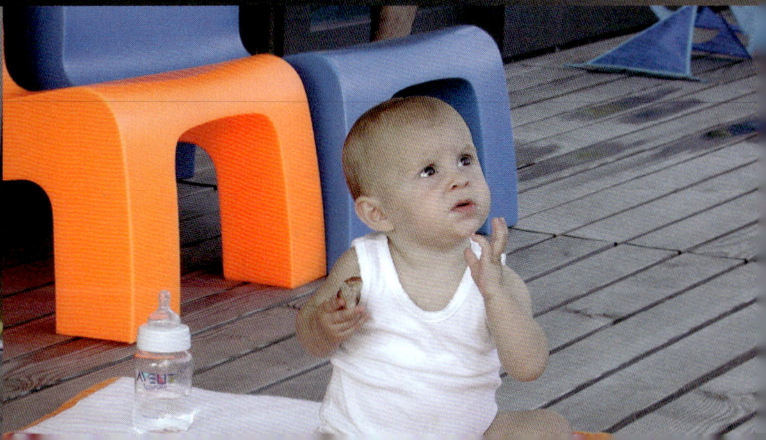

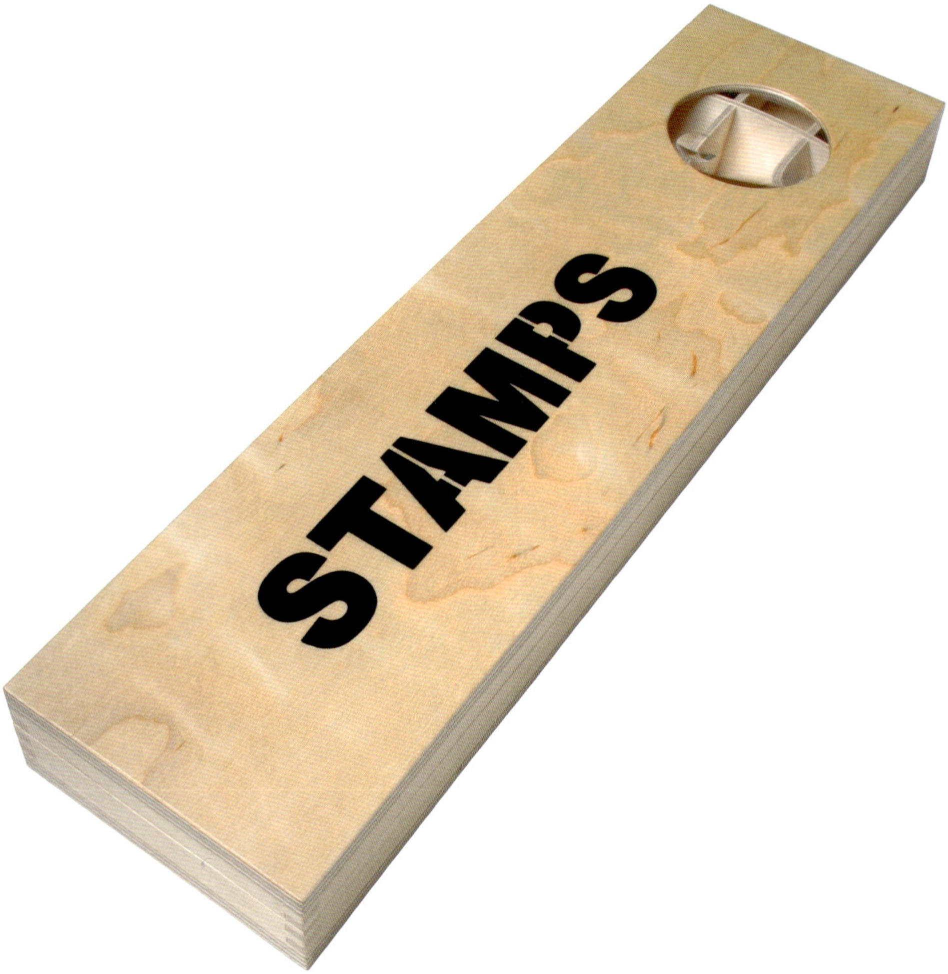

Stamps box — 2001

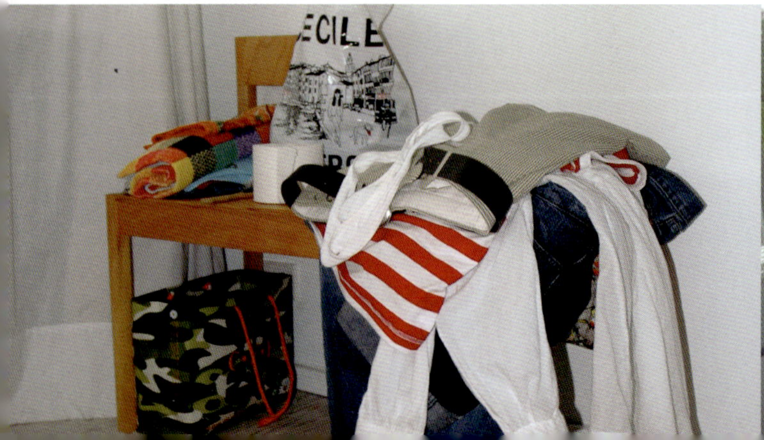

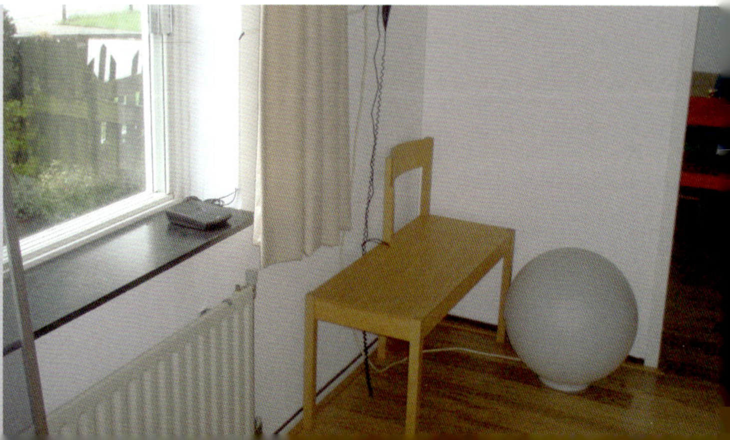

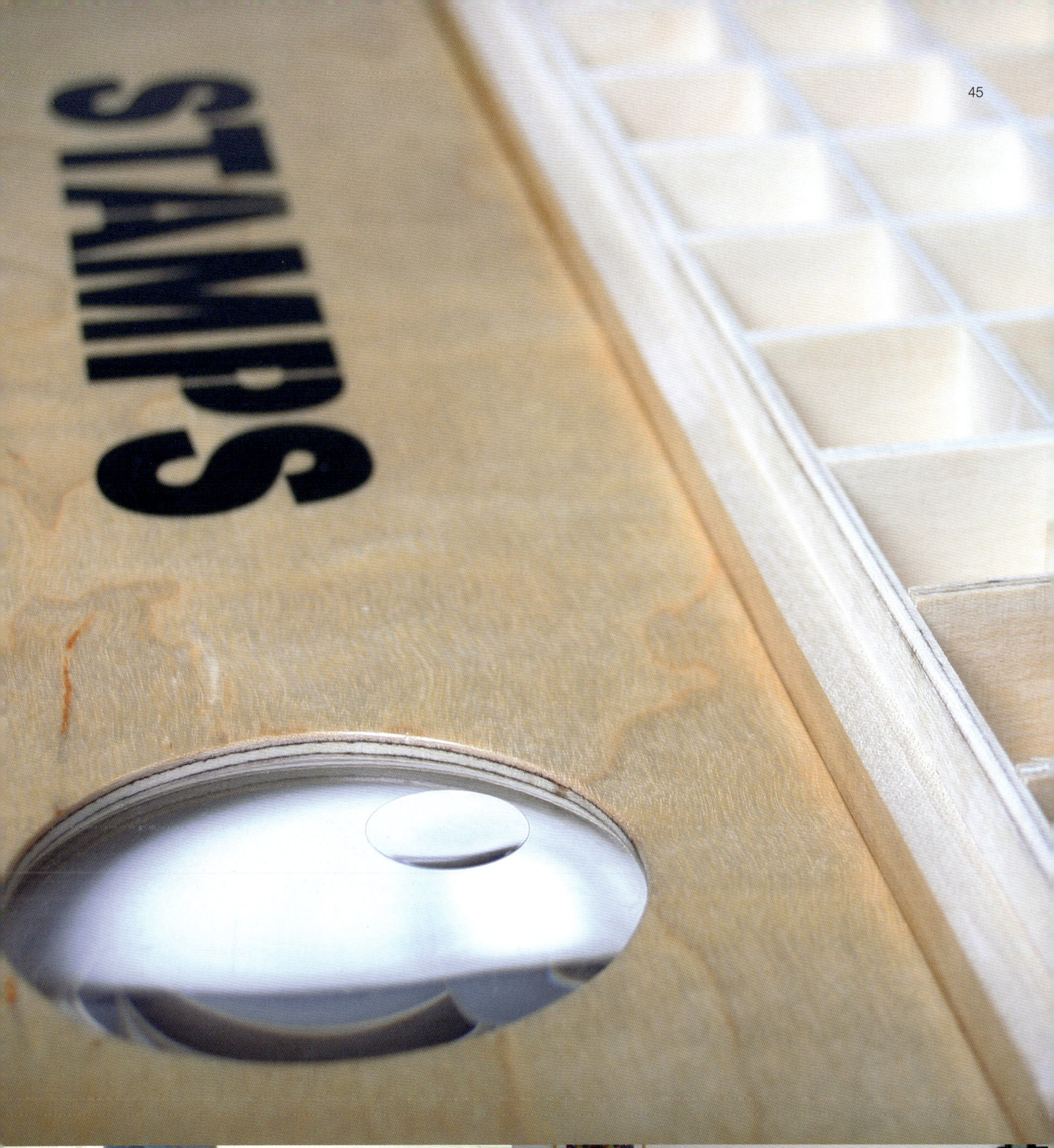
STAMPS

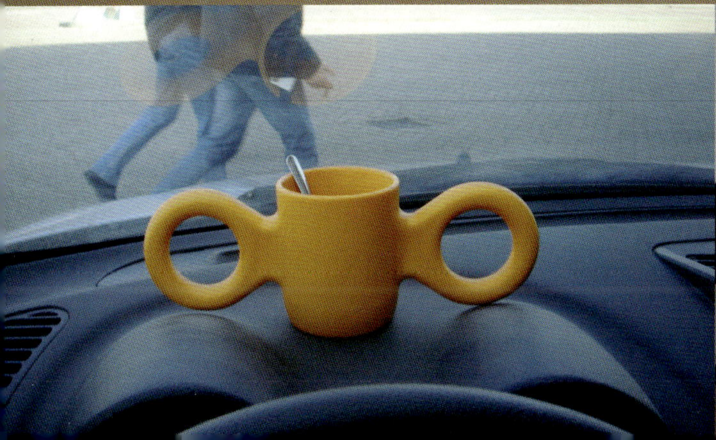

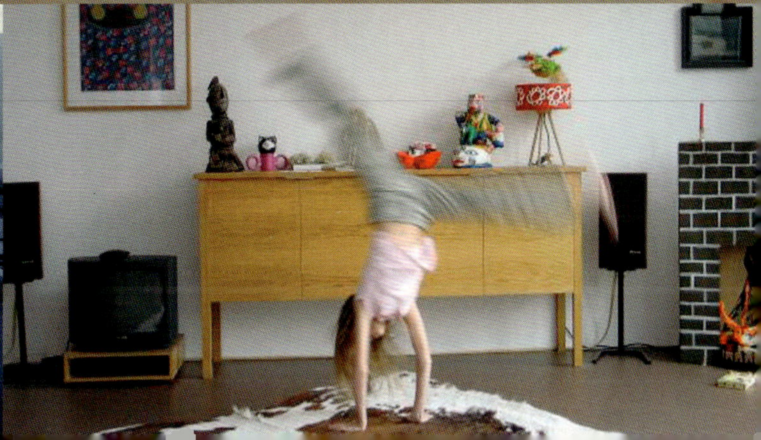

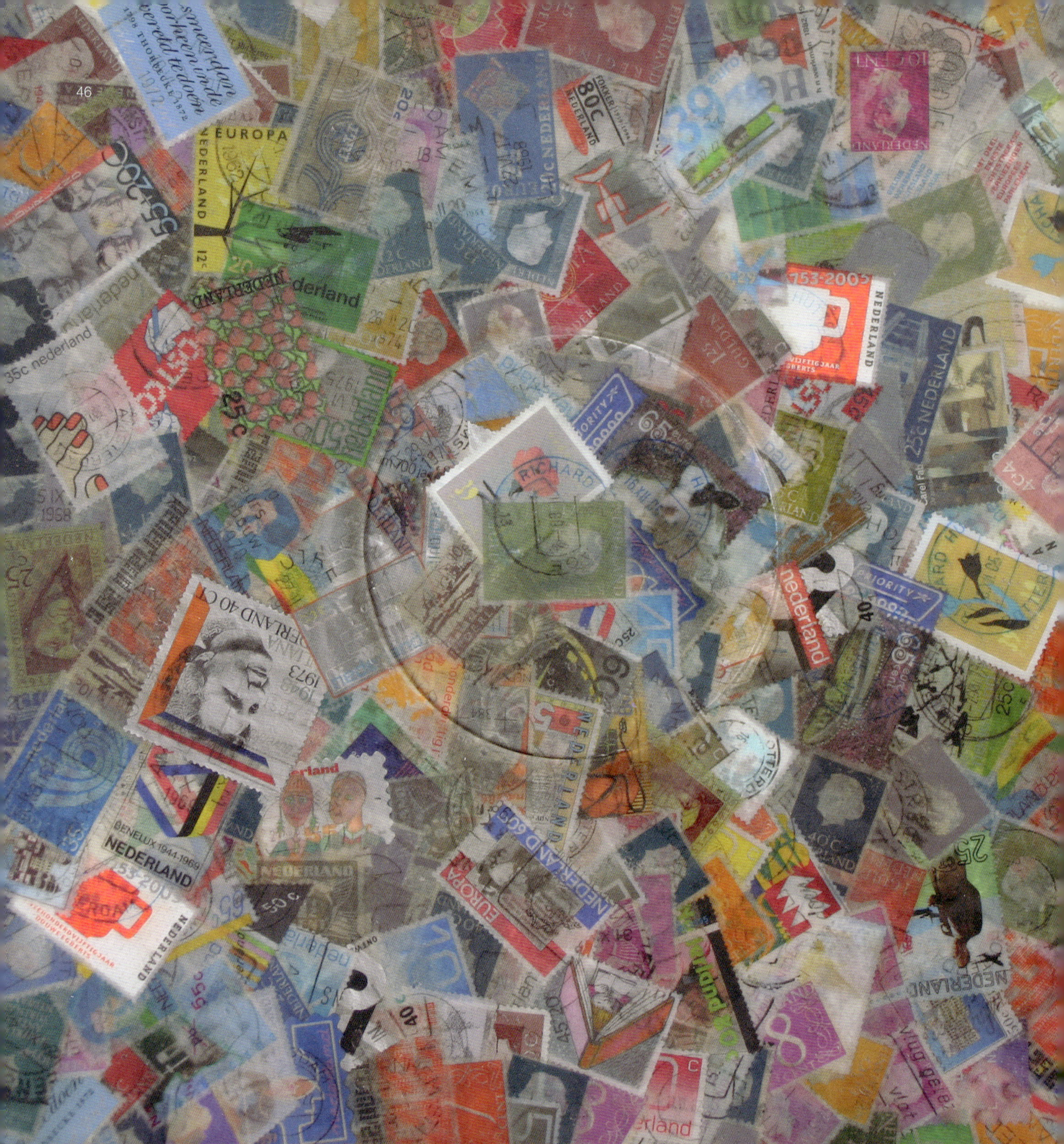
EUROPA
NEDERLAND
35c nederland
nederland
25c NEDERLAND
BENELUX 1944-1969
NEDERLAND
EUROPA
NEDERLAND
PRIORITY
NEDERLAND

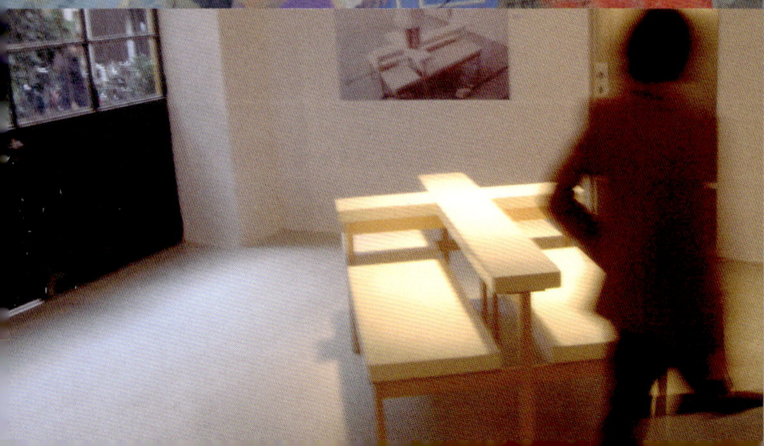

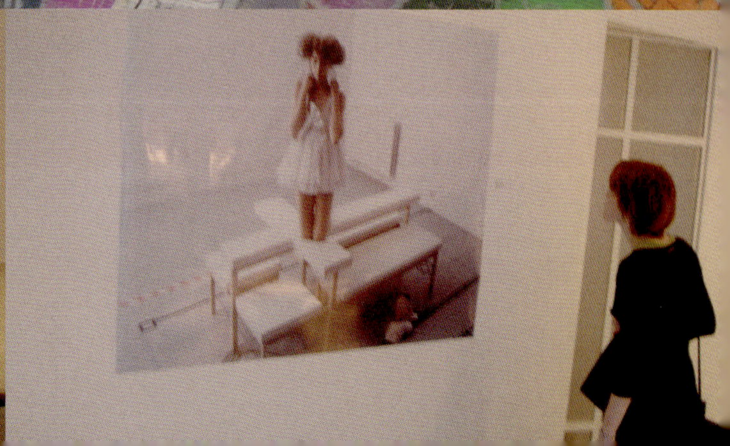

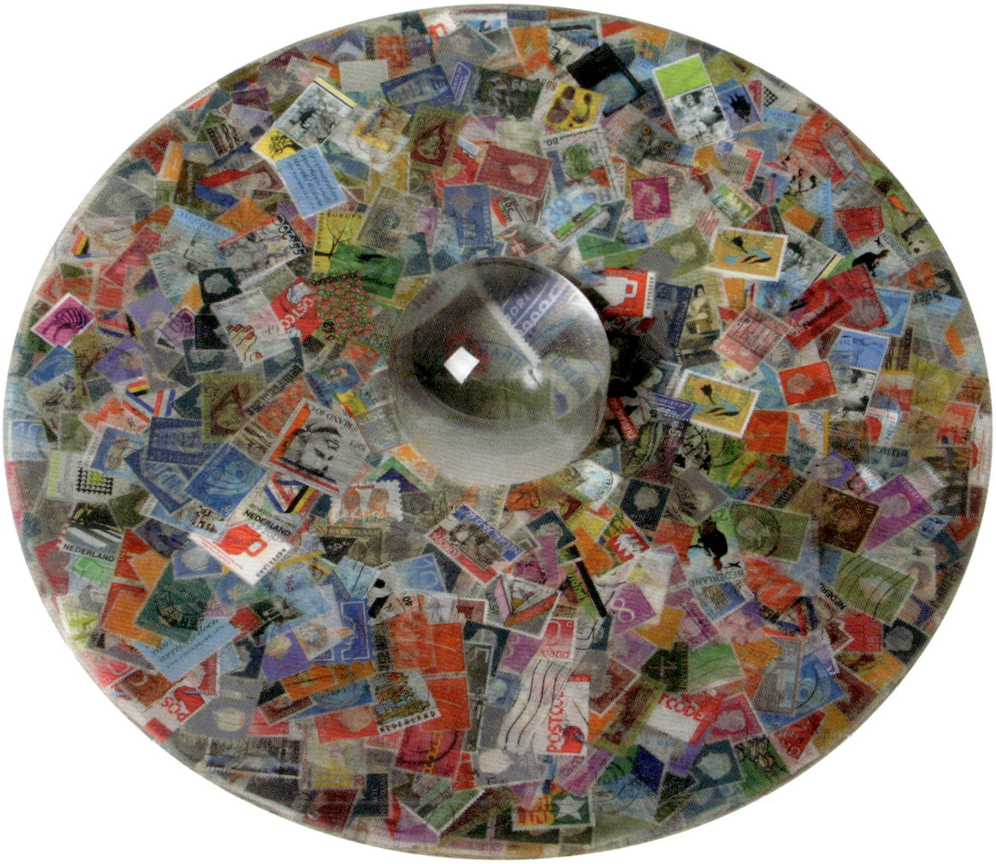

Stamp-plate — 2005

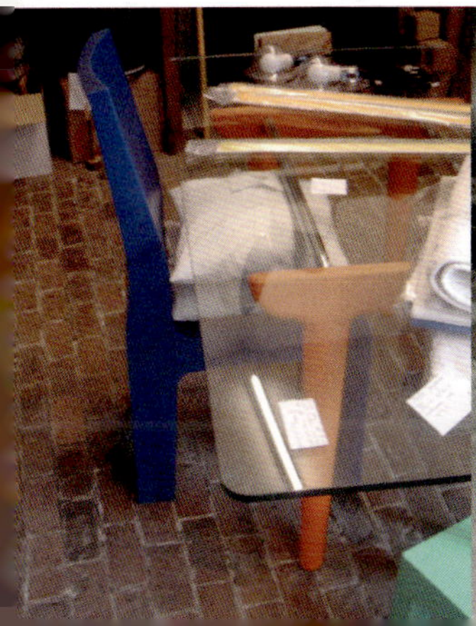

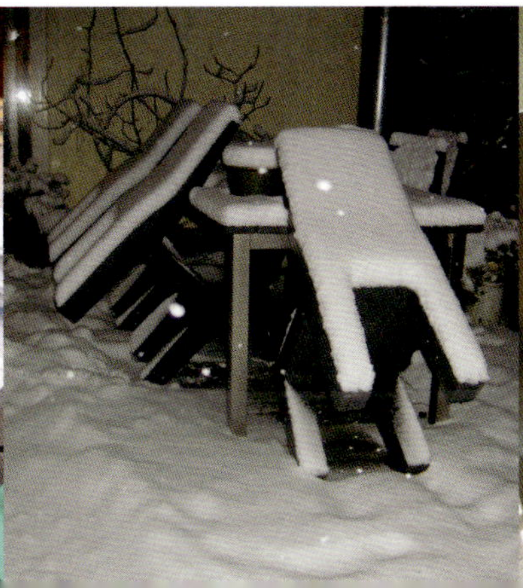

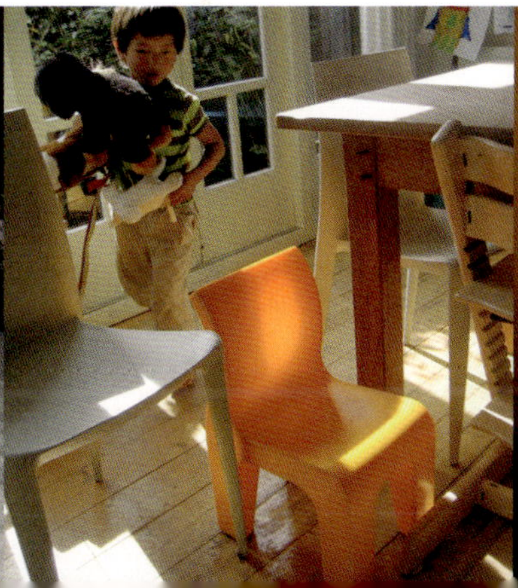

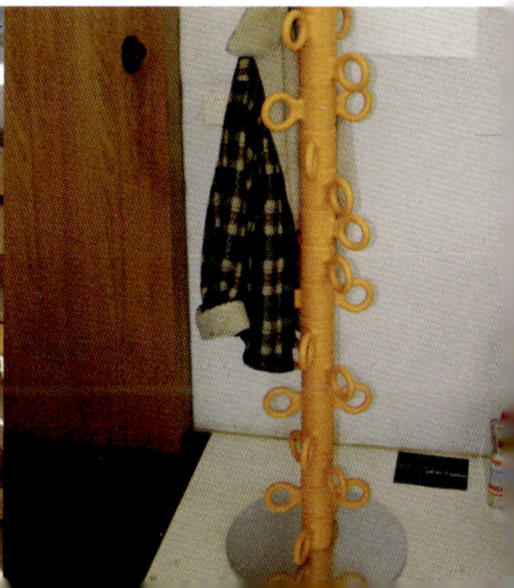

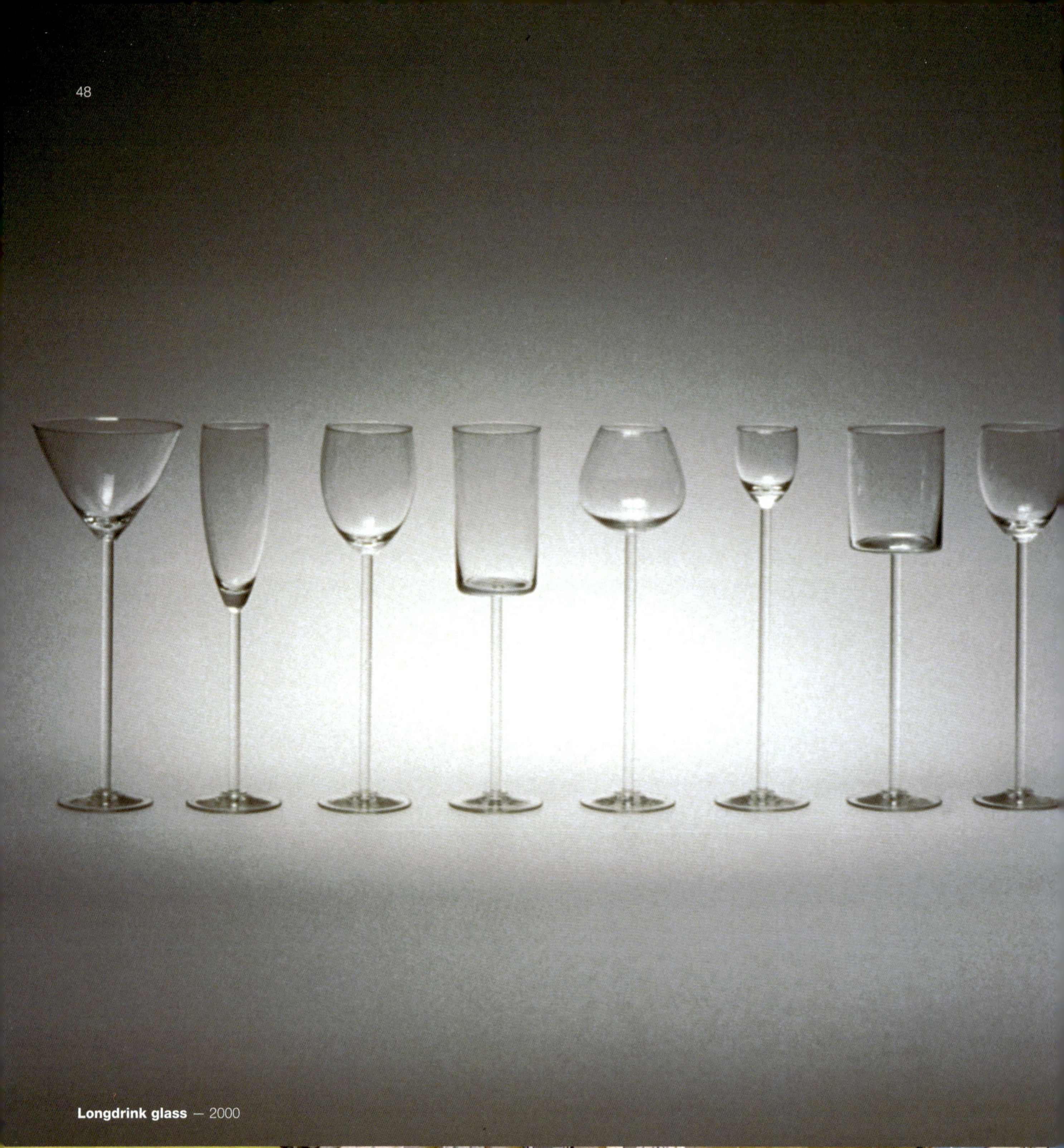

Longdrink glass — 2000

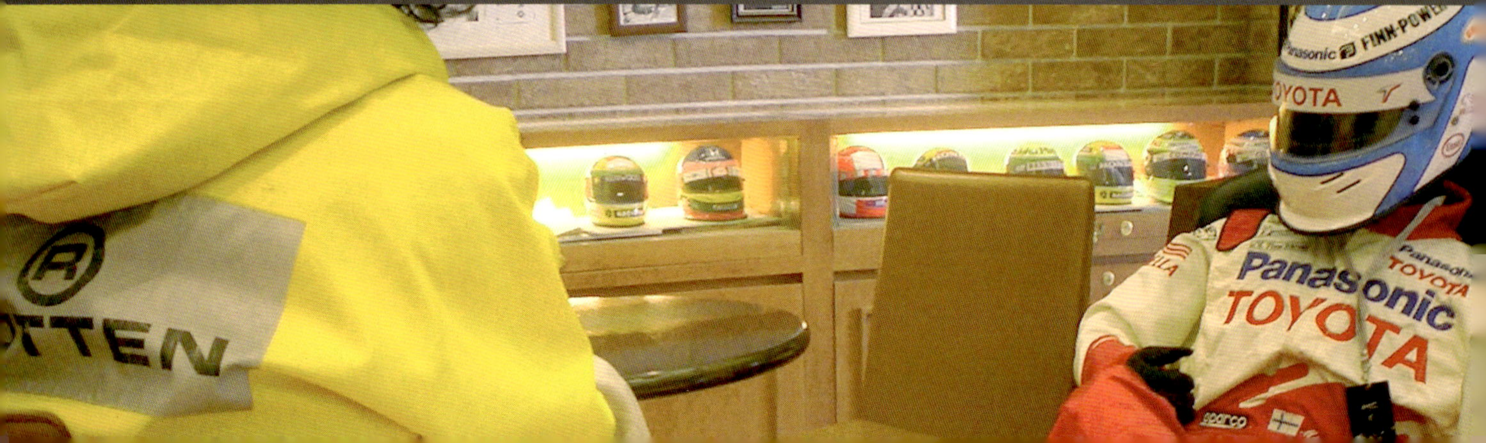

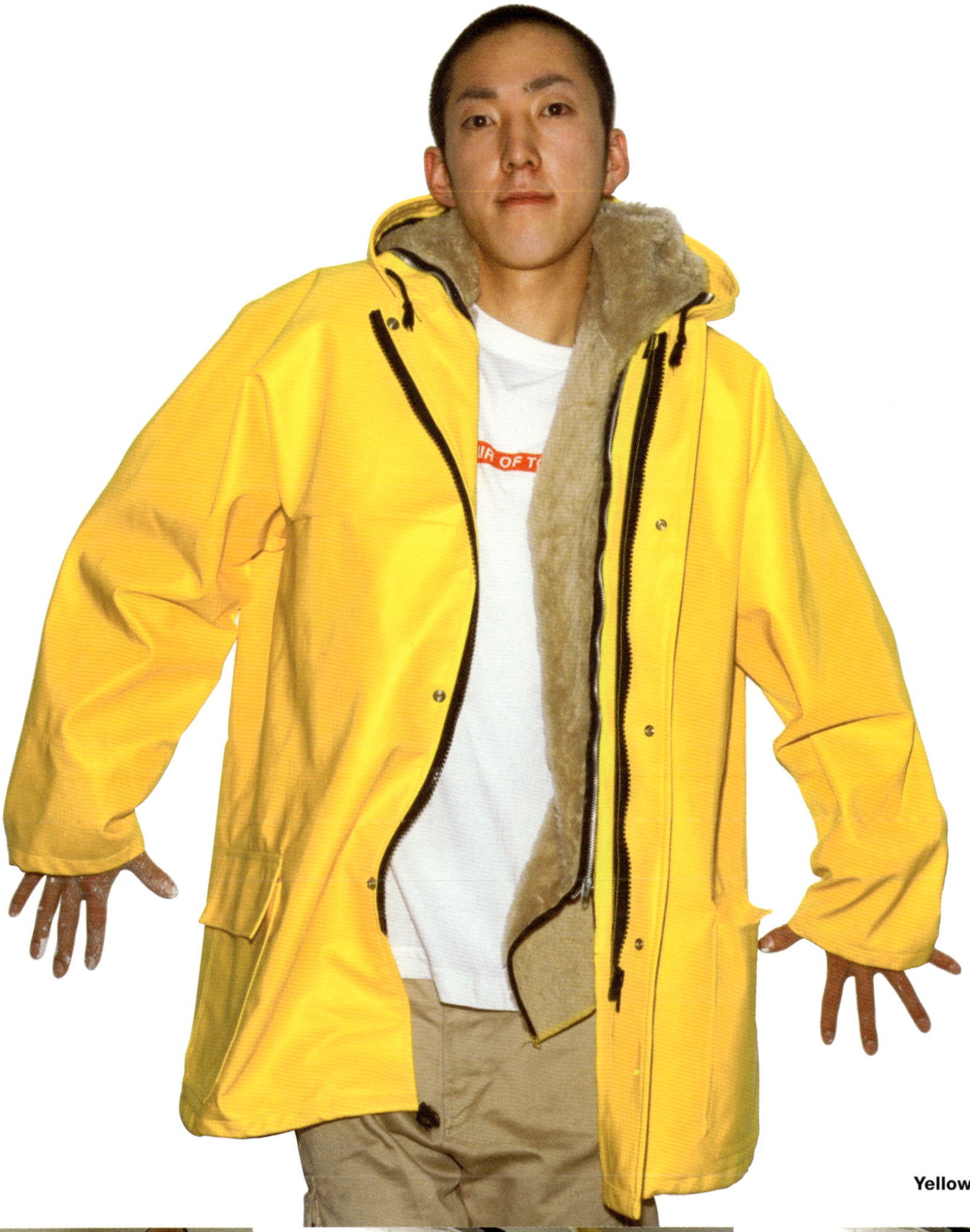

Yellow Jacket — 1996

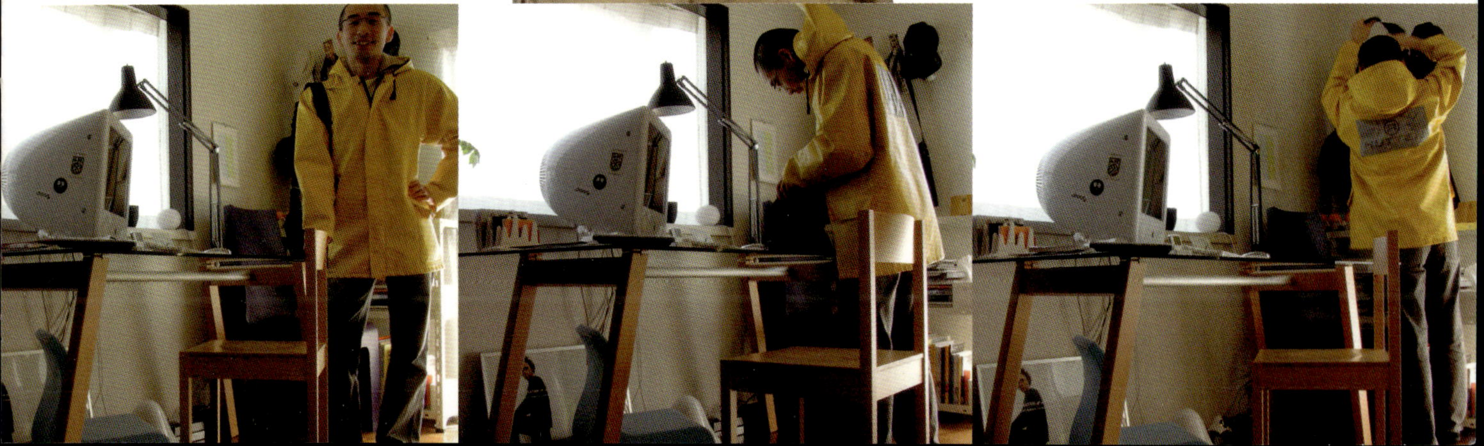

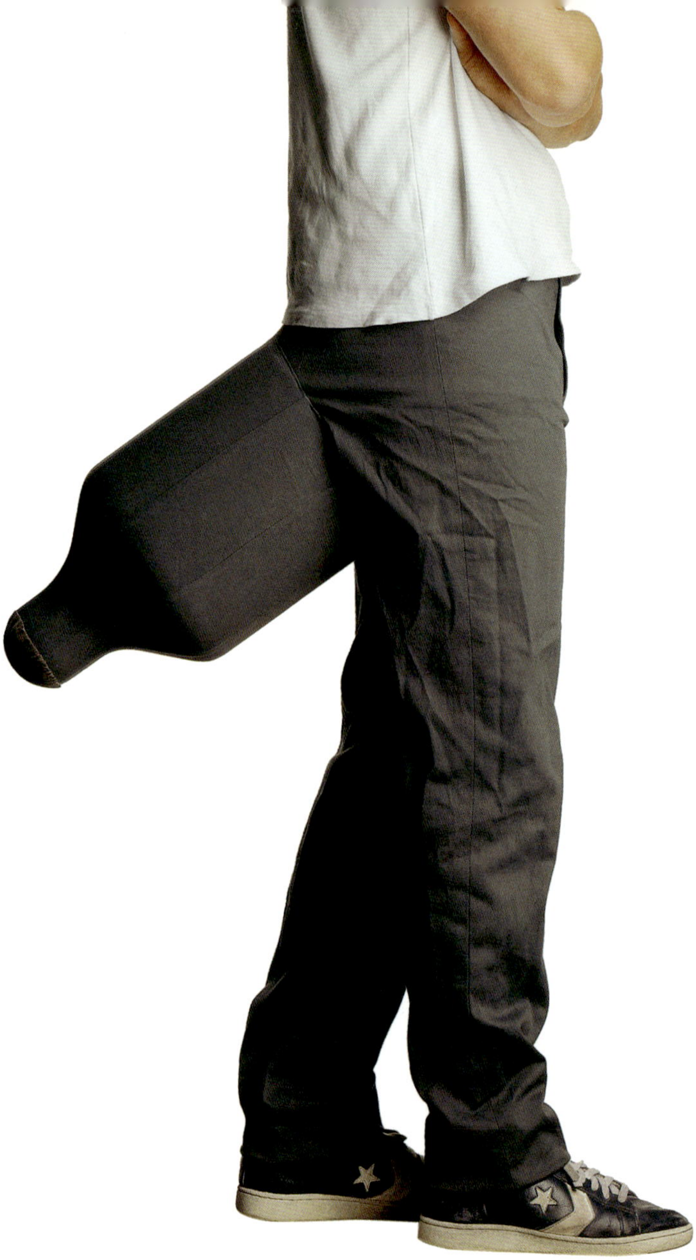

Stoolpants — 2002

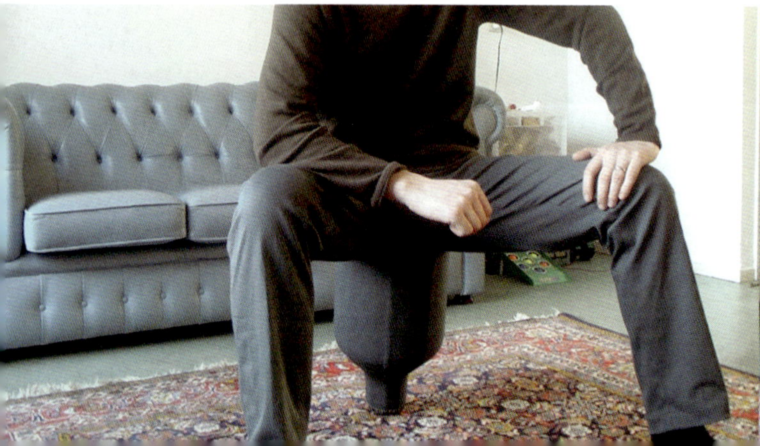

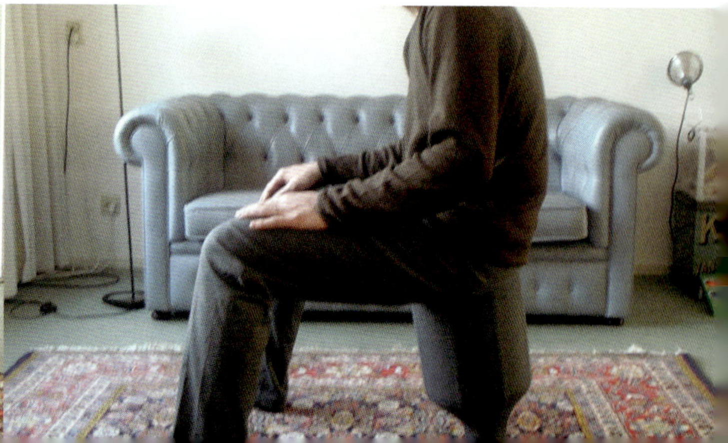

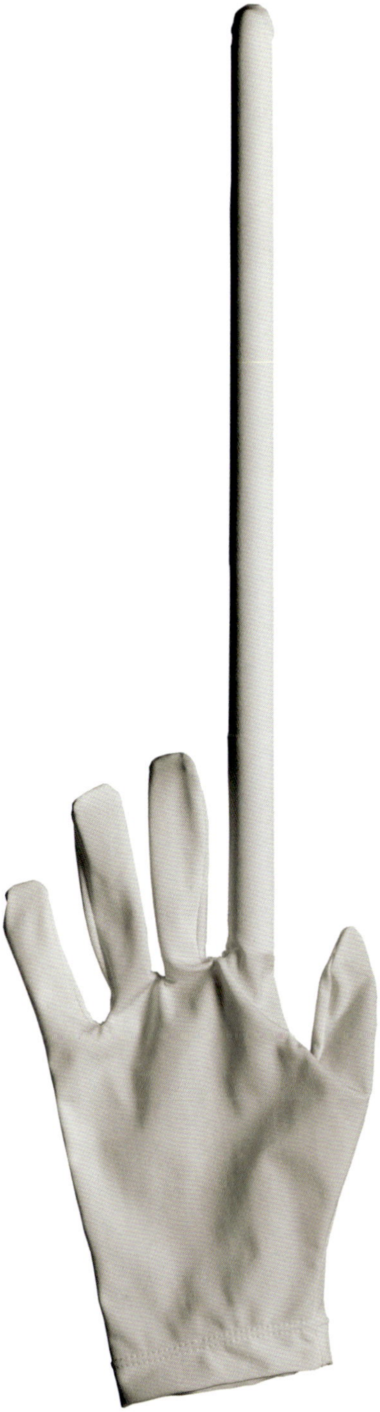

Glove-batton — 2003

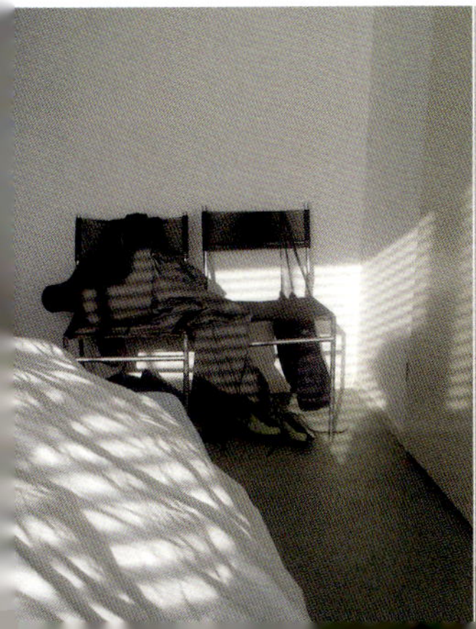

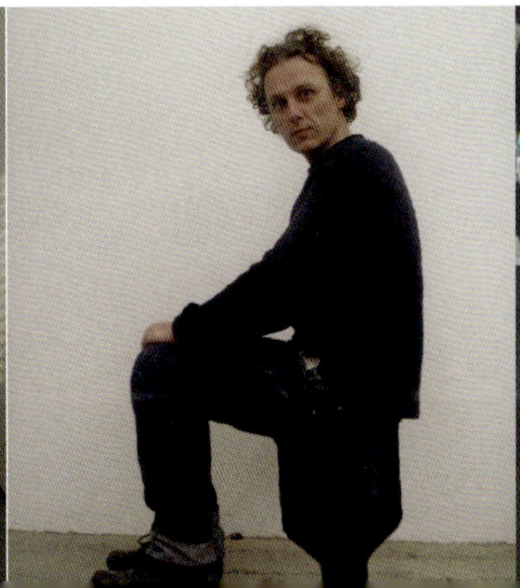

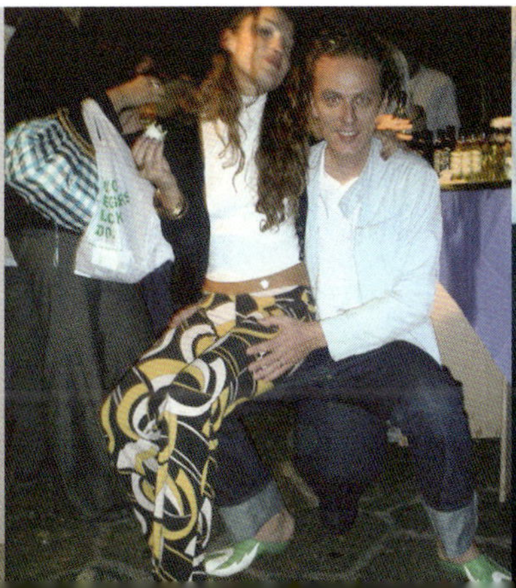

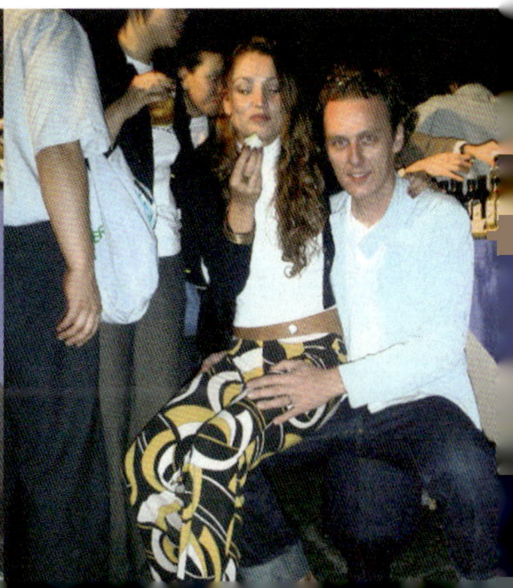

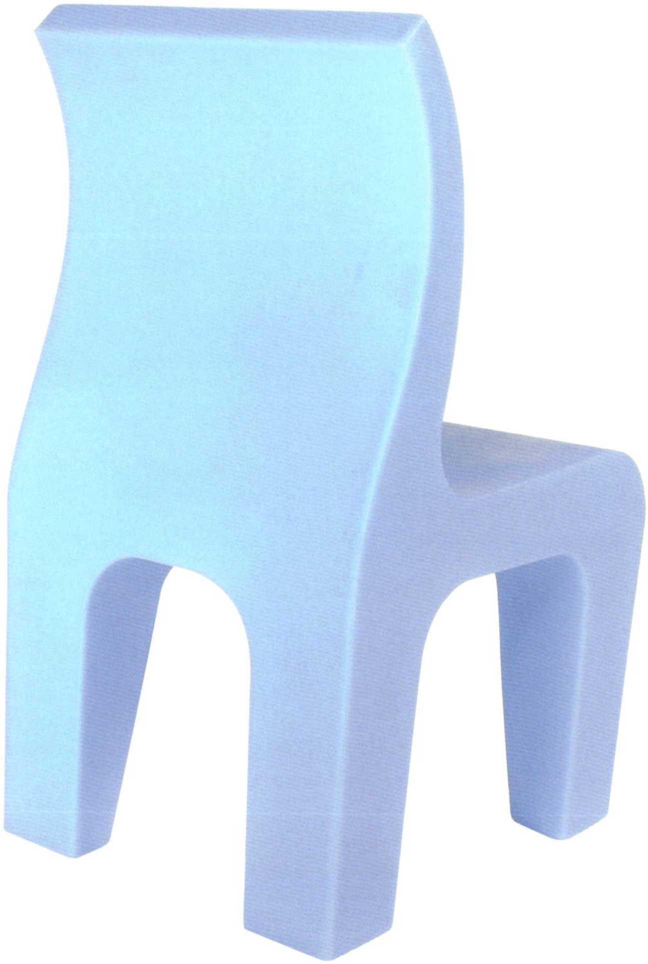

Bronto chair — 2004

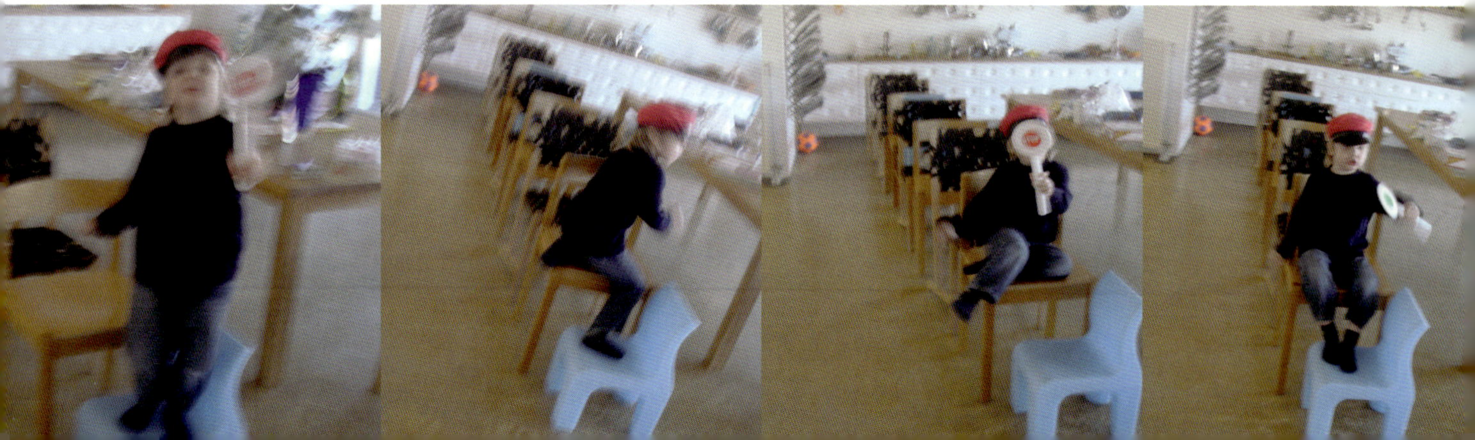

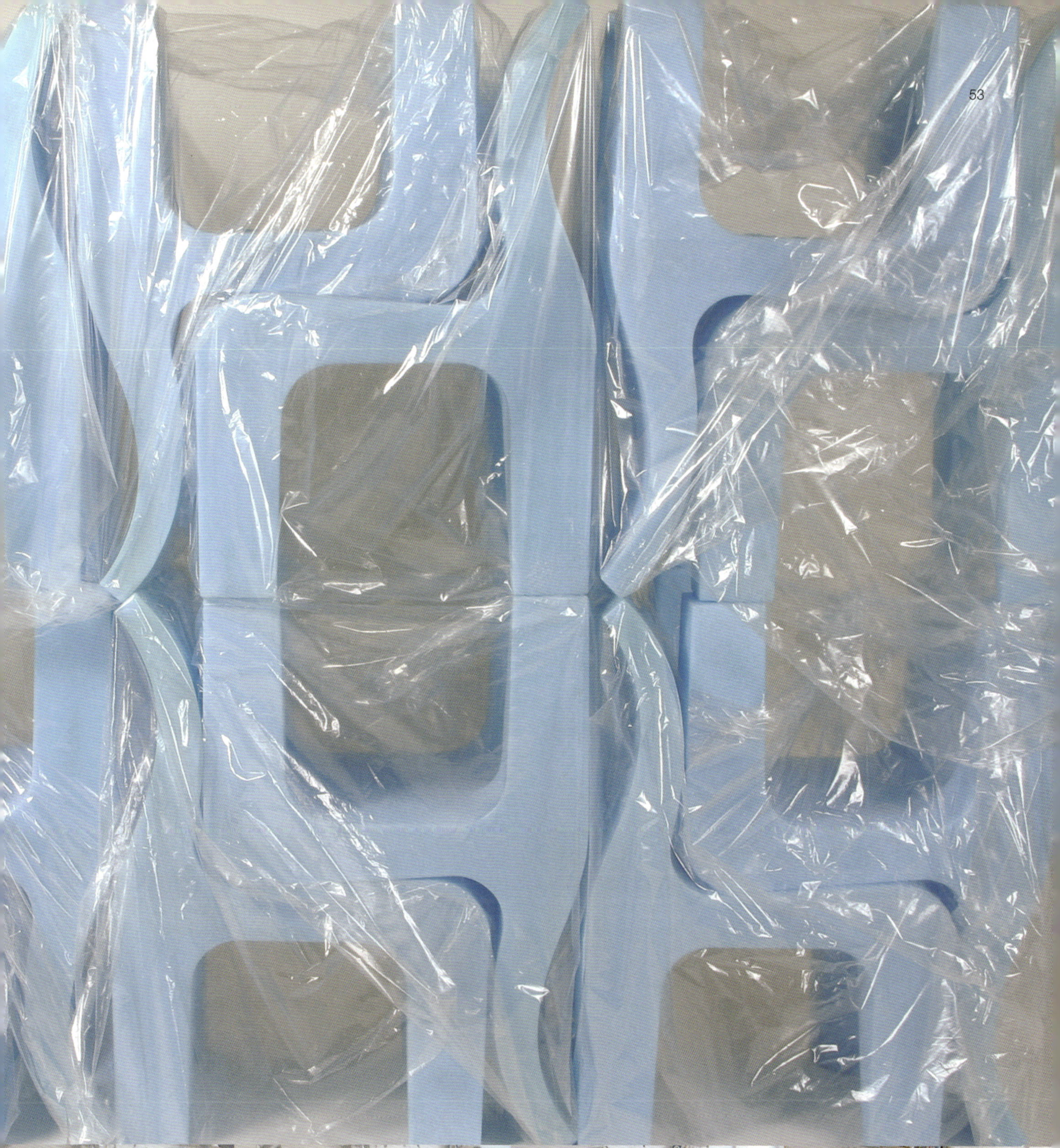

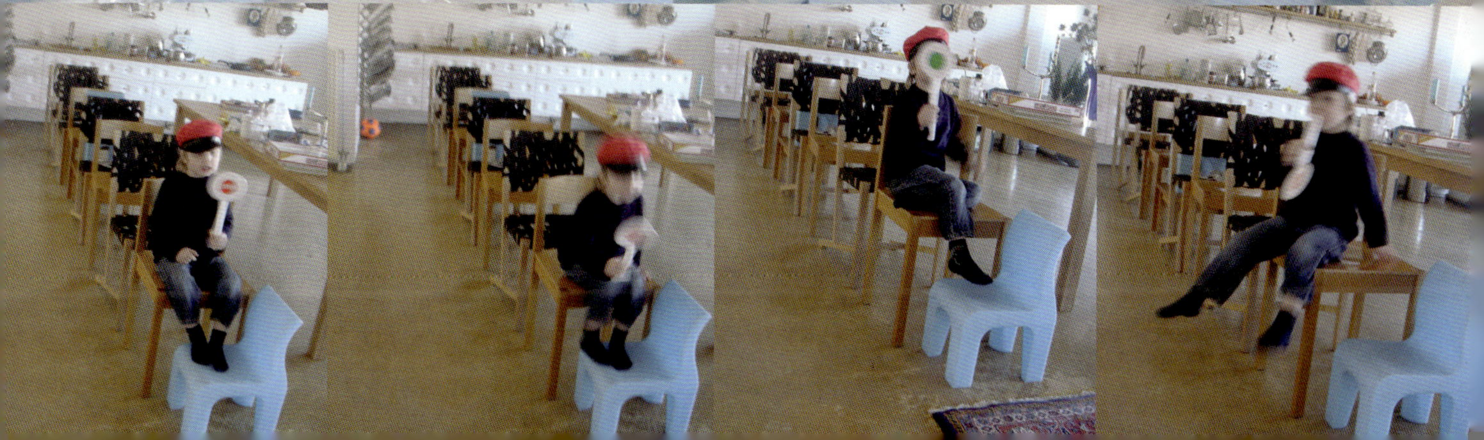

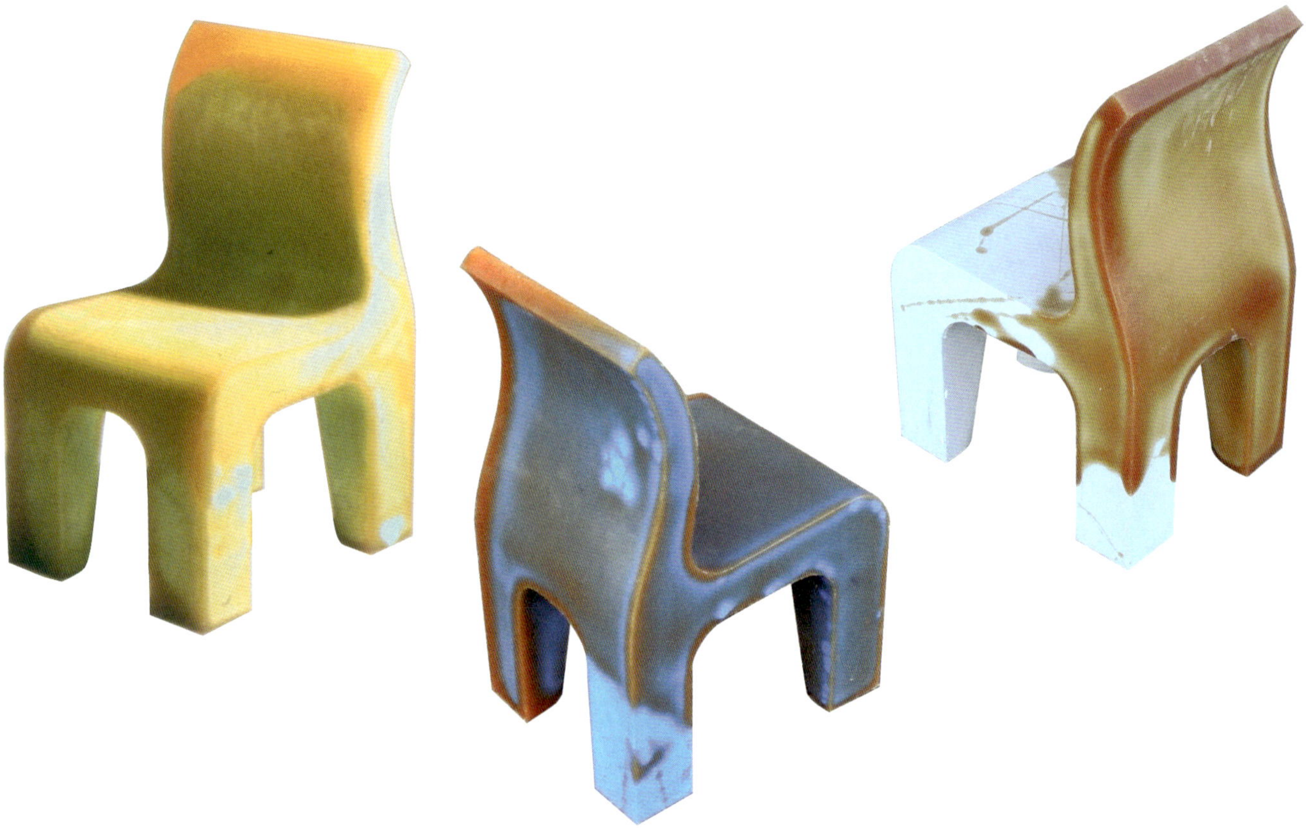

Bronto constructionless chair — 1996

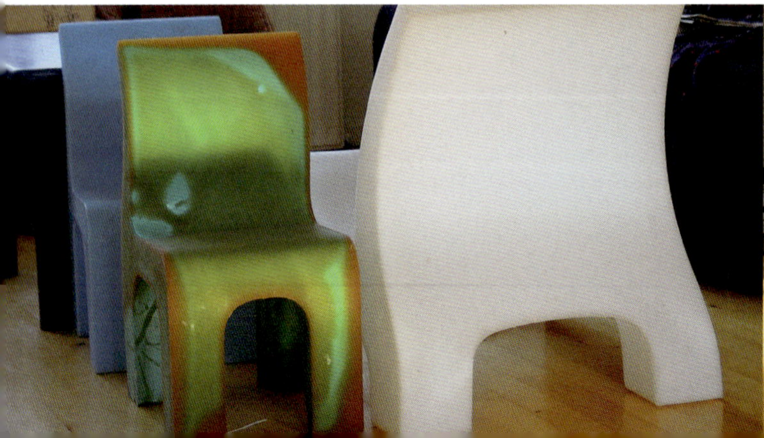

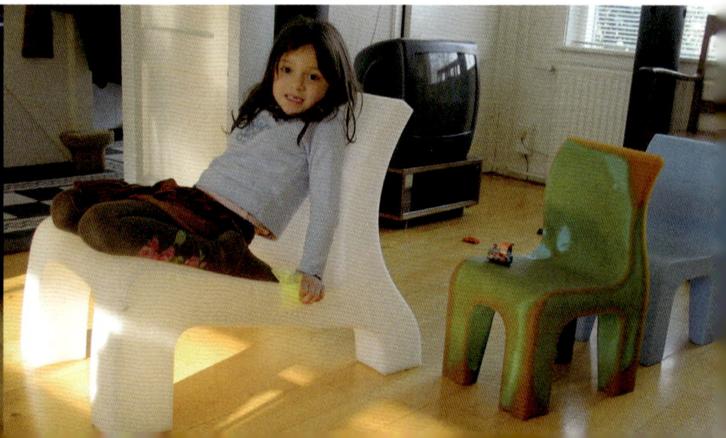

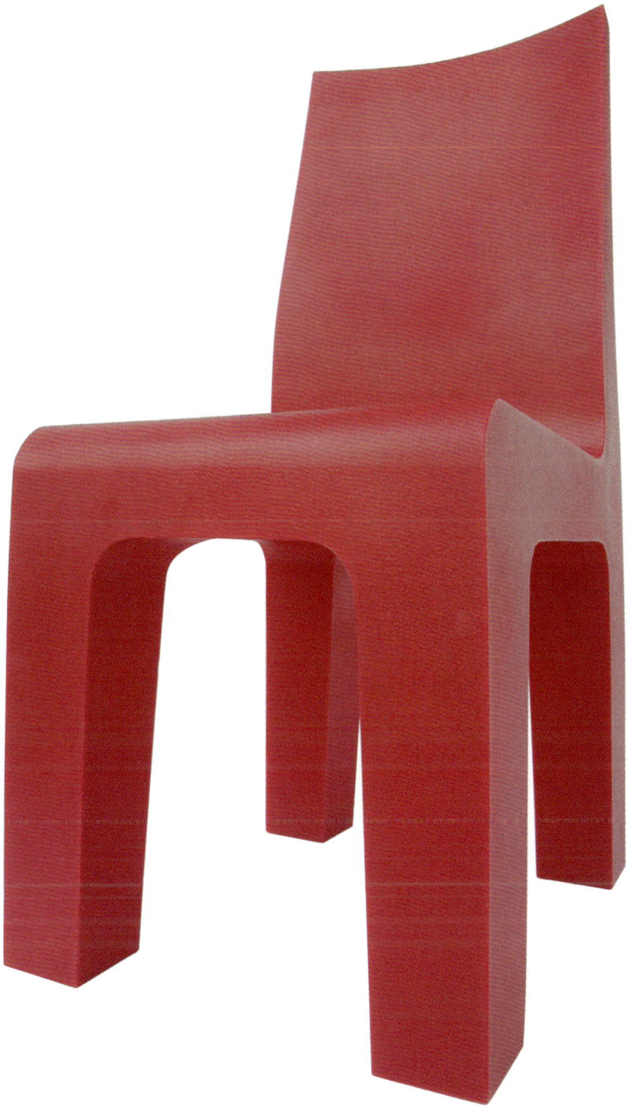

Centraal Museum chair — 1998

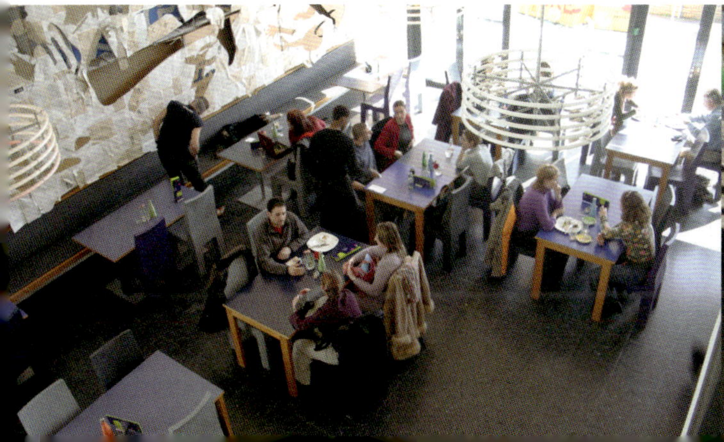

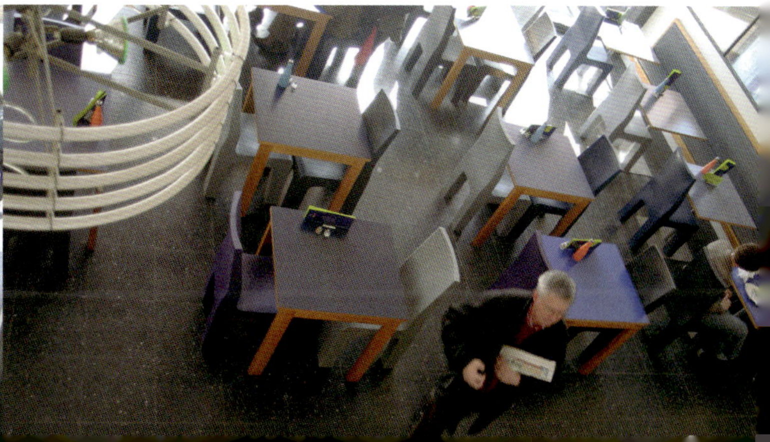

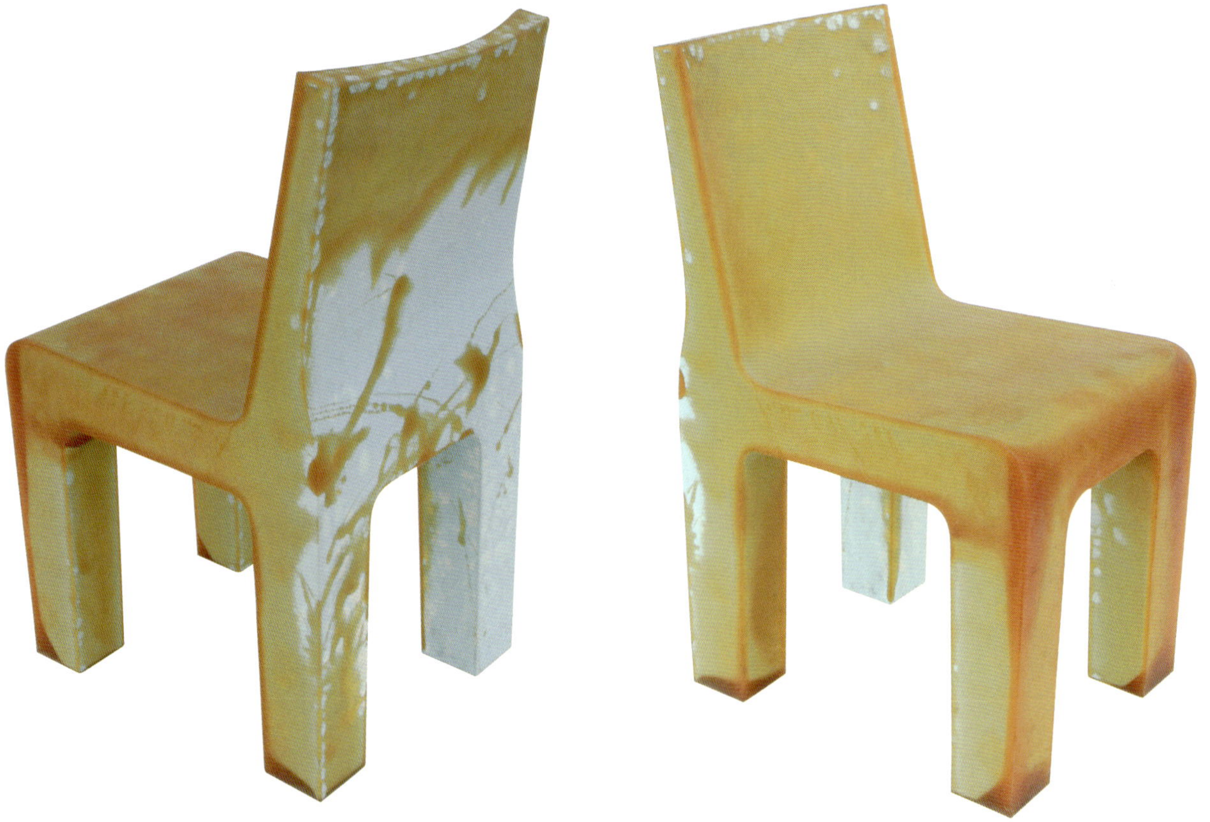

Constructionless chair — 1996

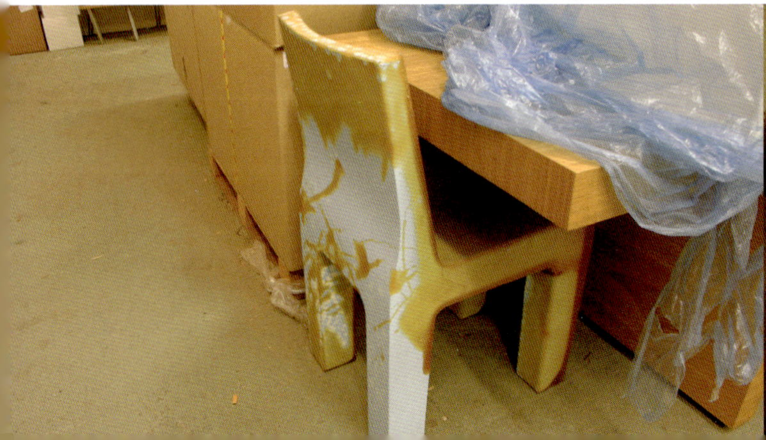

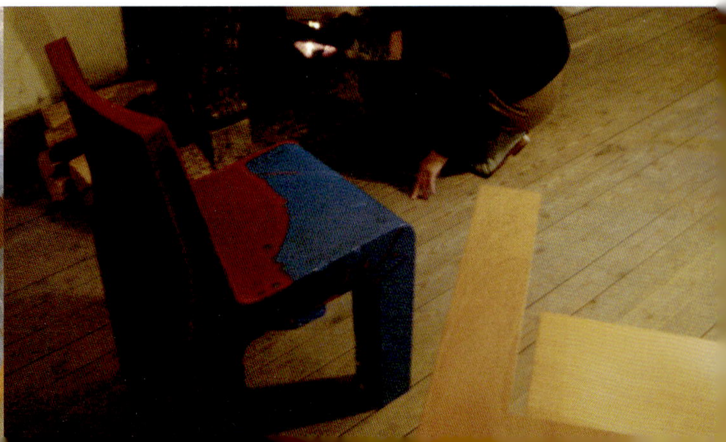

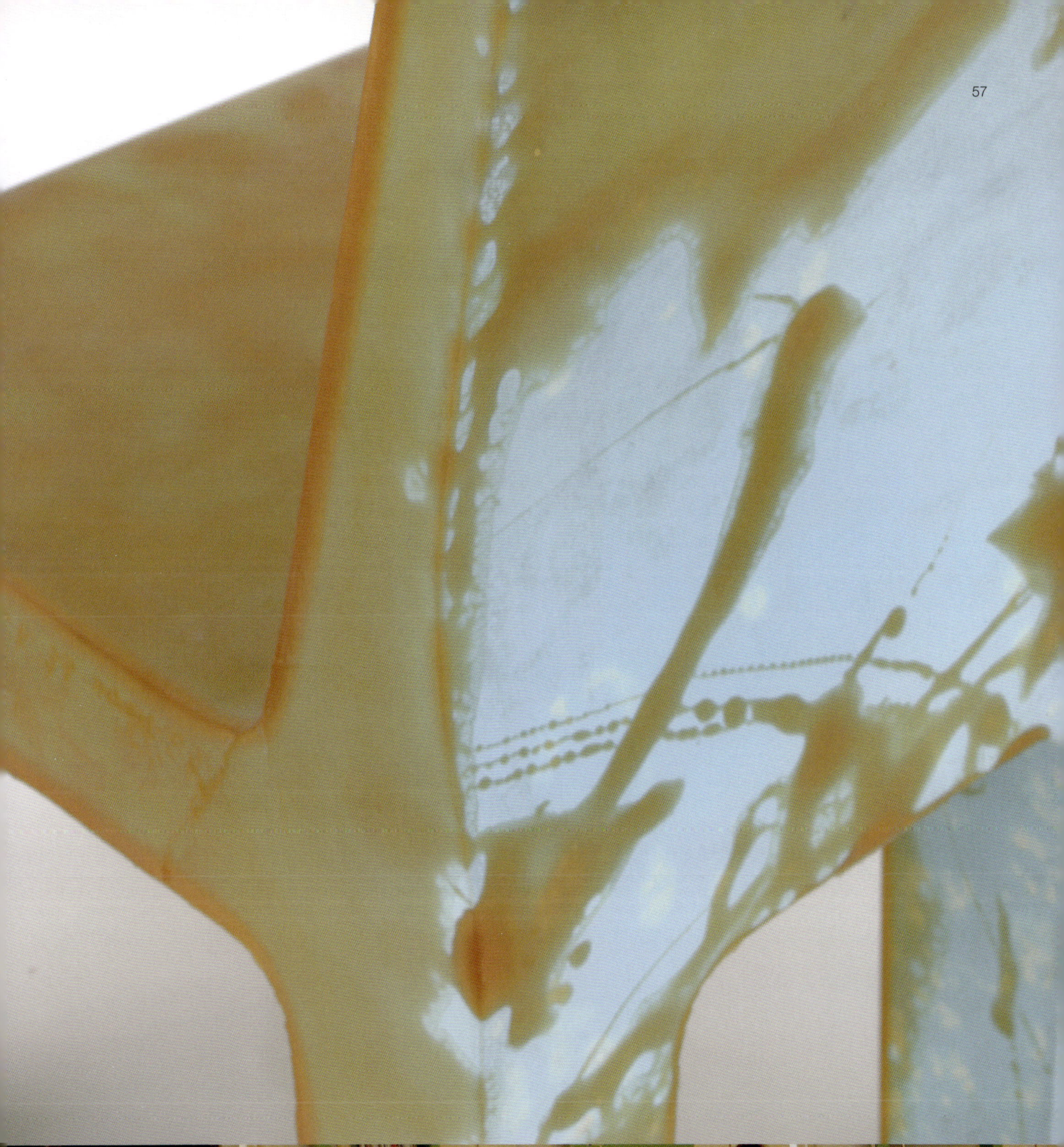

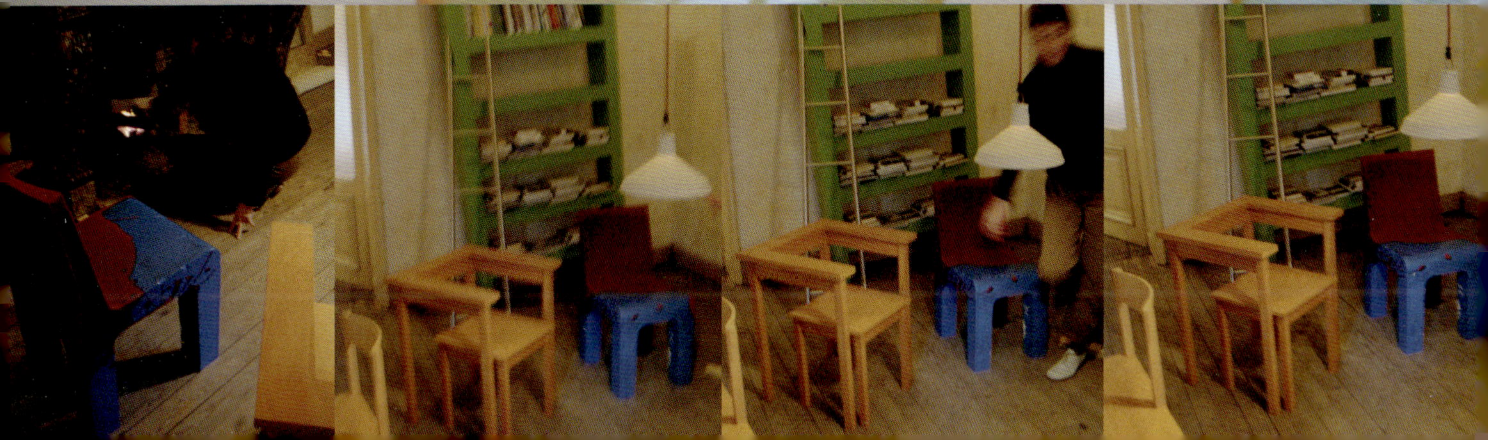

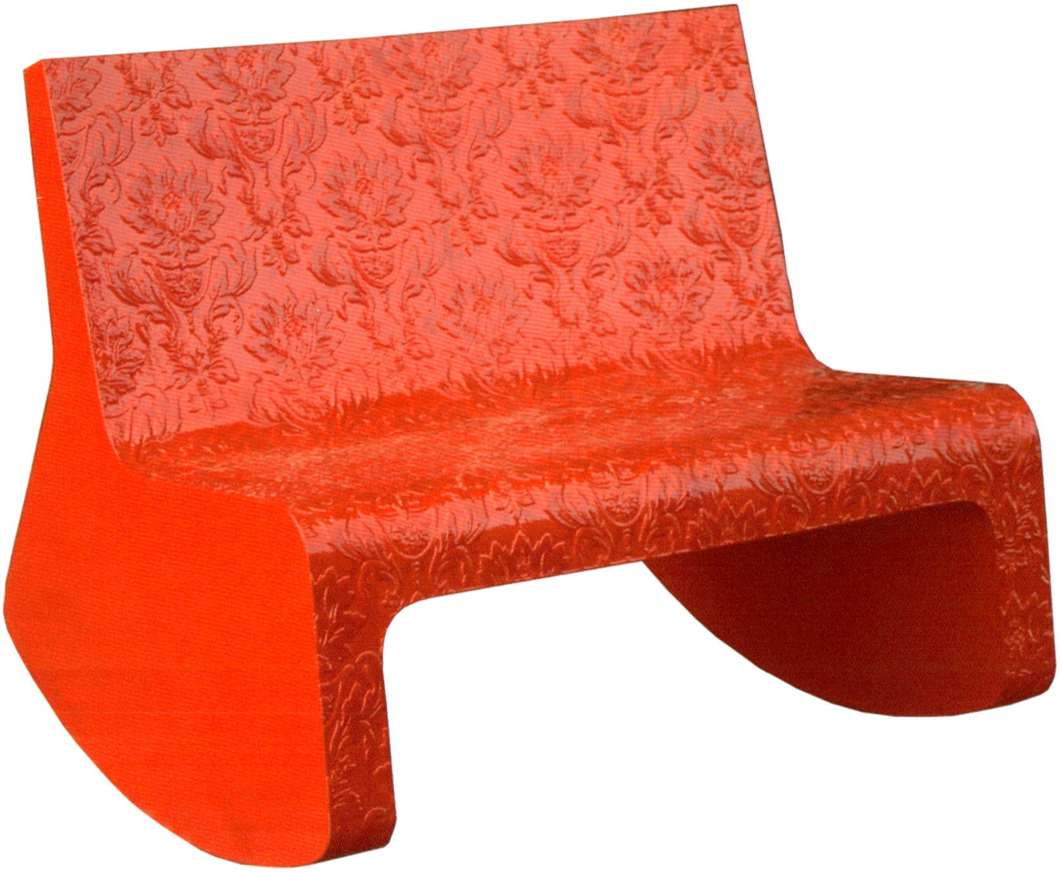

2 Loveseat — 2003

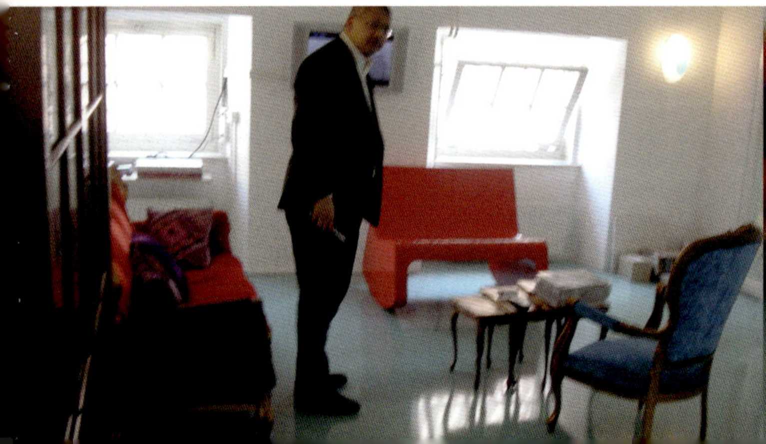

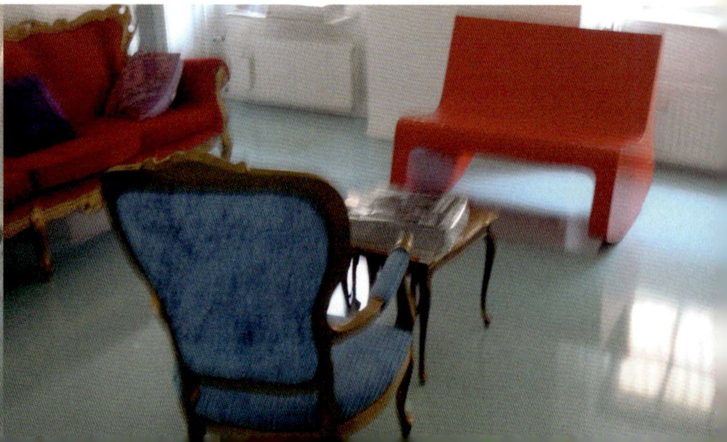

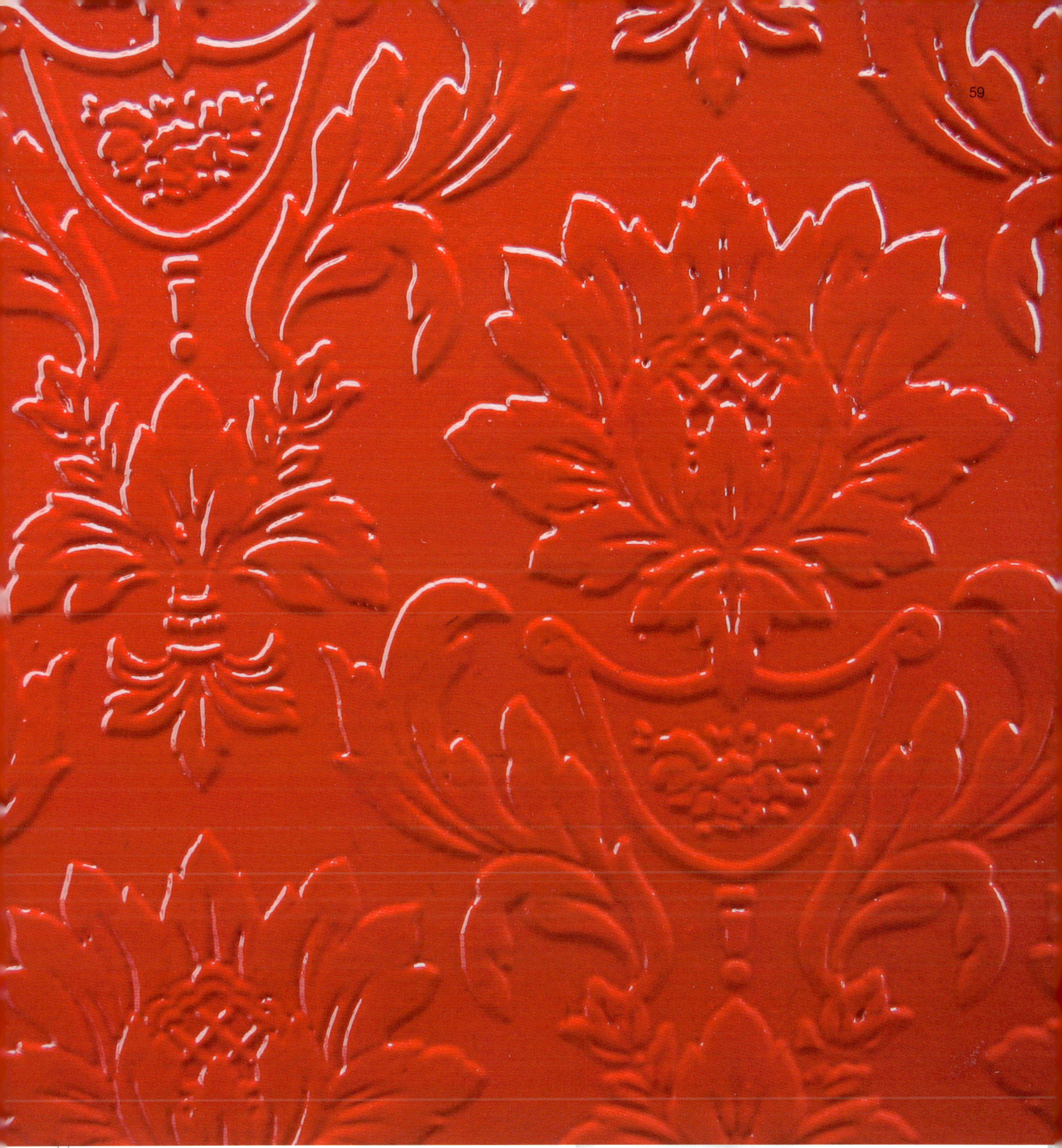

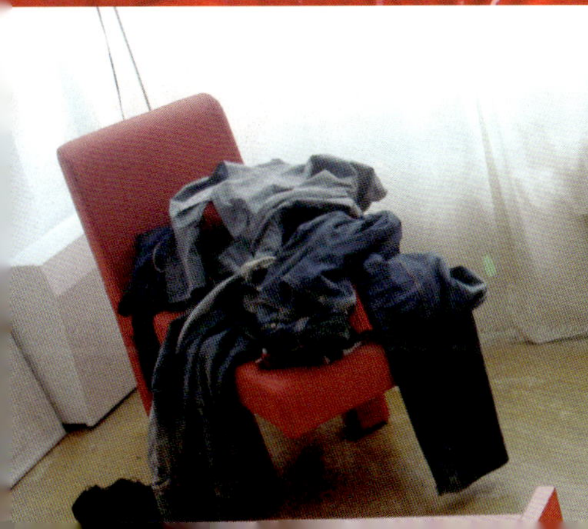

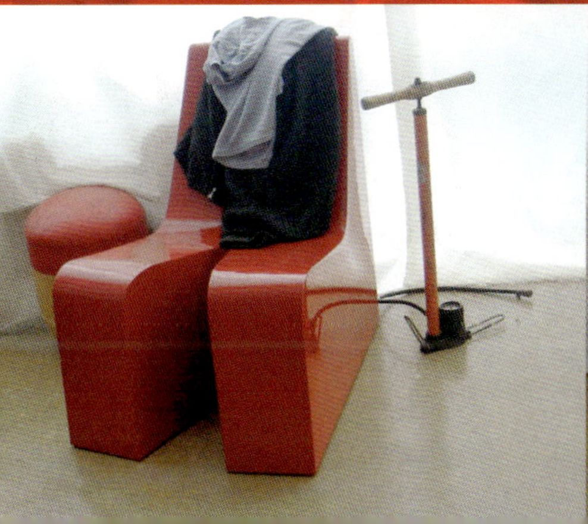

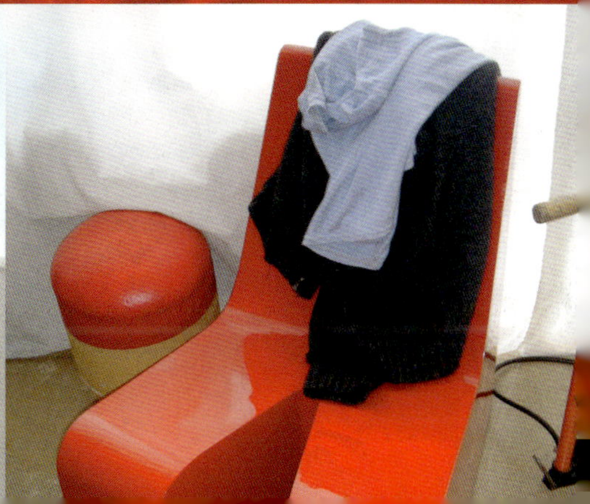

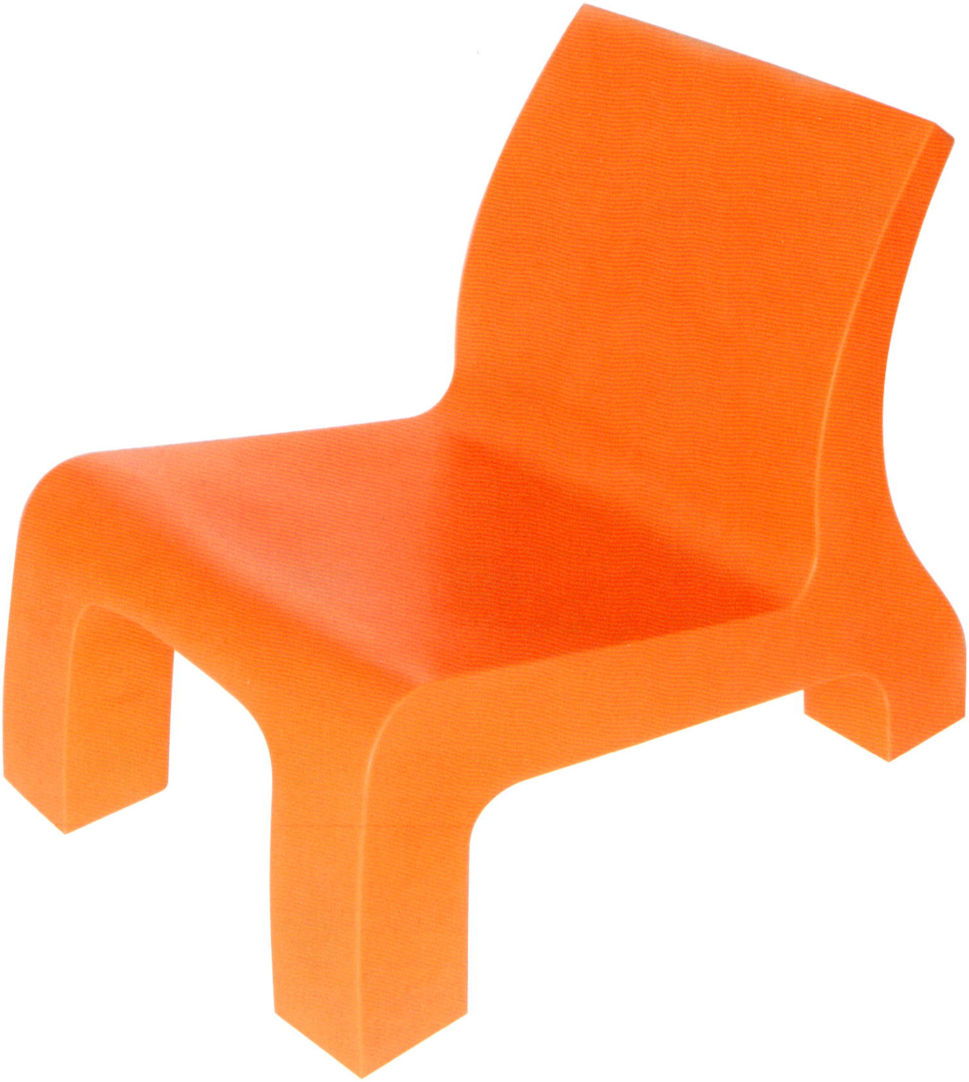

Rhino chair — 2001

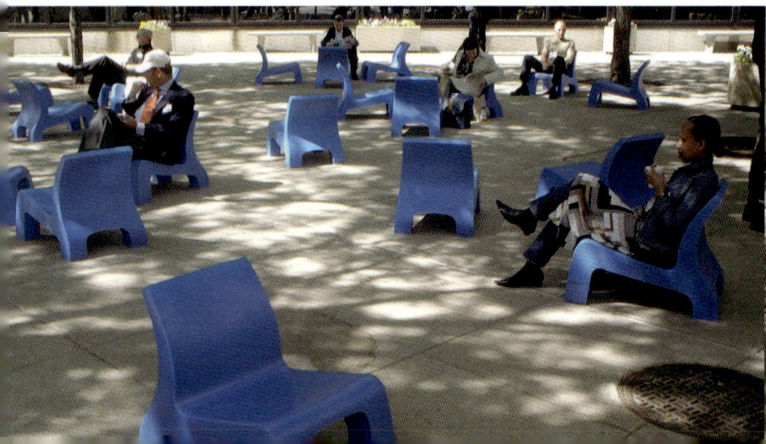

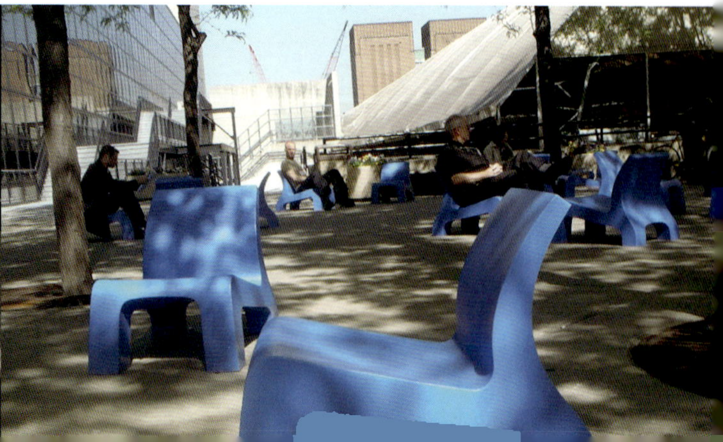

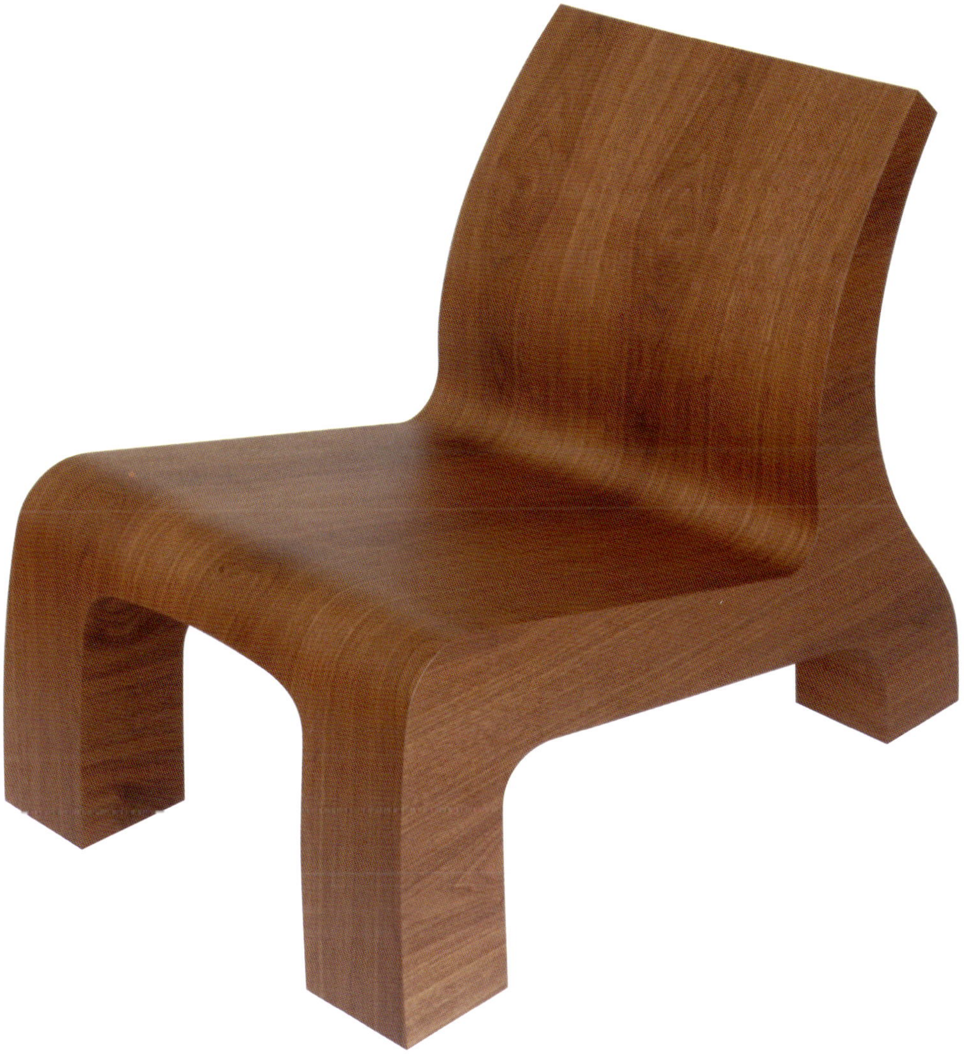

Rhino walnut — 2001

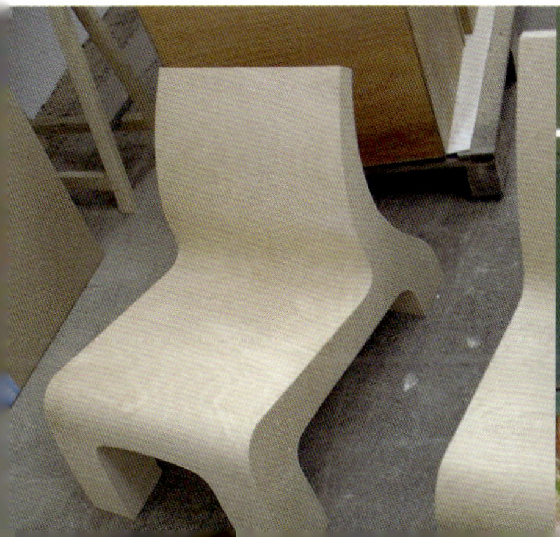

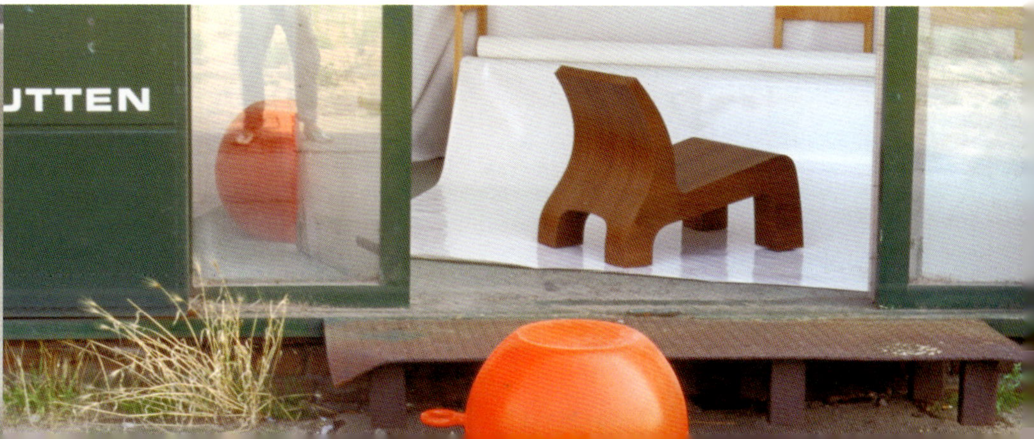

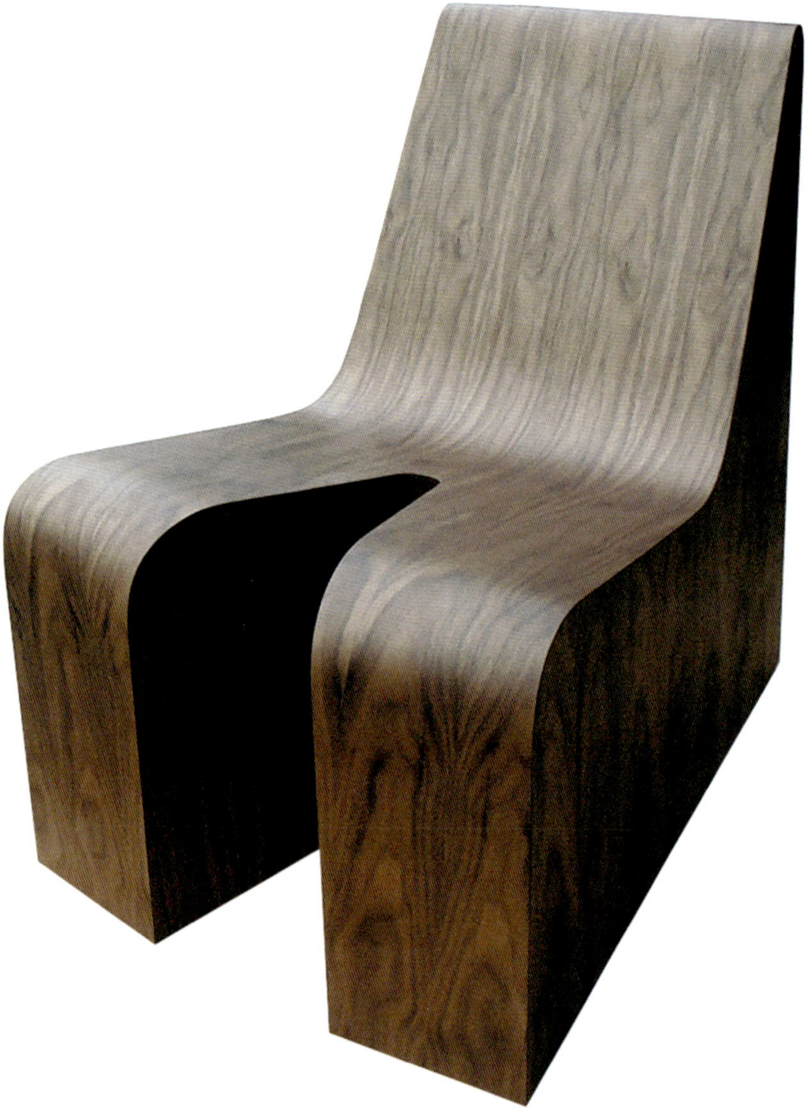

Sexy Relaxy walnut — 2001

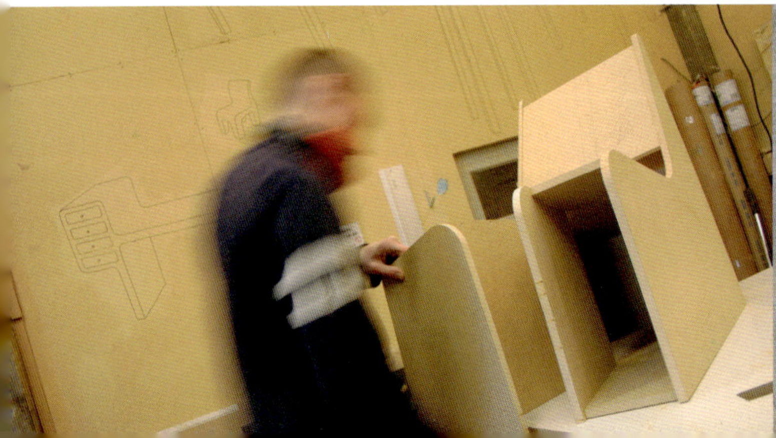

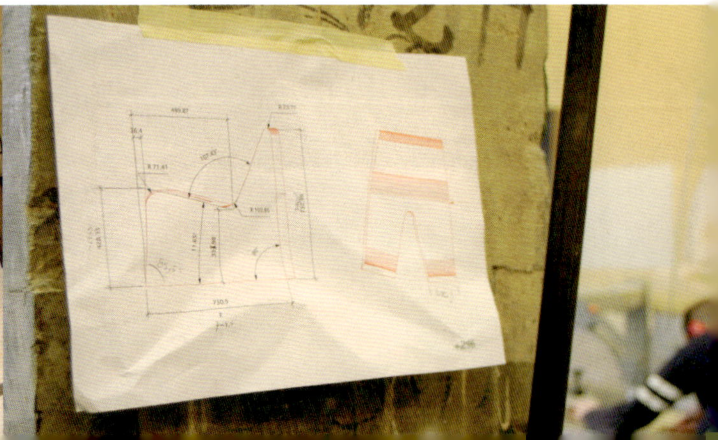

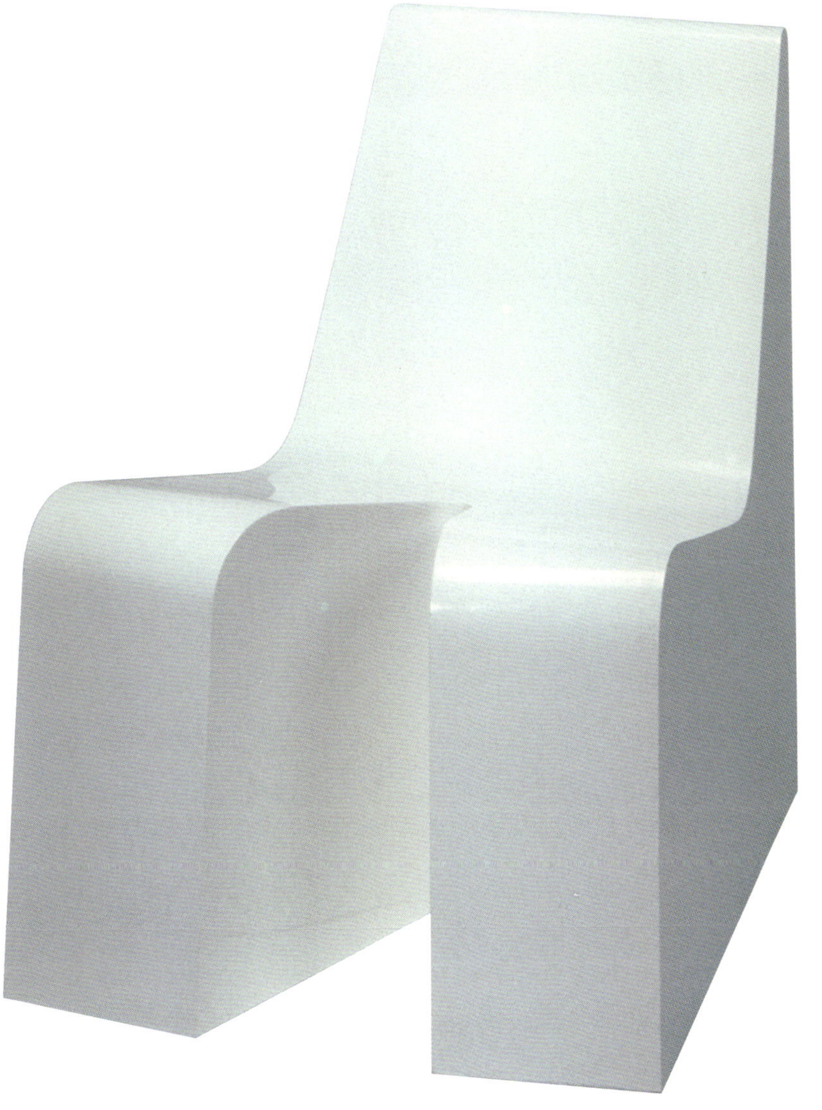

Sexy Relaxy polyester — 2000

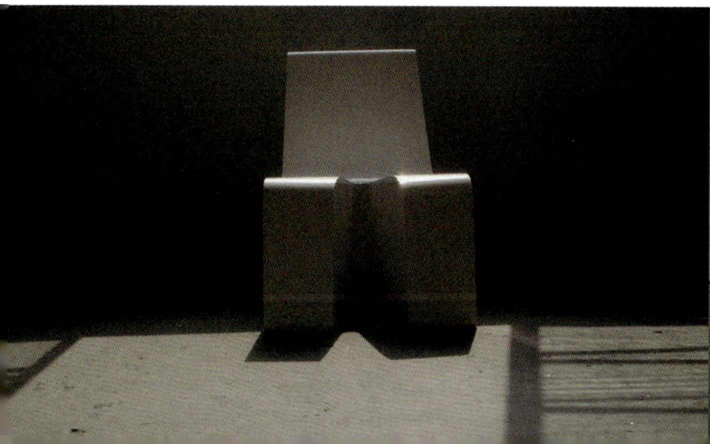

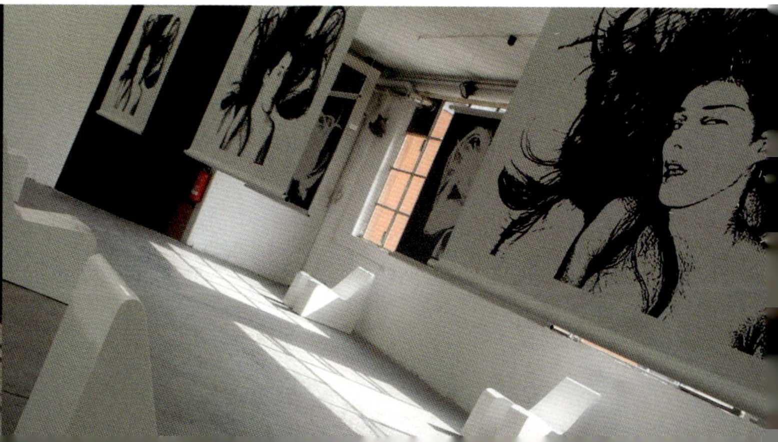

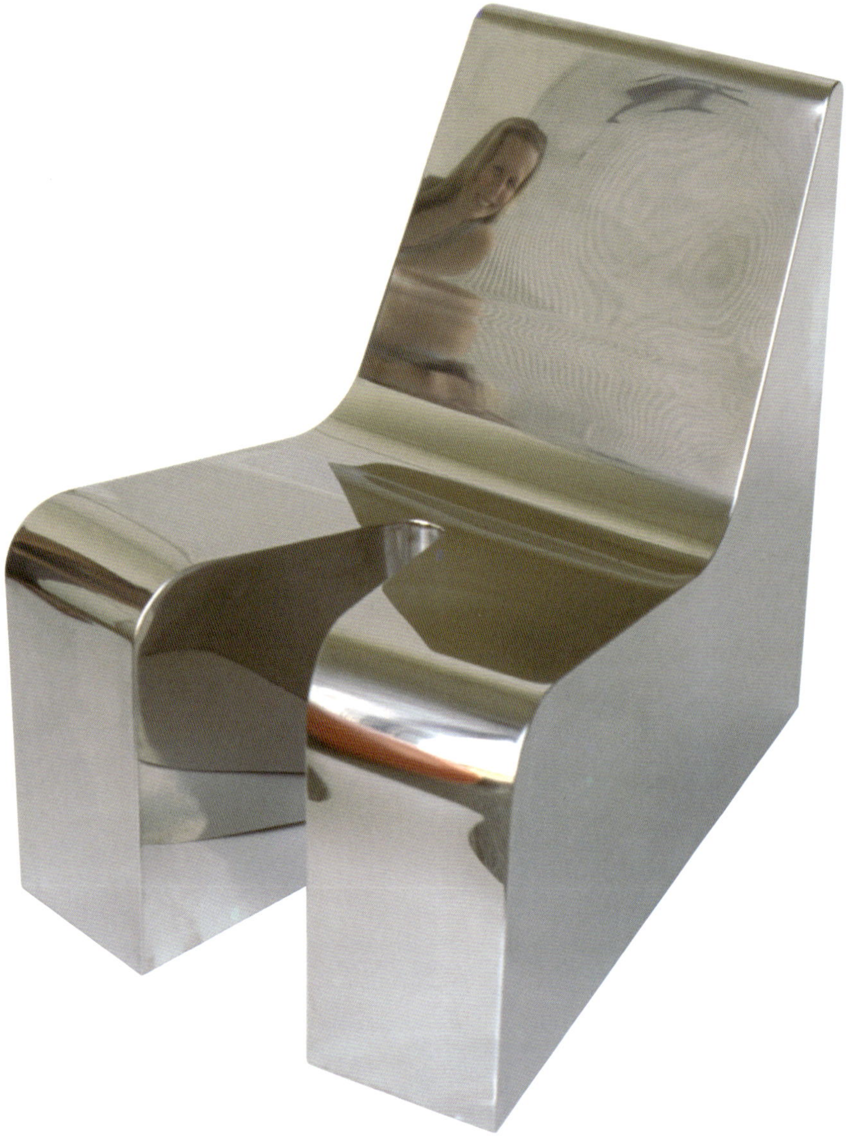

Sexy Relaxy mirrorsteel — 2001

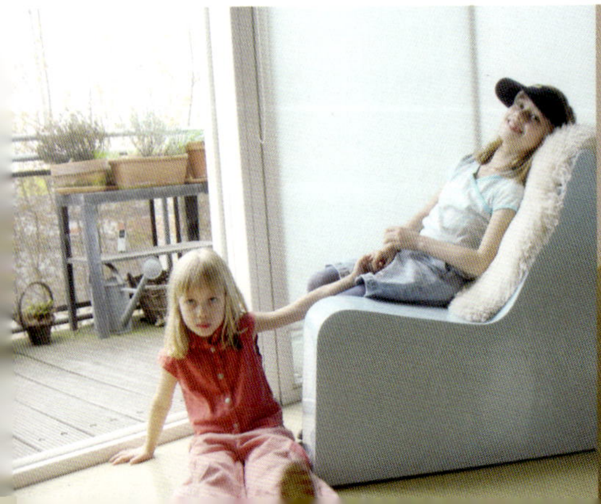

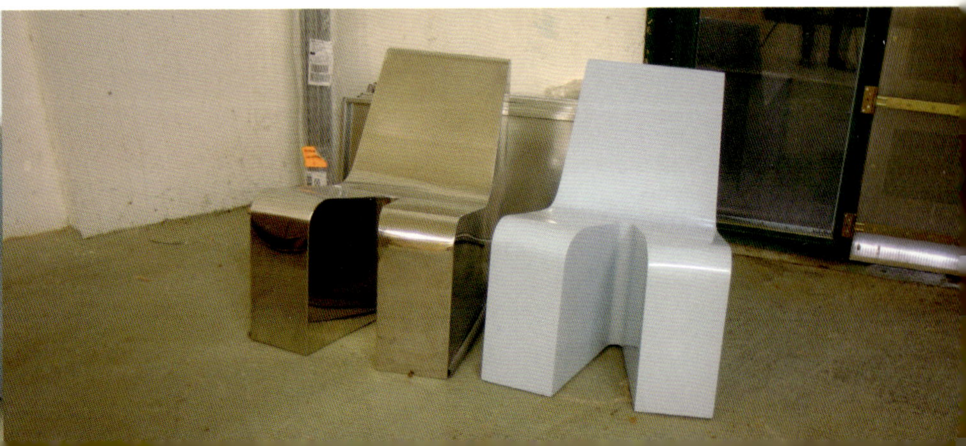

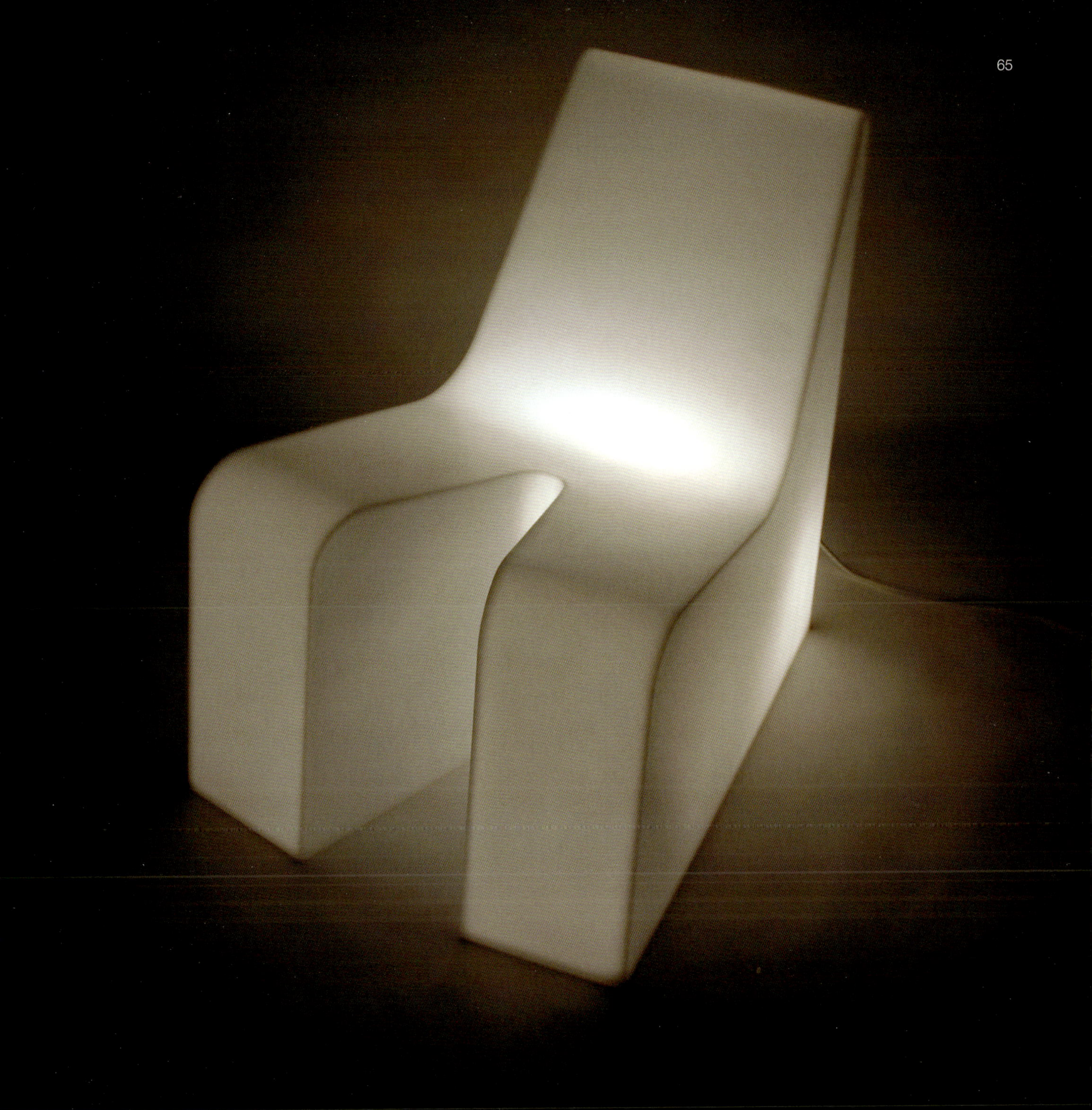

Sexy Relaxy rotation moulded with light — 2004

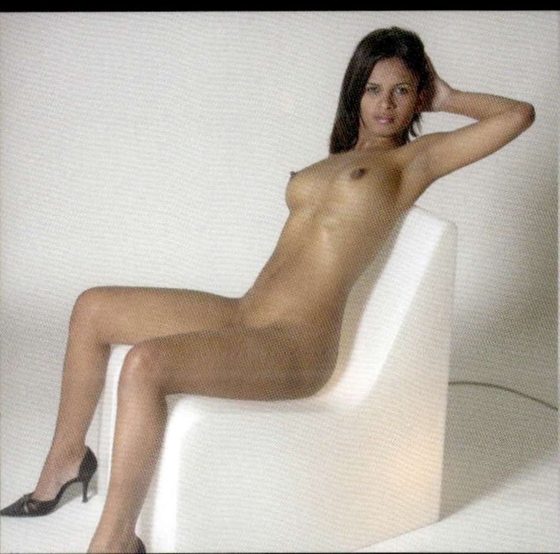

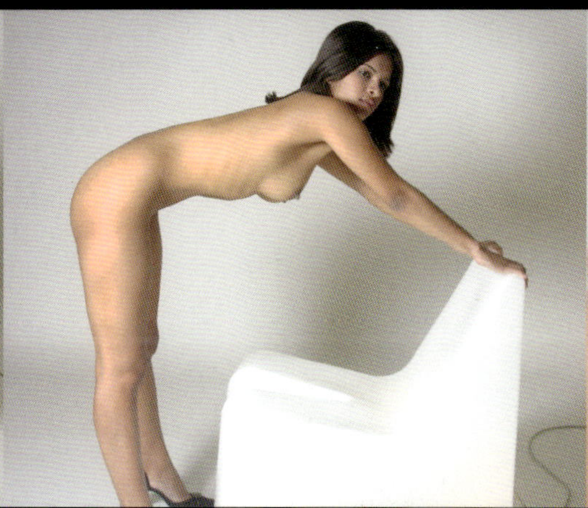

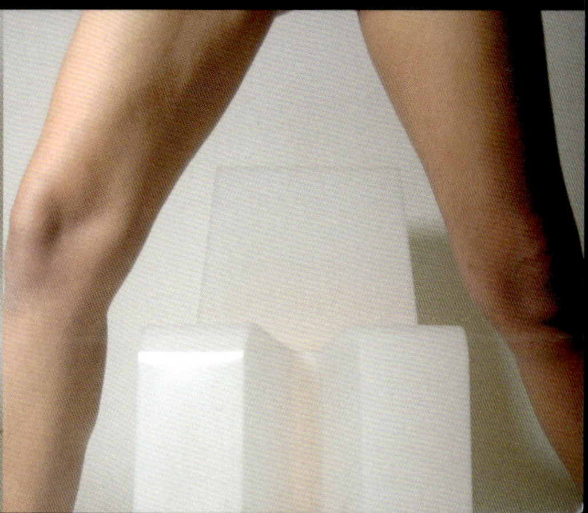

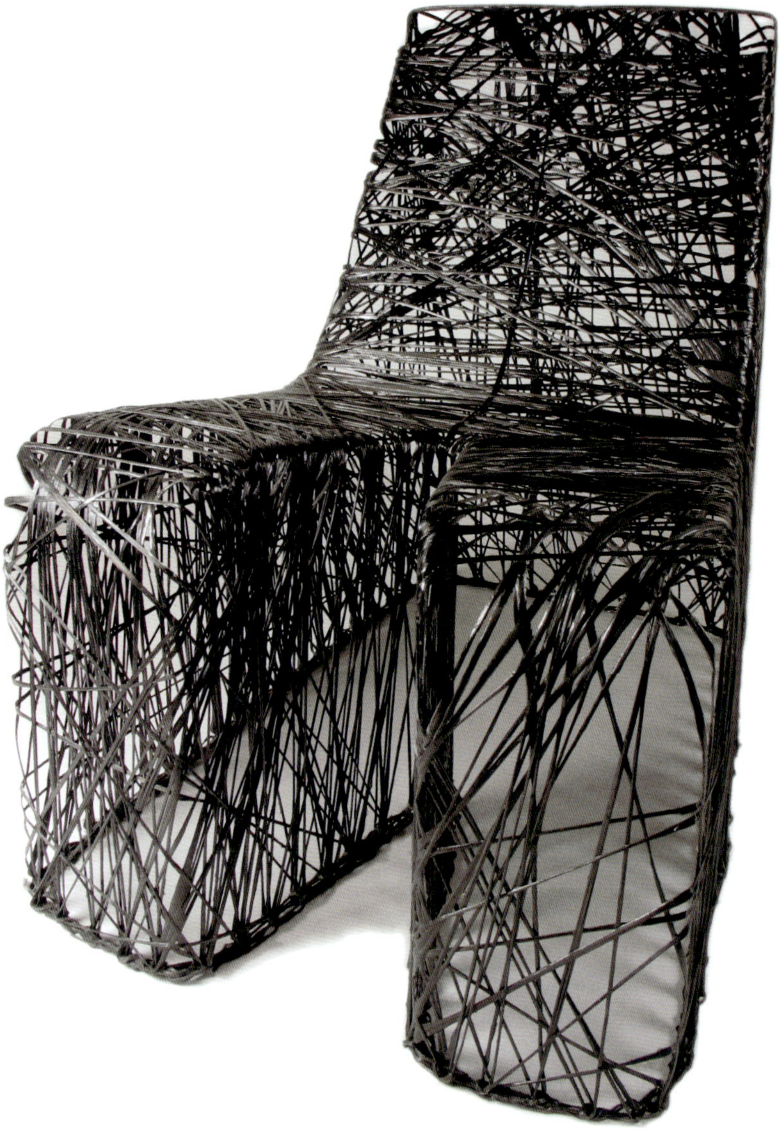

Sexy Relaxy random carbon copy — 2005

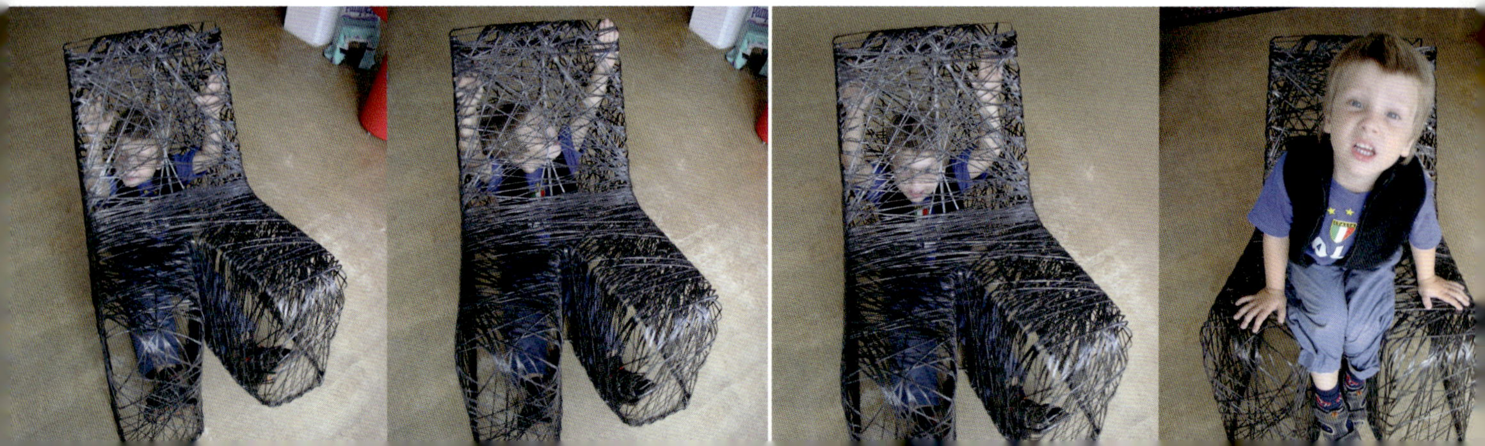

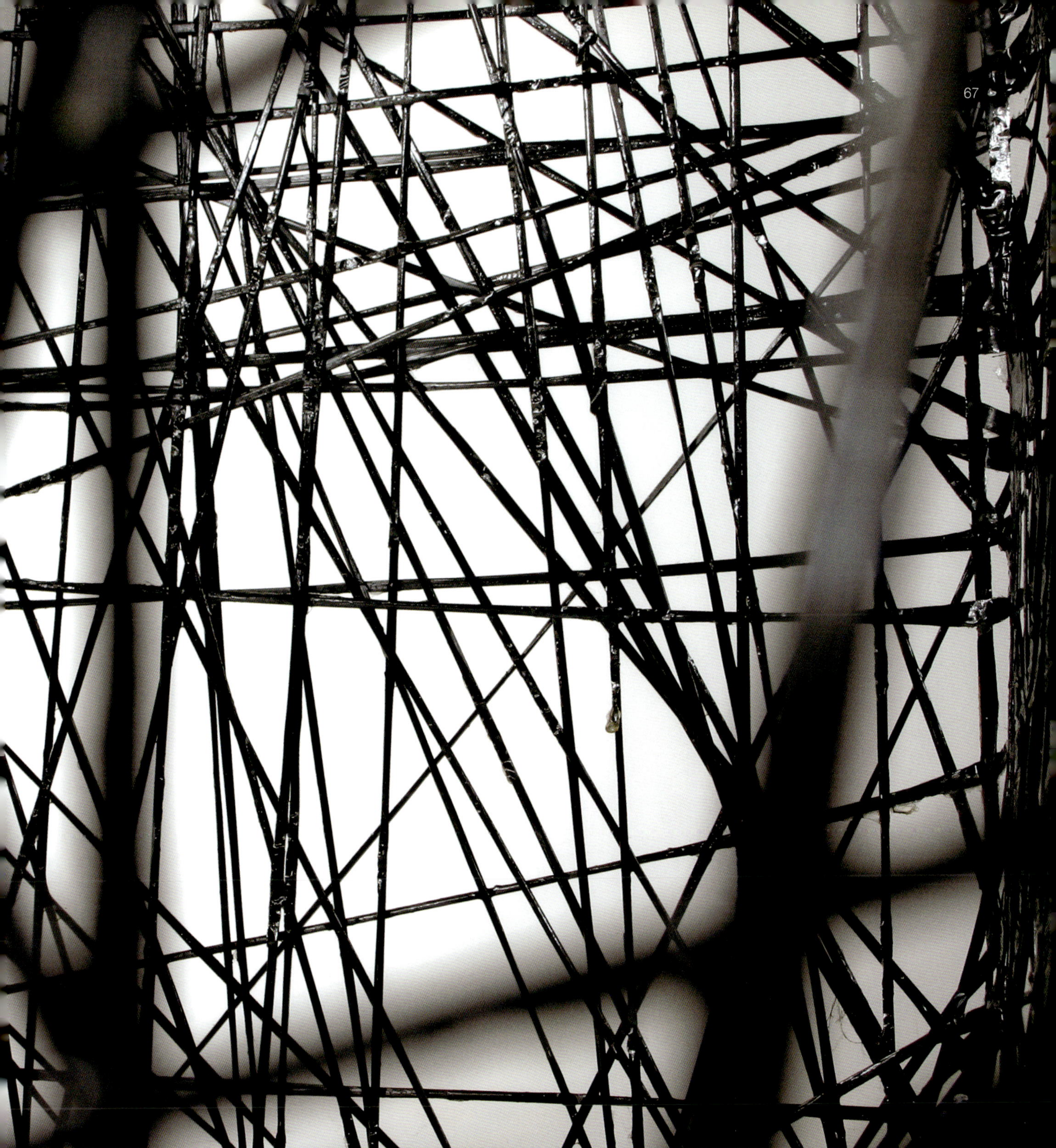

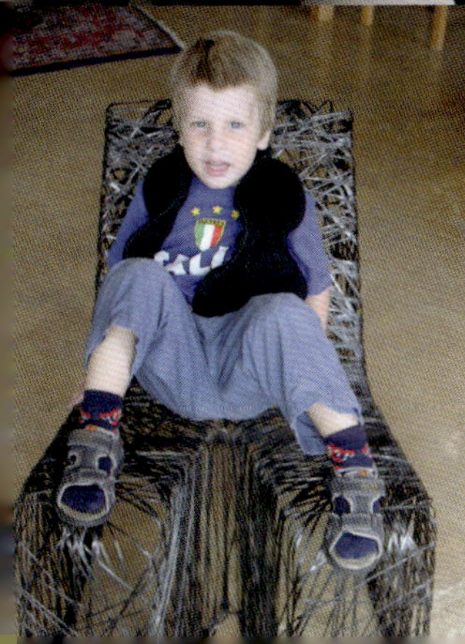

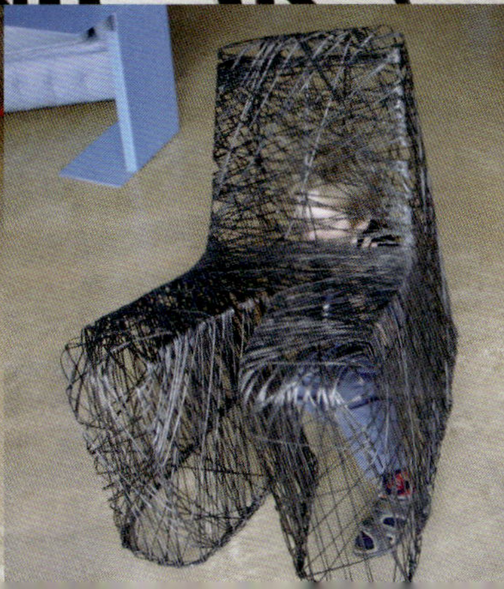

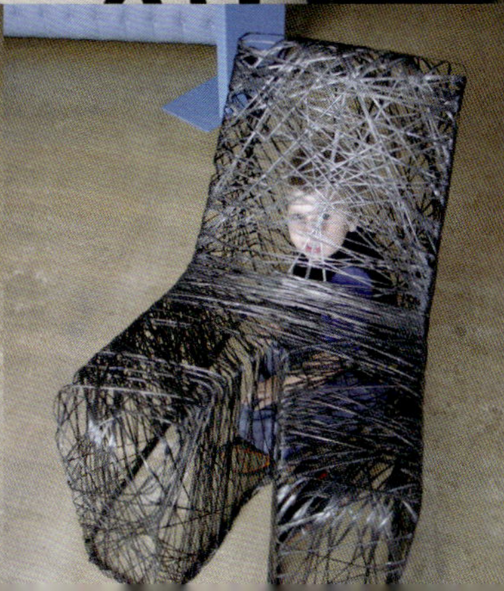

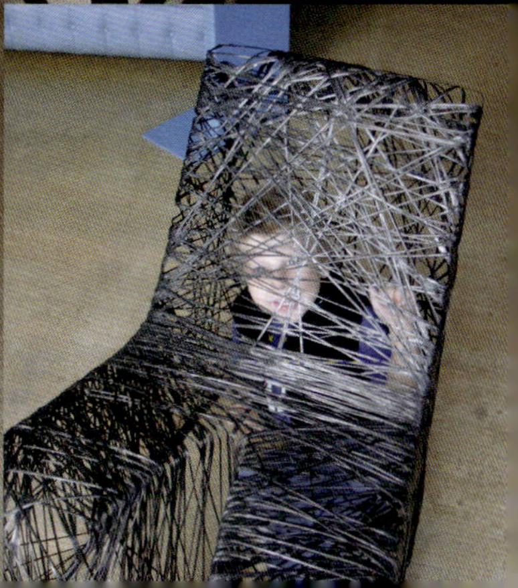

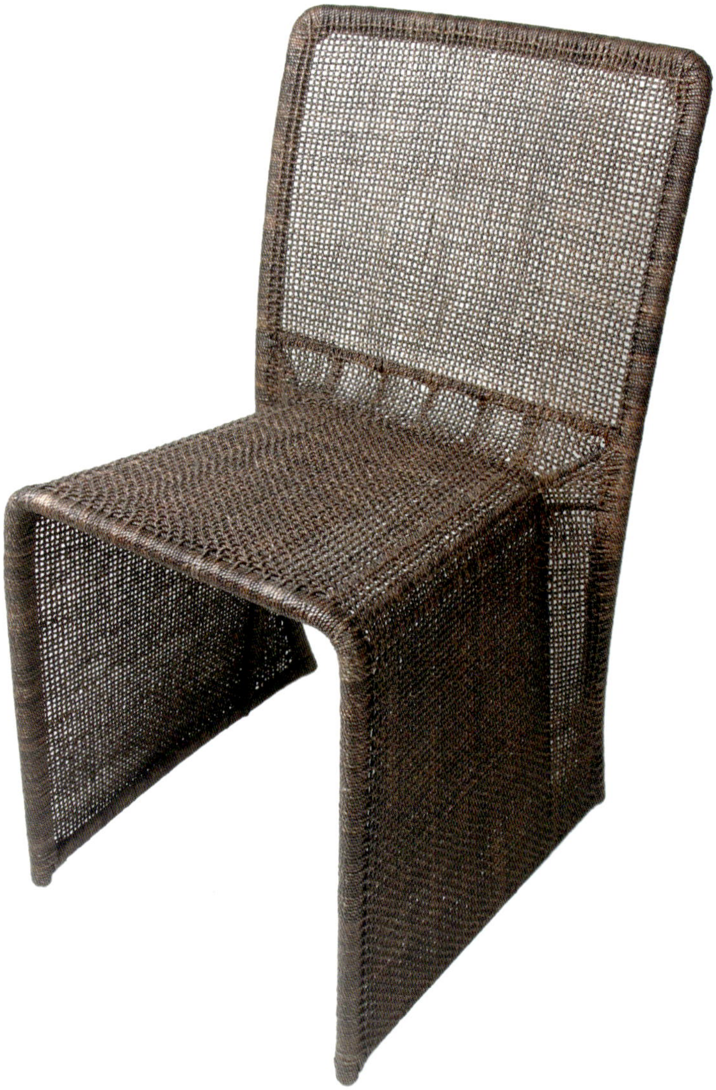

Back-up chair — 2000

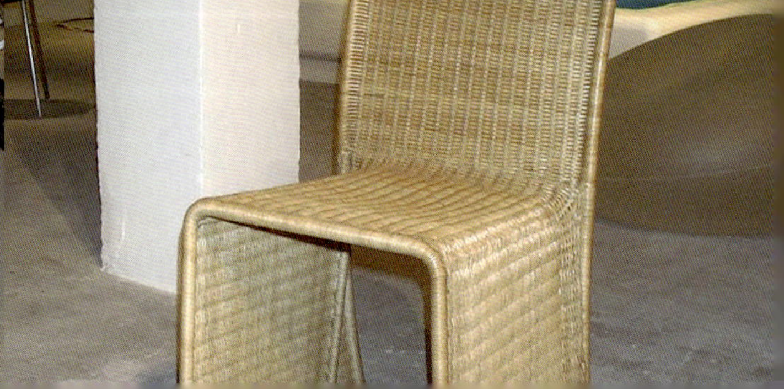

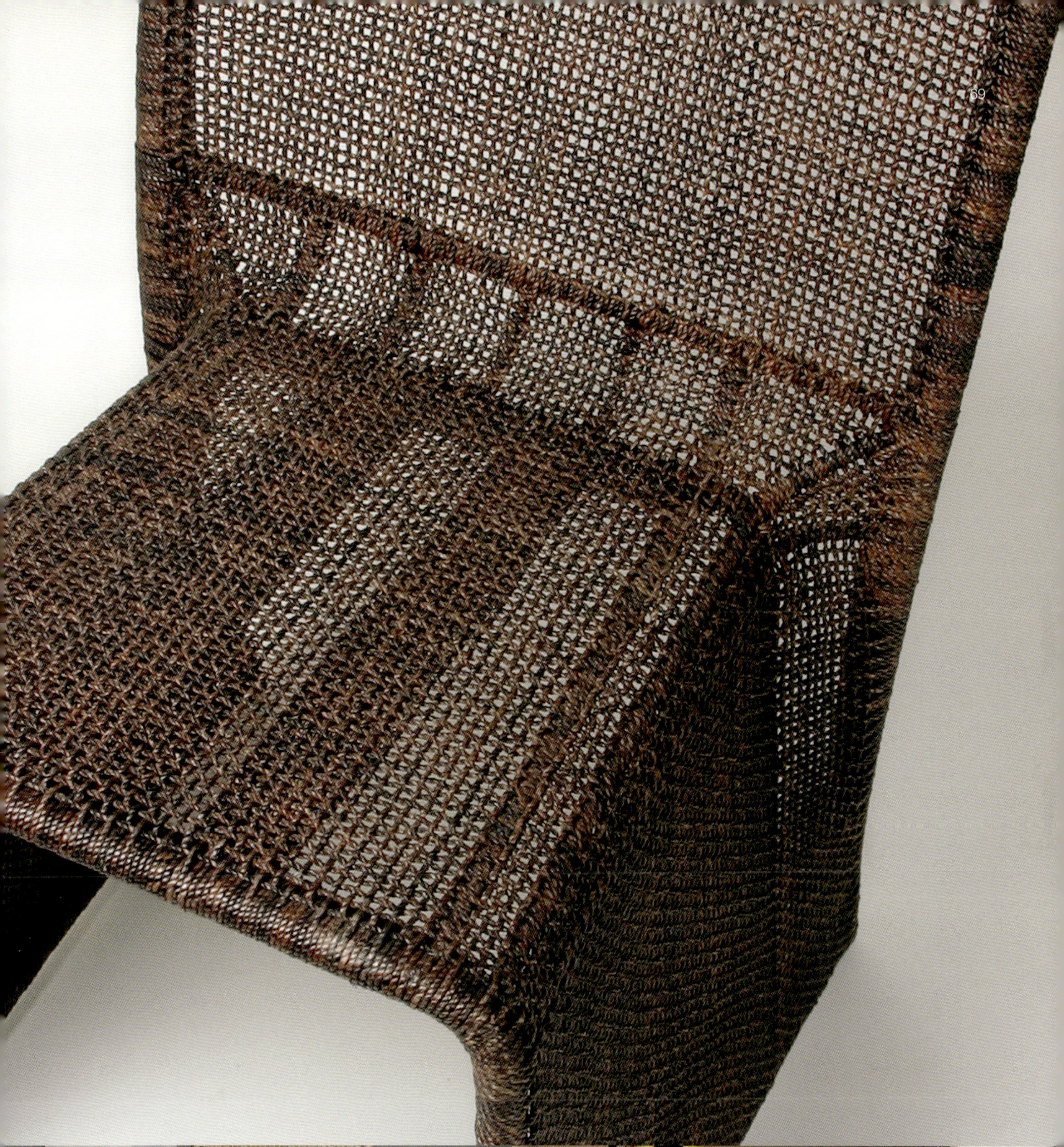

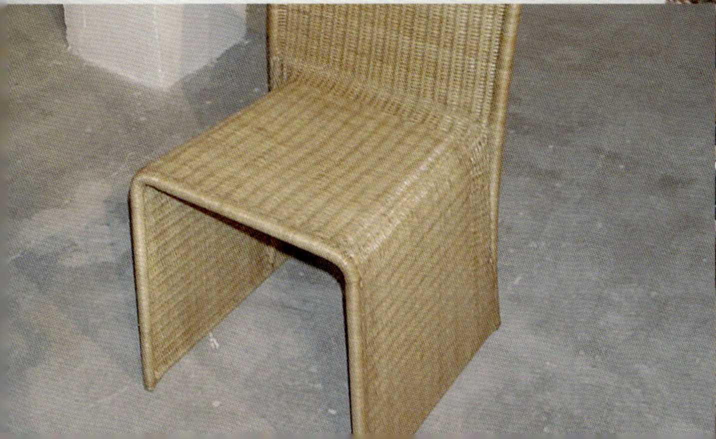

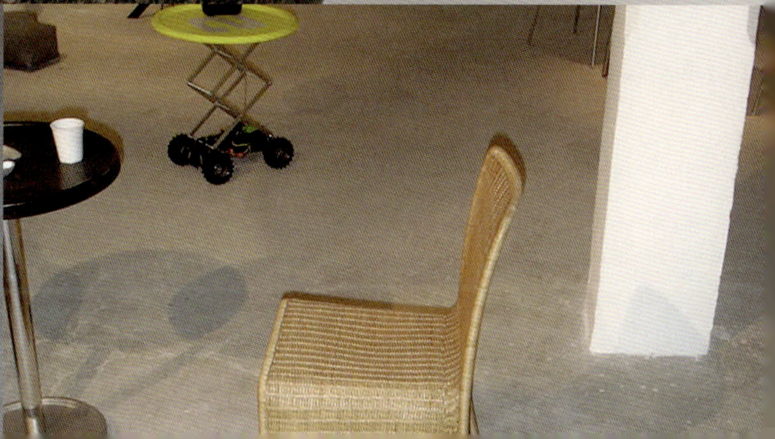

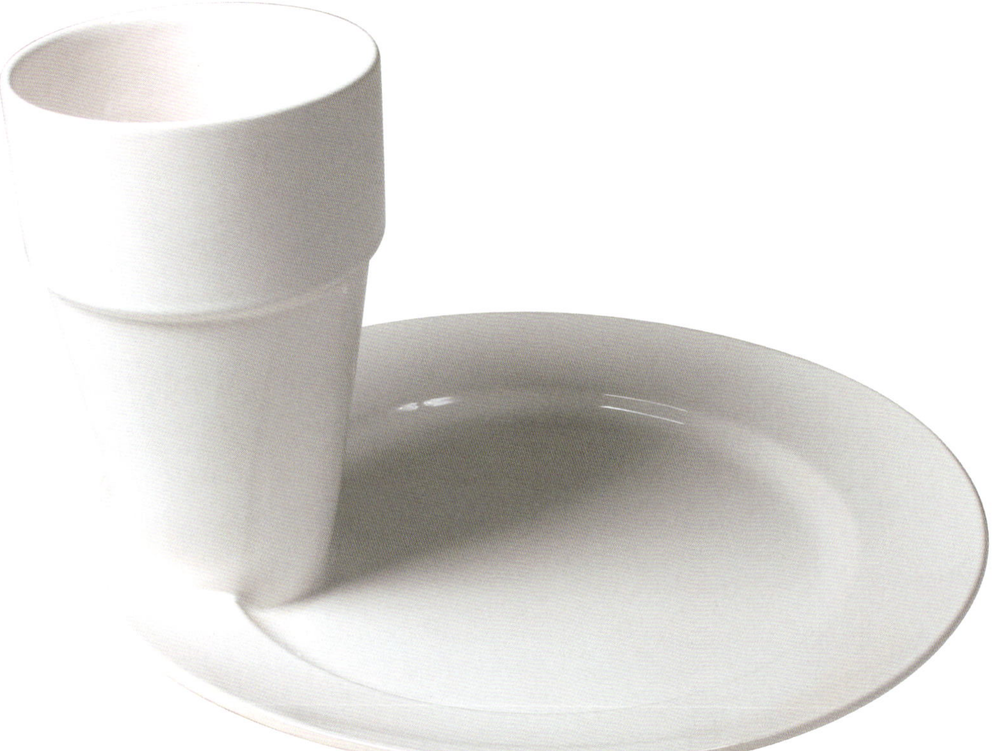

Cup-plate — 2003

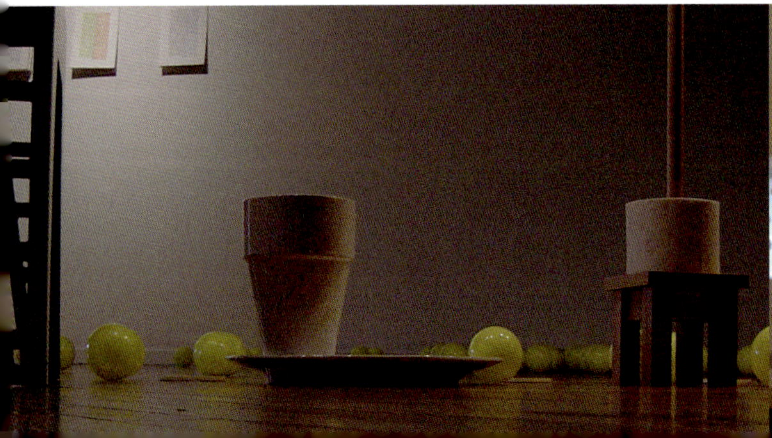

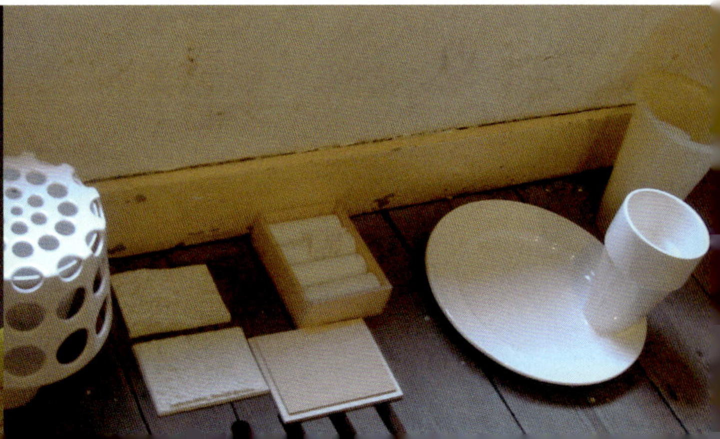

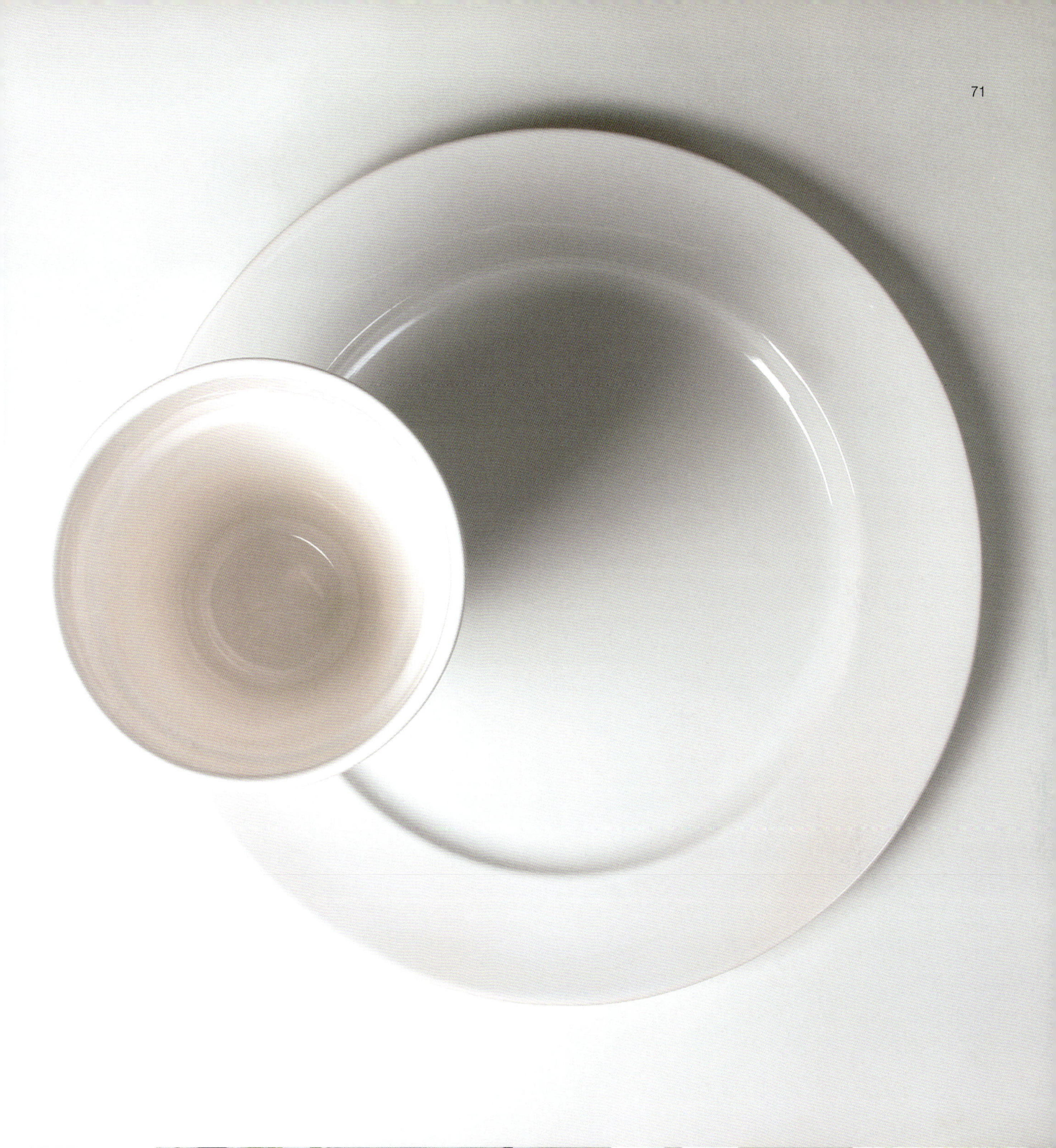

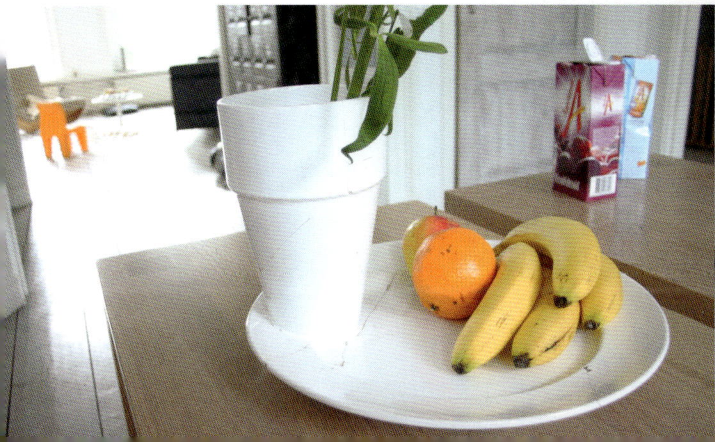

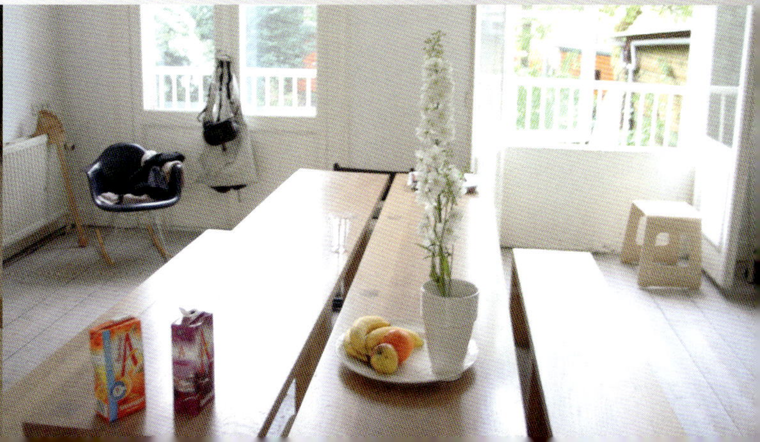

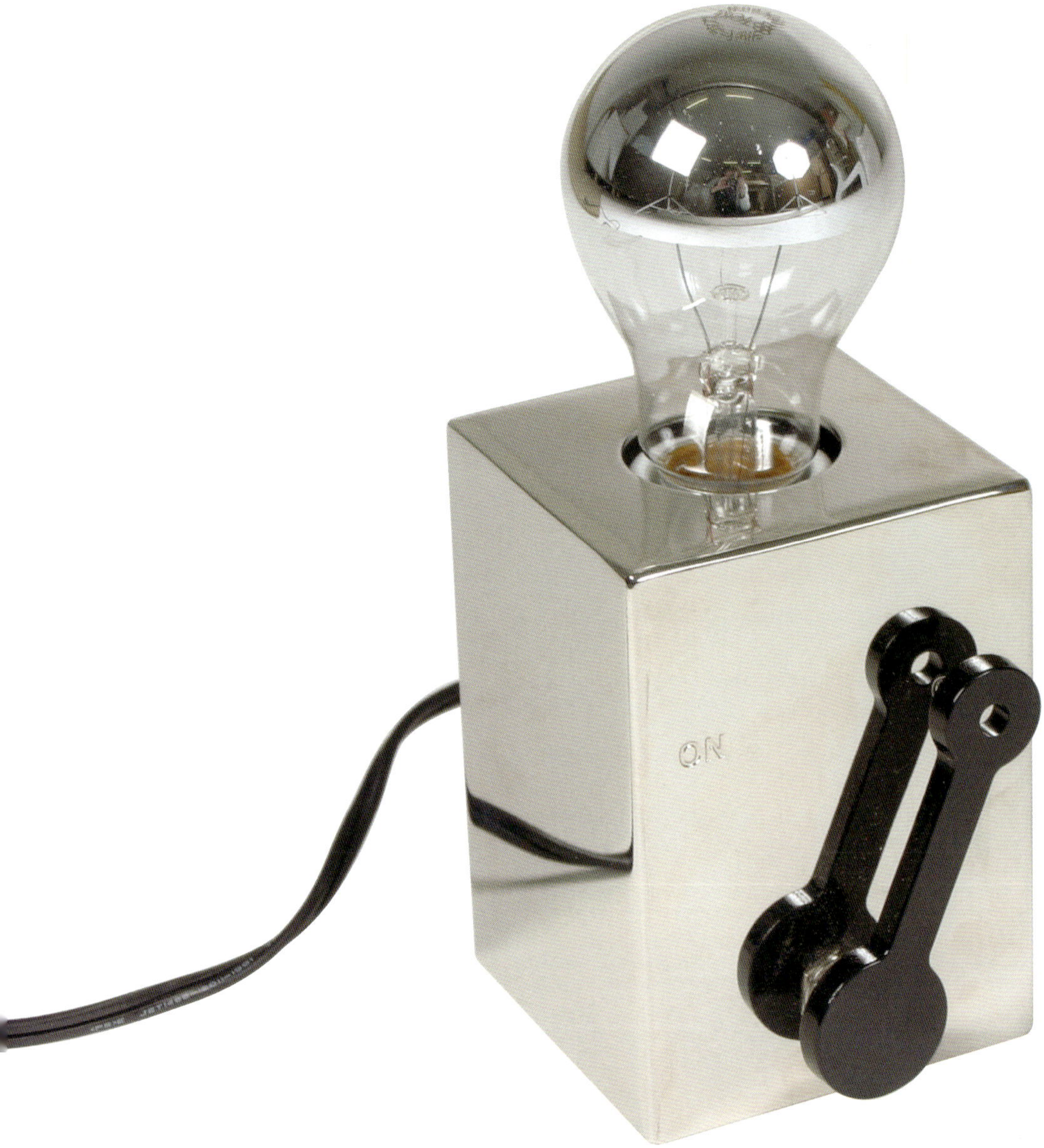

The switch — 1998

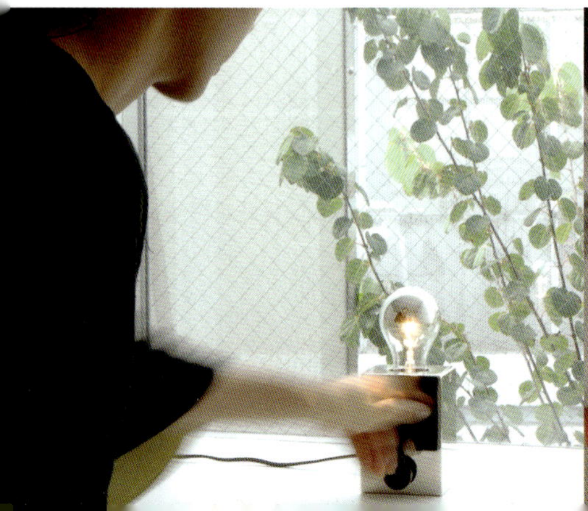

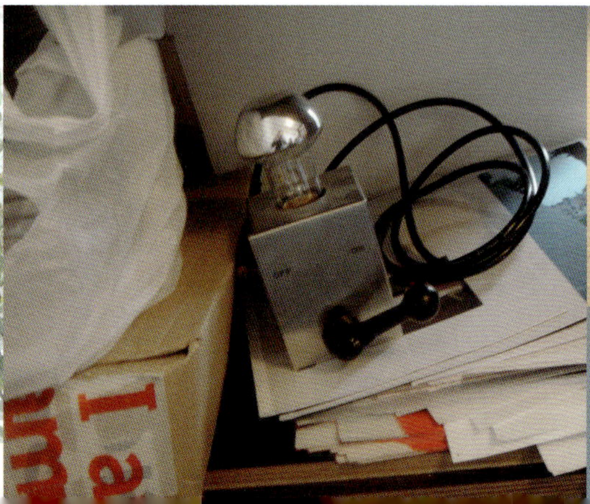

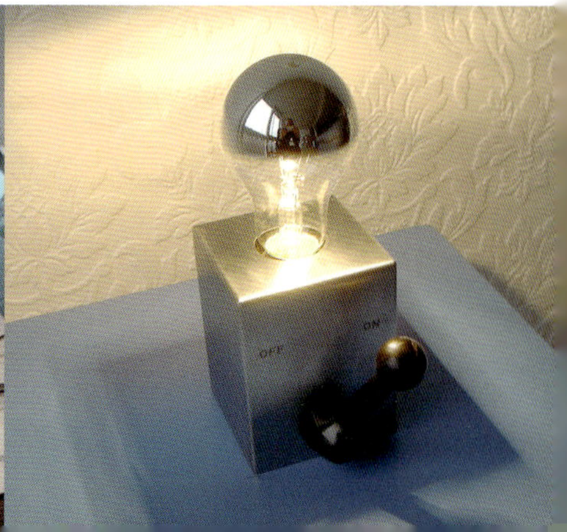

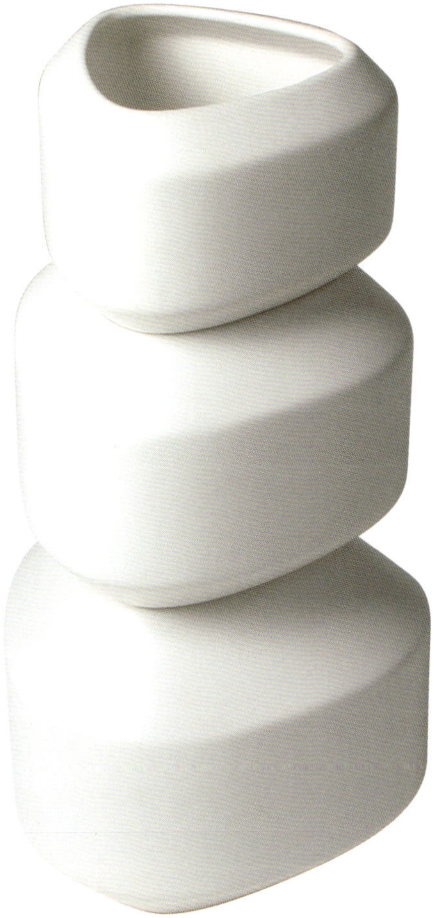

Rocket vase — 2004

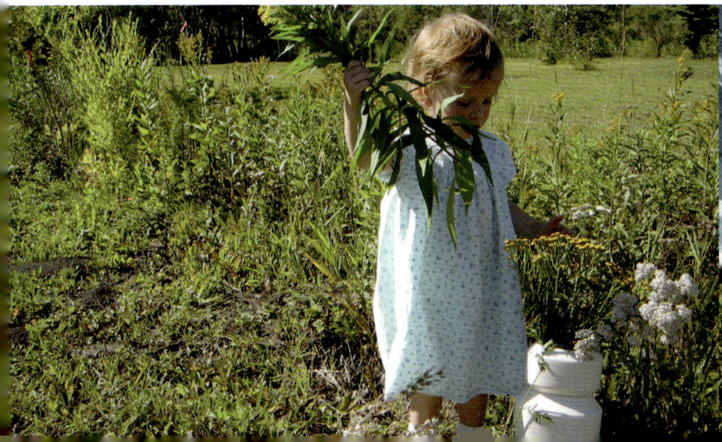

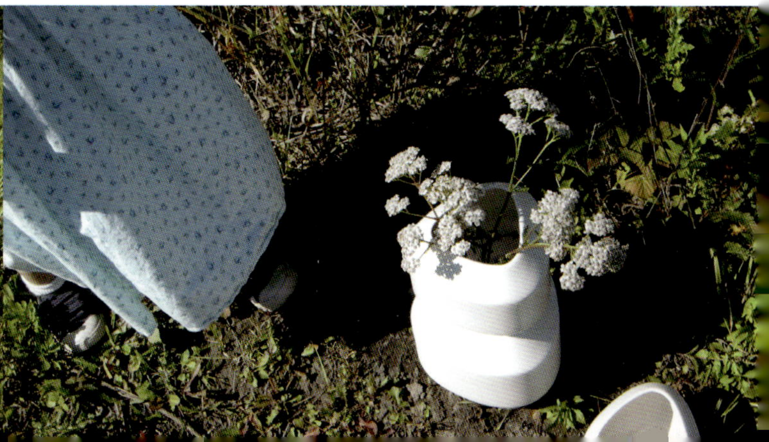

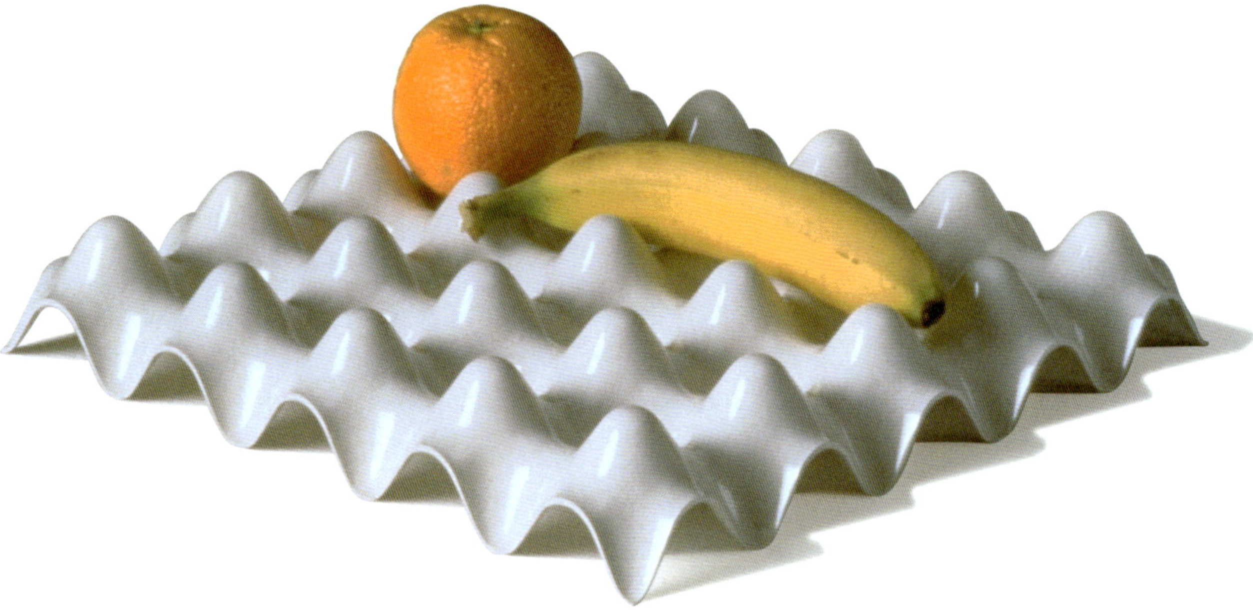

Fruittray — 1996

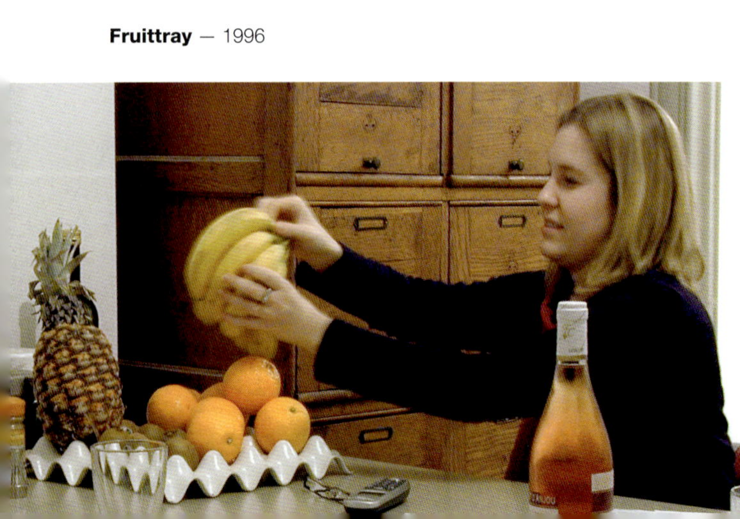

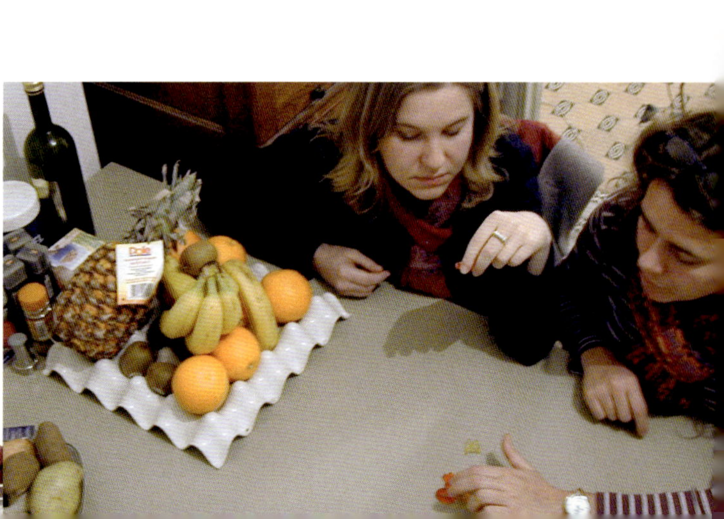

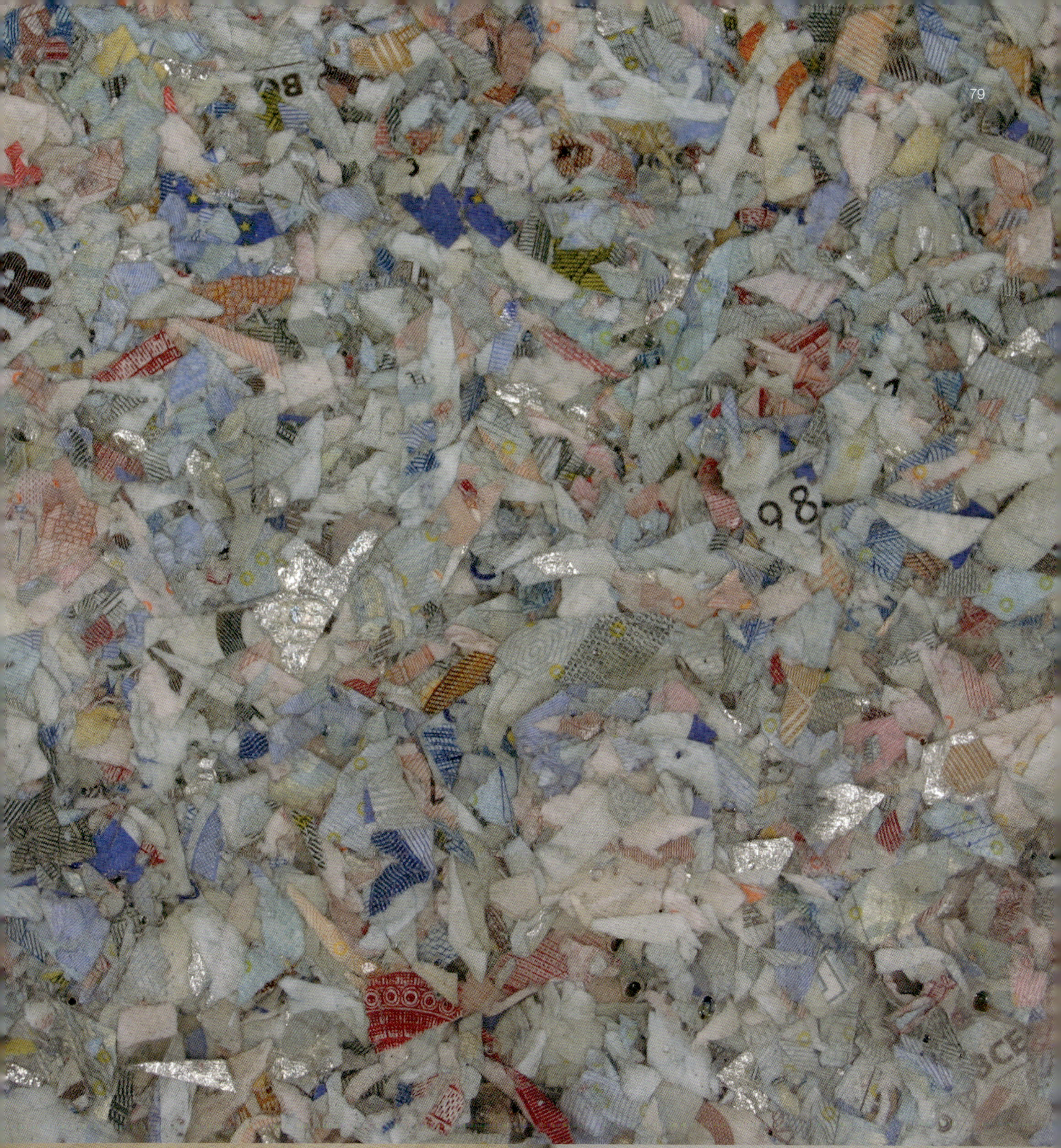

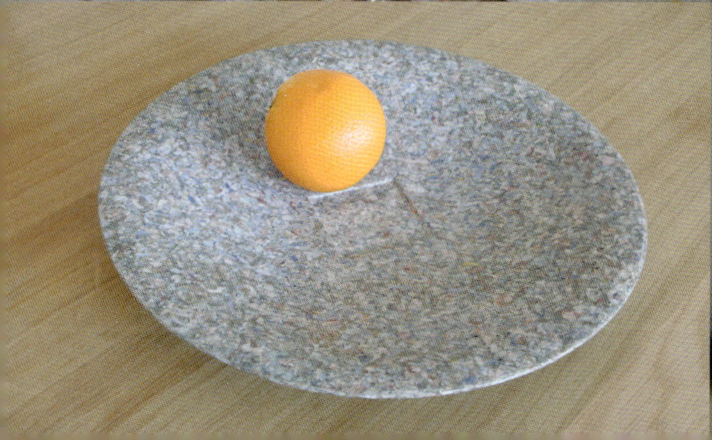

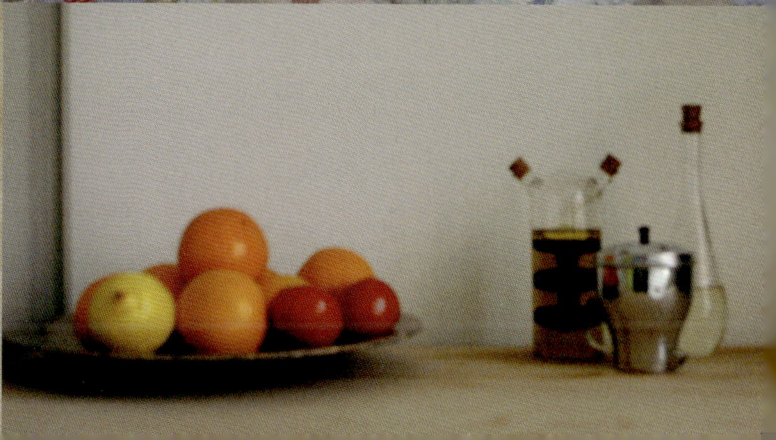

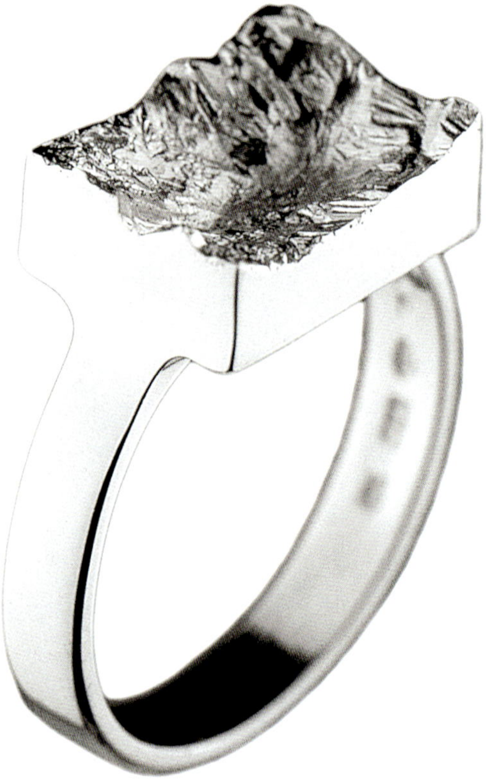

Ring S.M.A.K. male — 1999

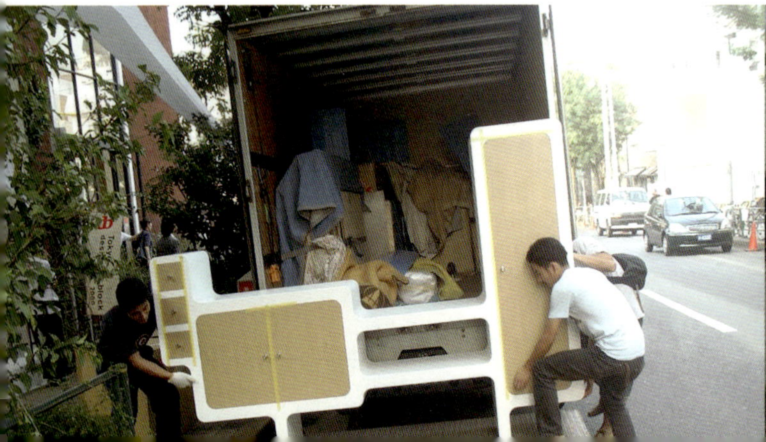

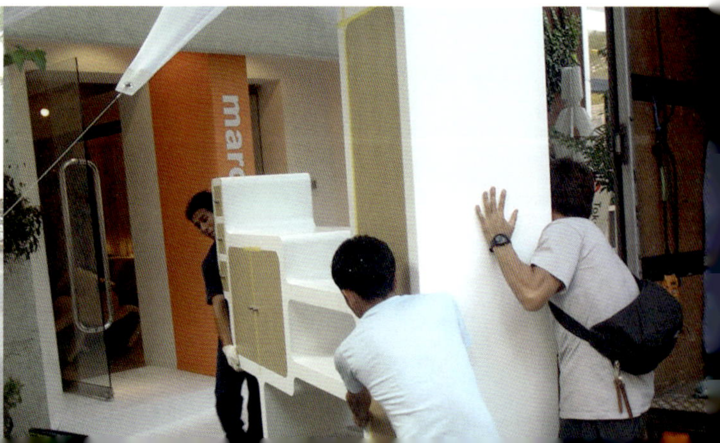

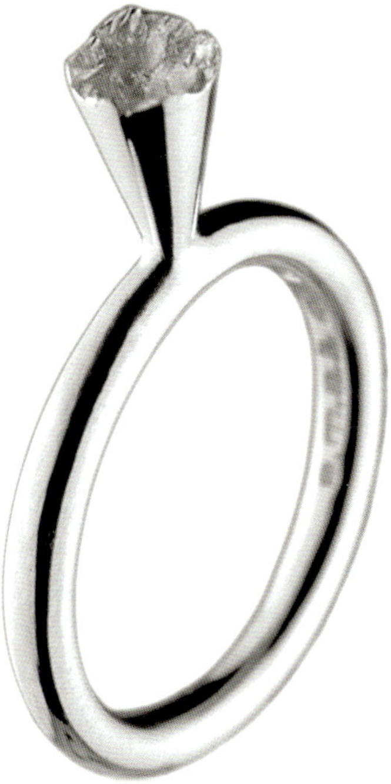

Ring S.M.A.K. female — 1999

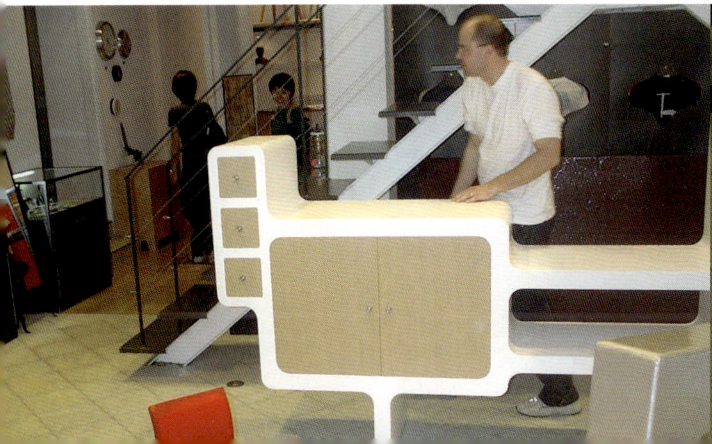

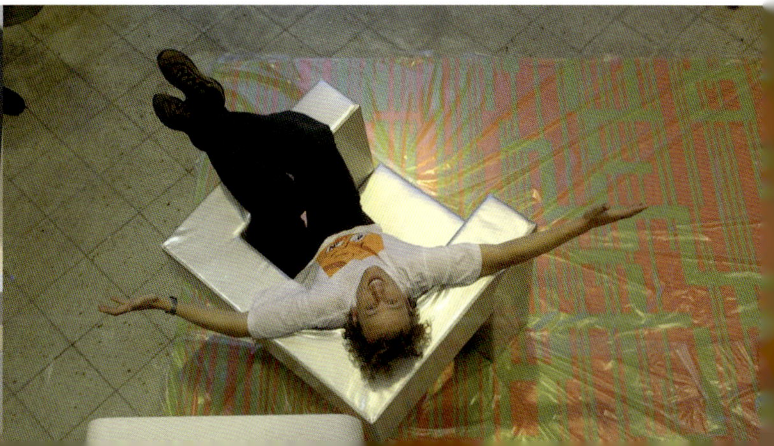

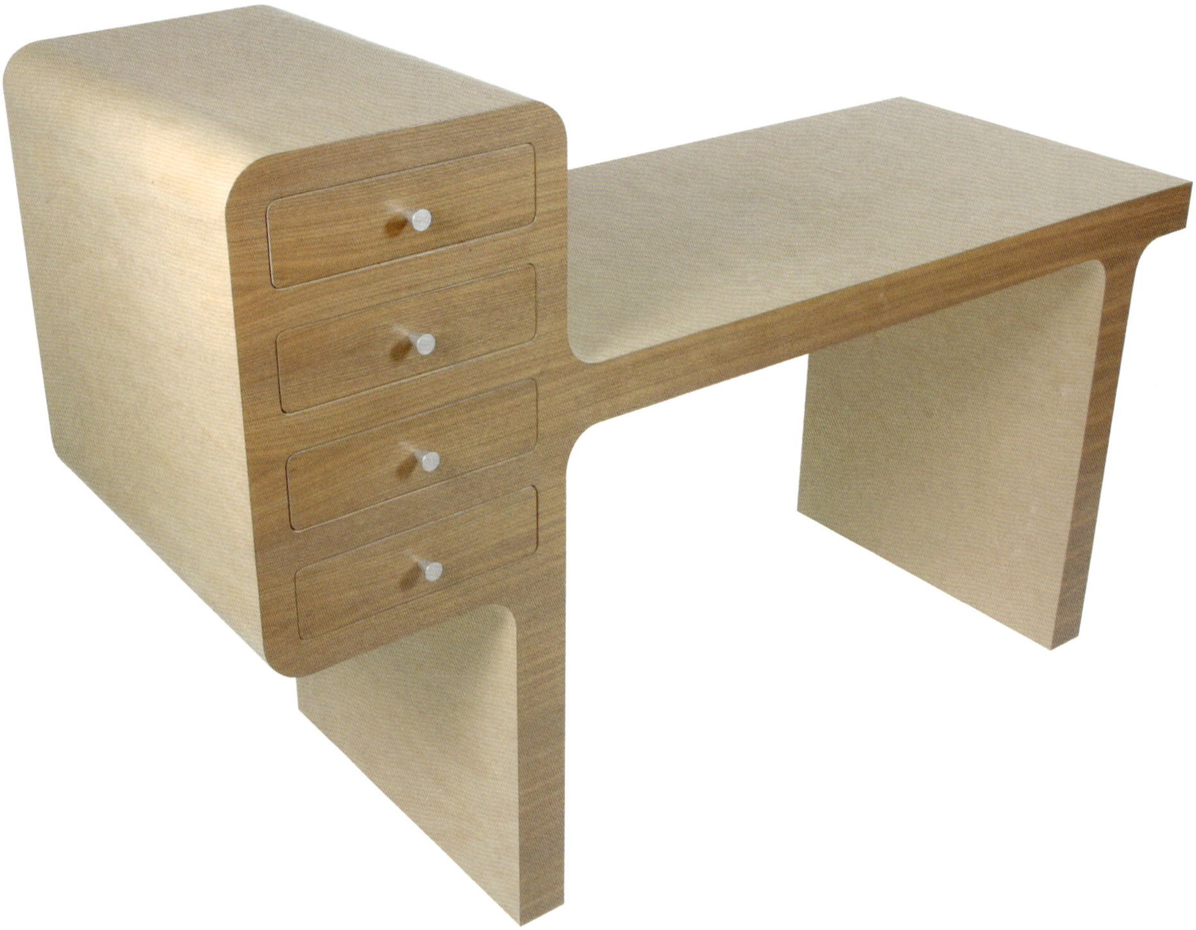

3 minus 1 desk — 1996

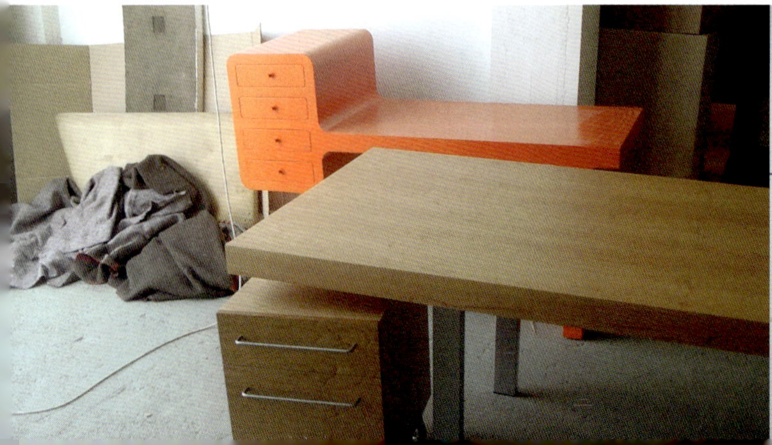

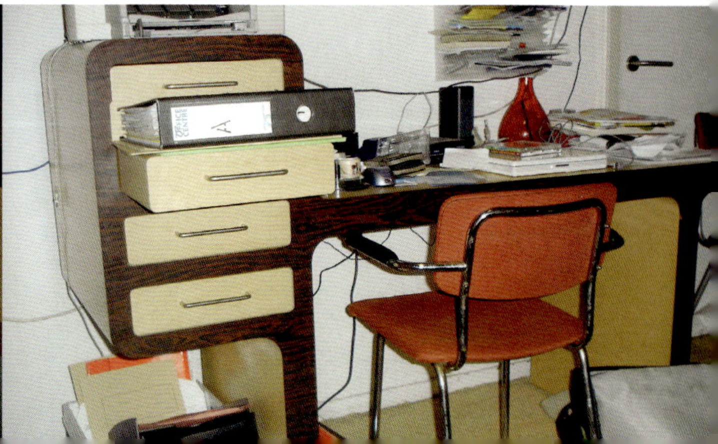

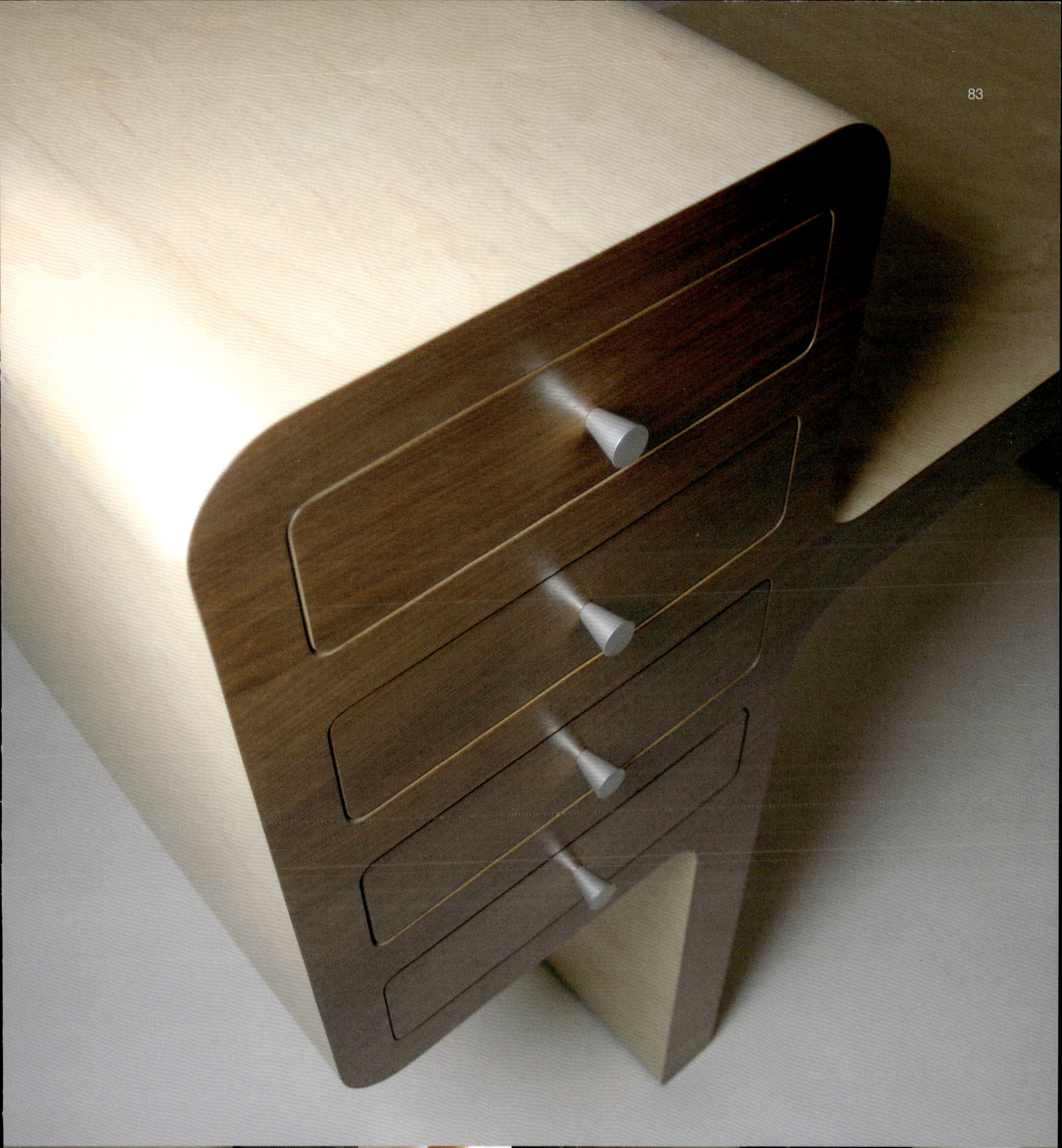

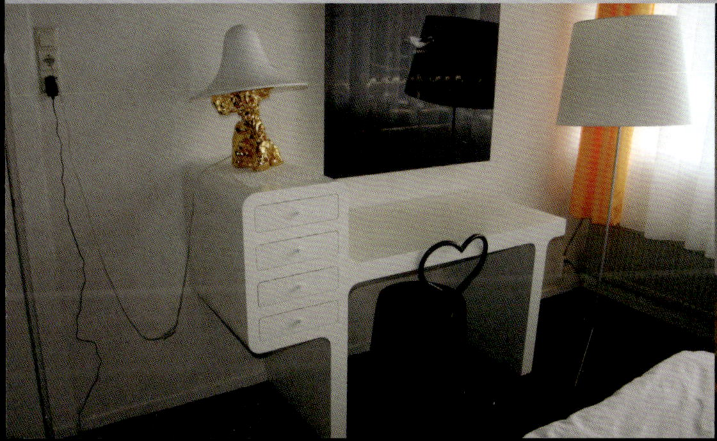

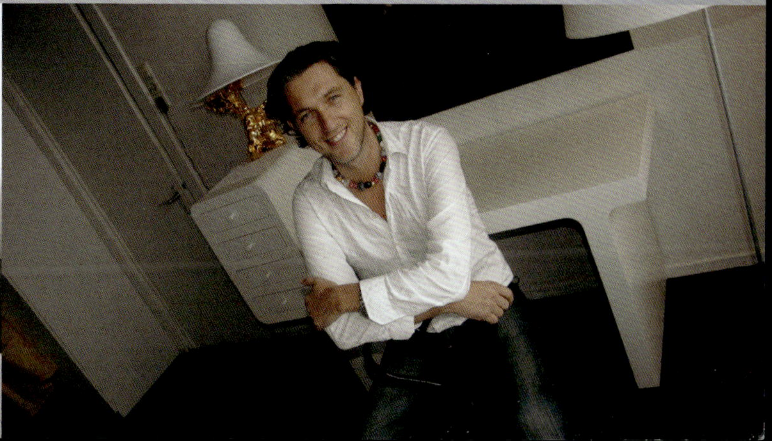

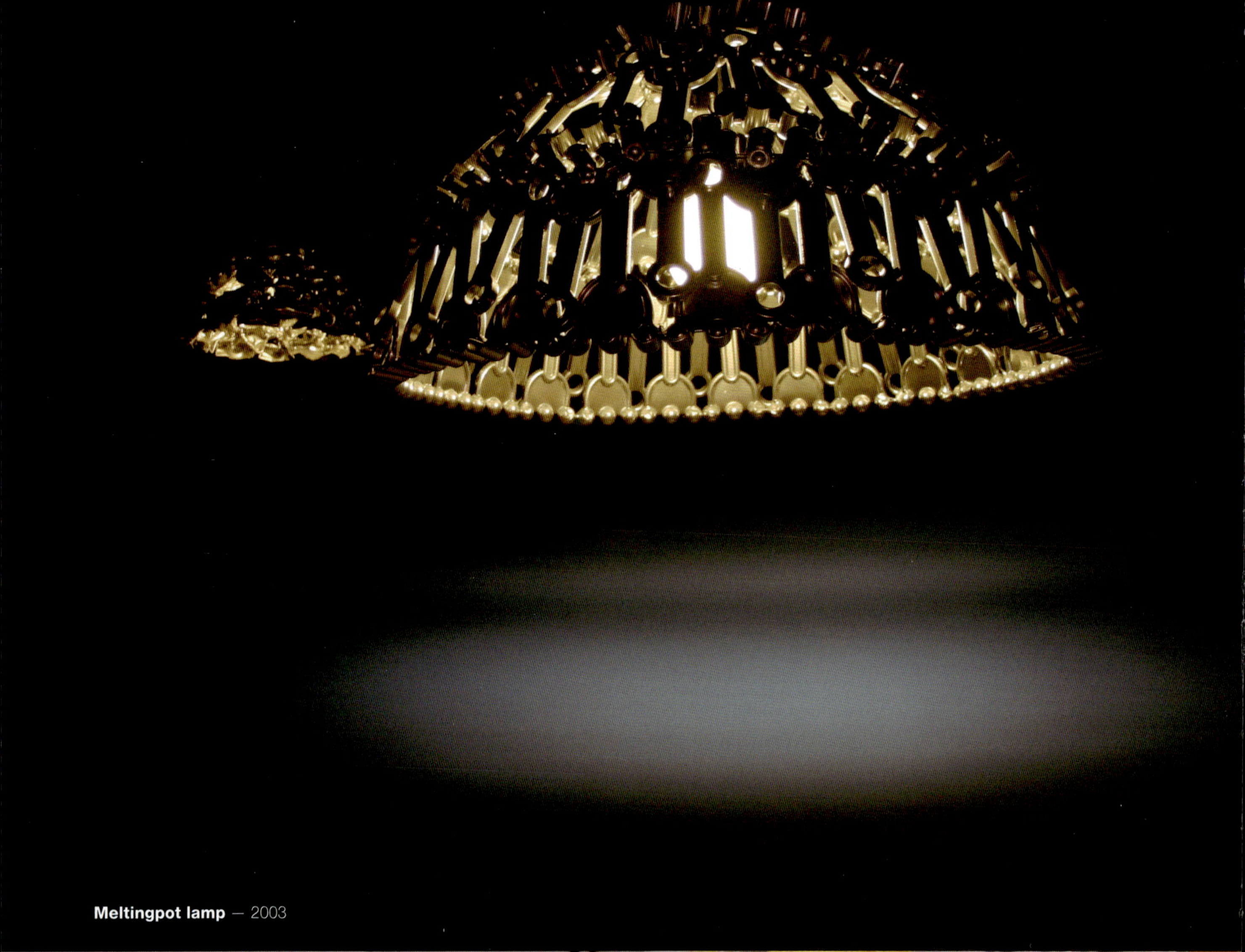

Meltingpot lamp — 2003

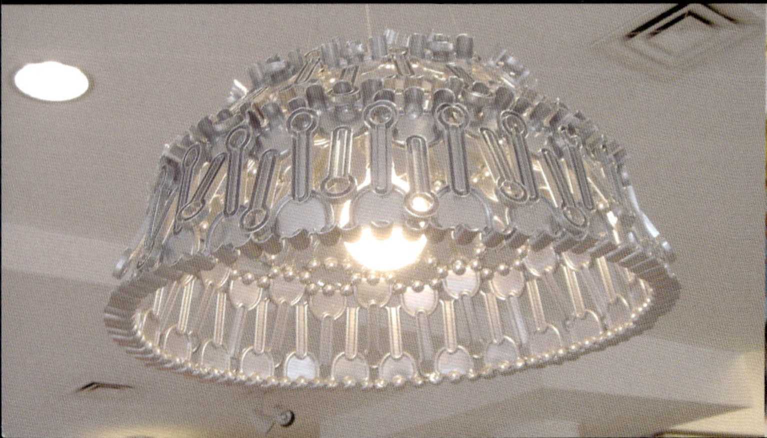

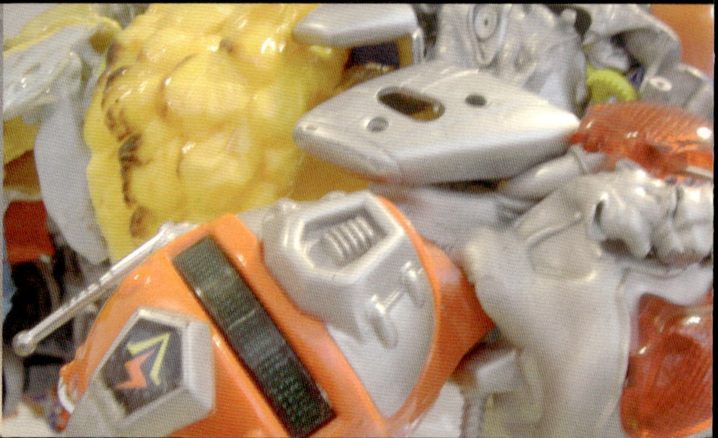

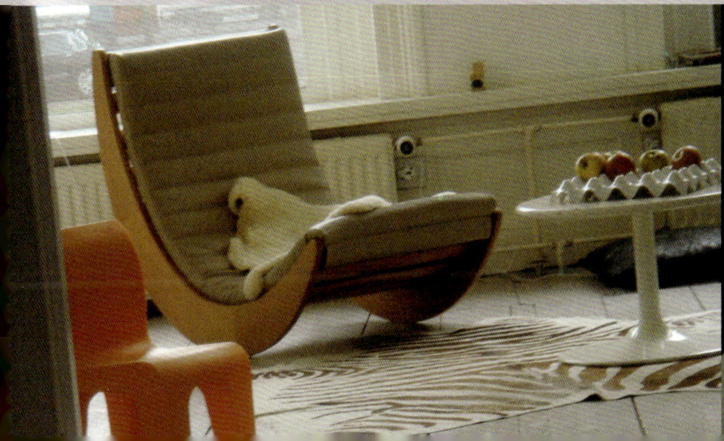

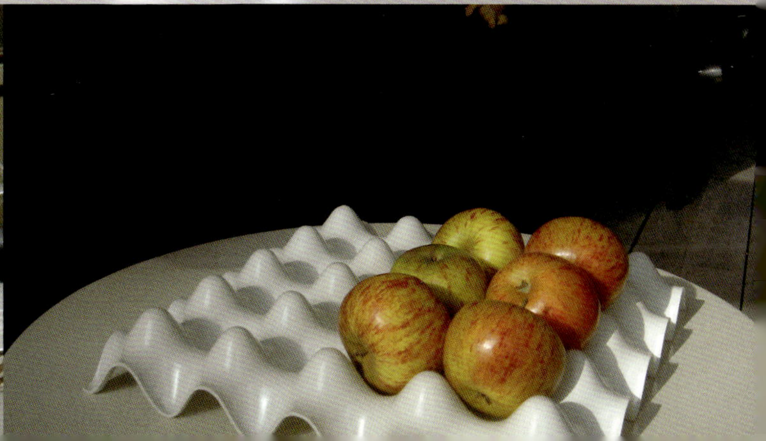

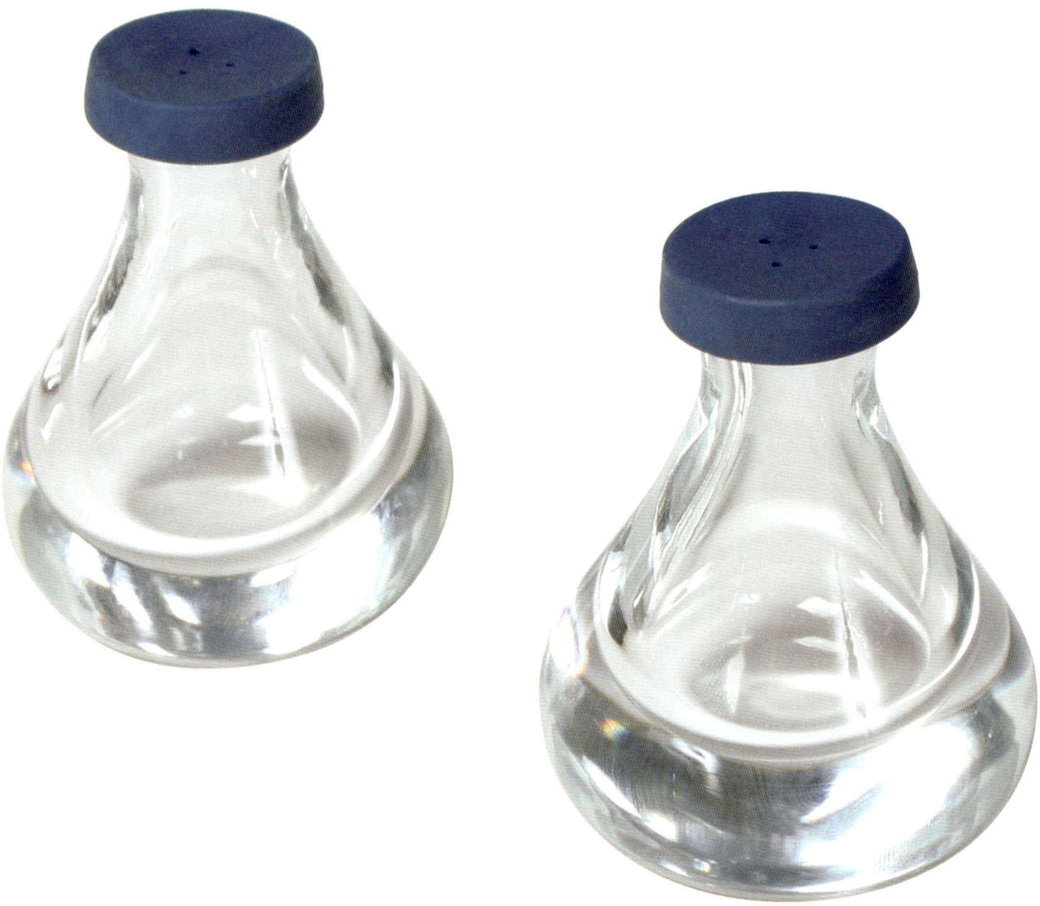

Table-tumblers — 1996

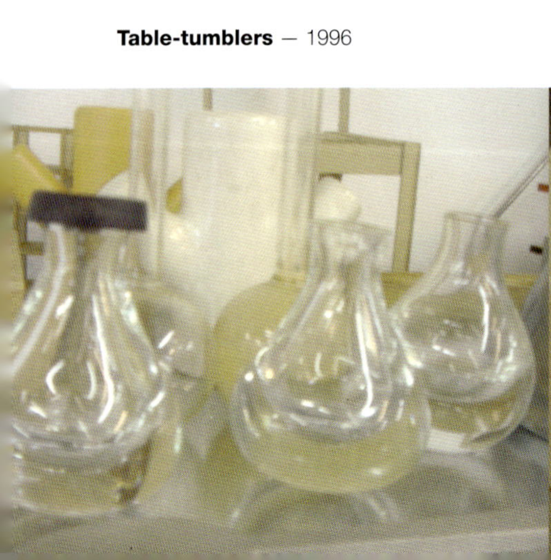

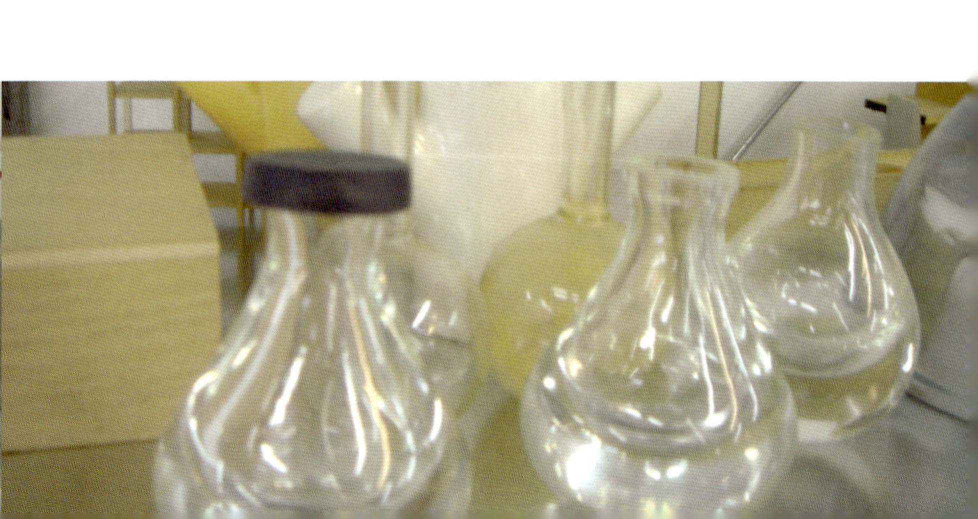

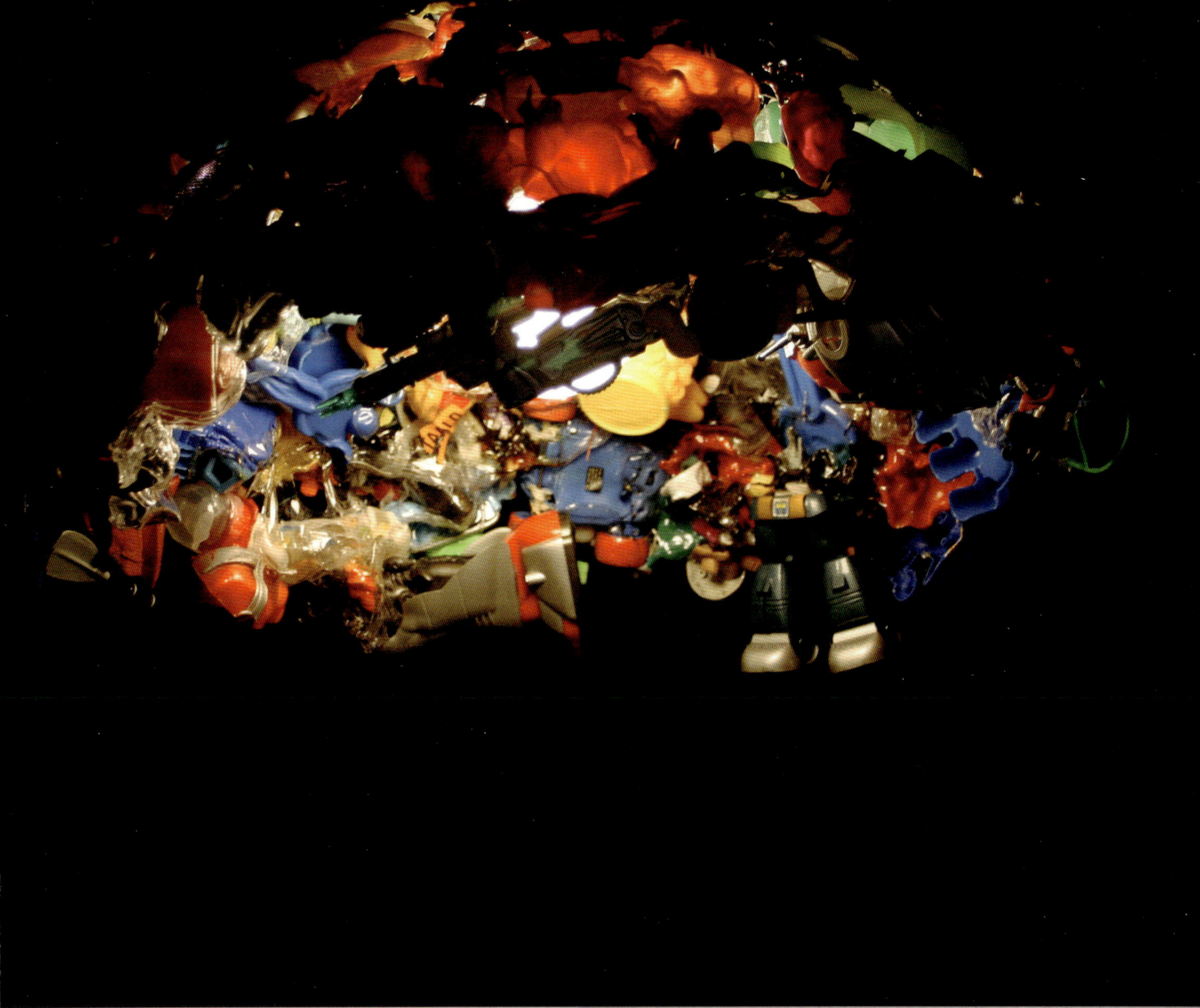

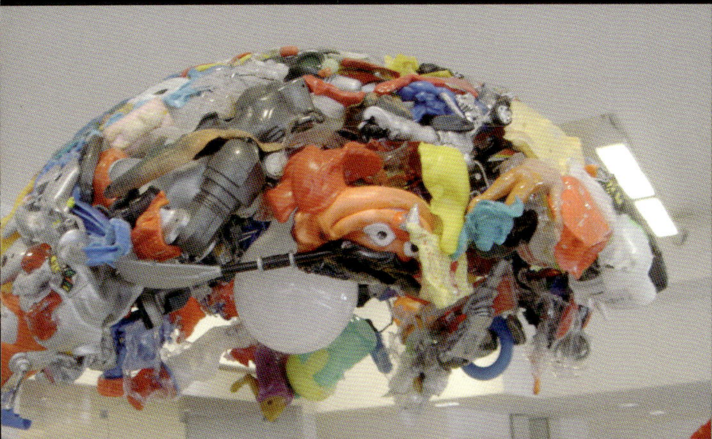

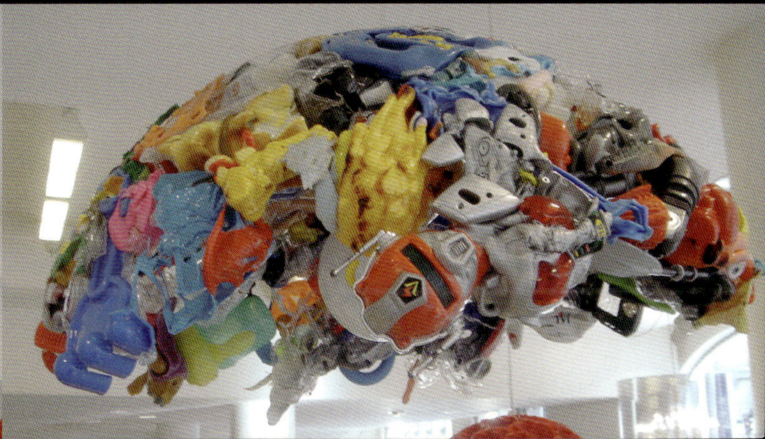

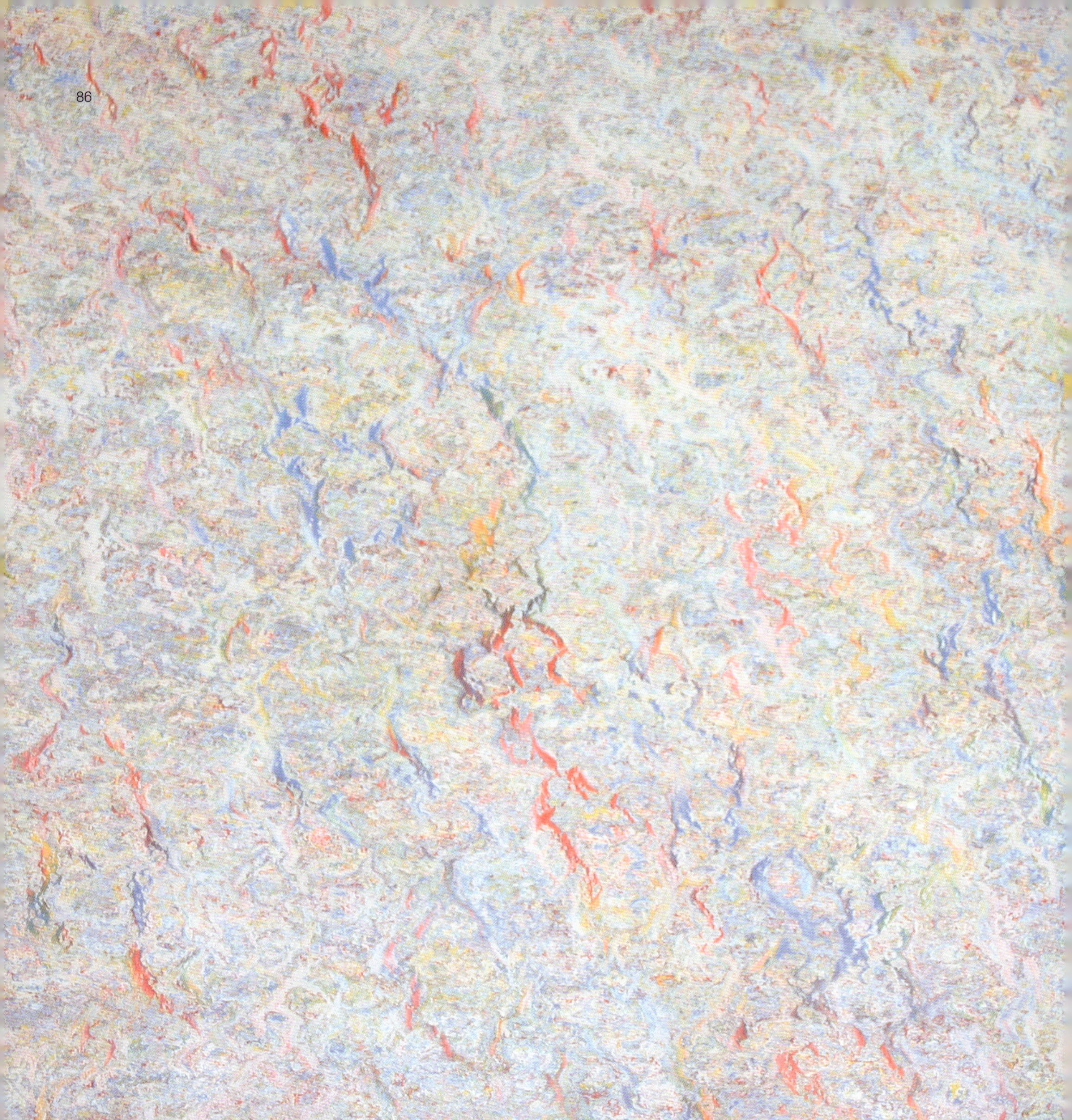

Marmoleum floor — 2006

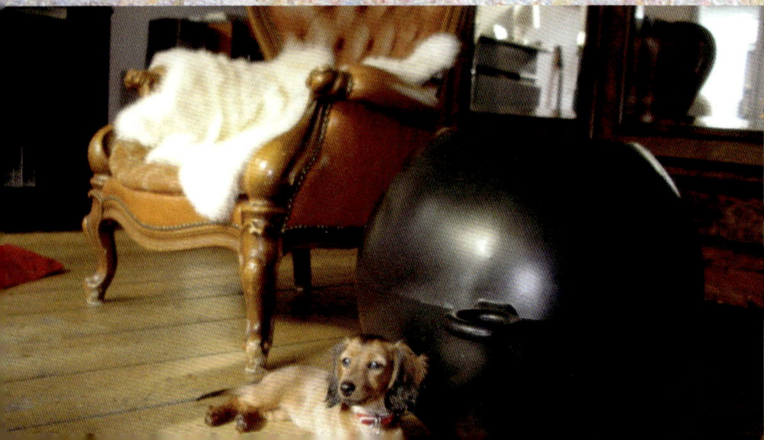

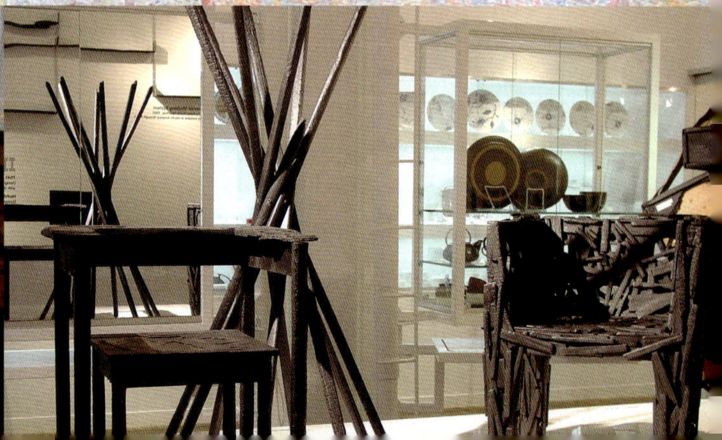

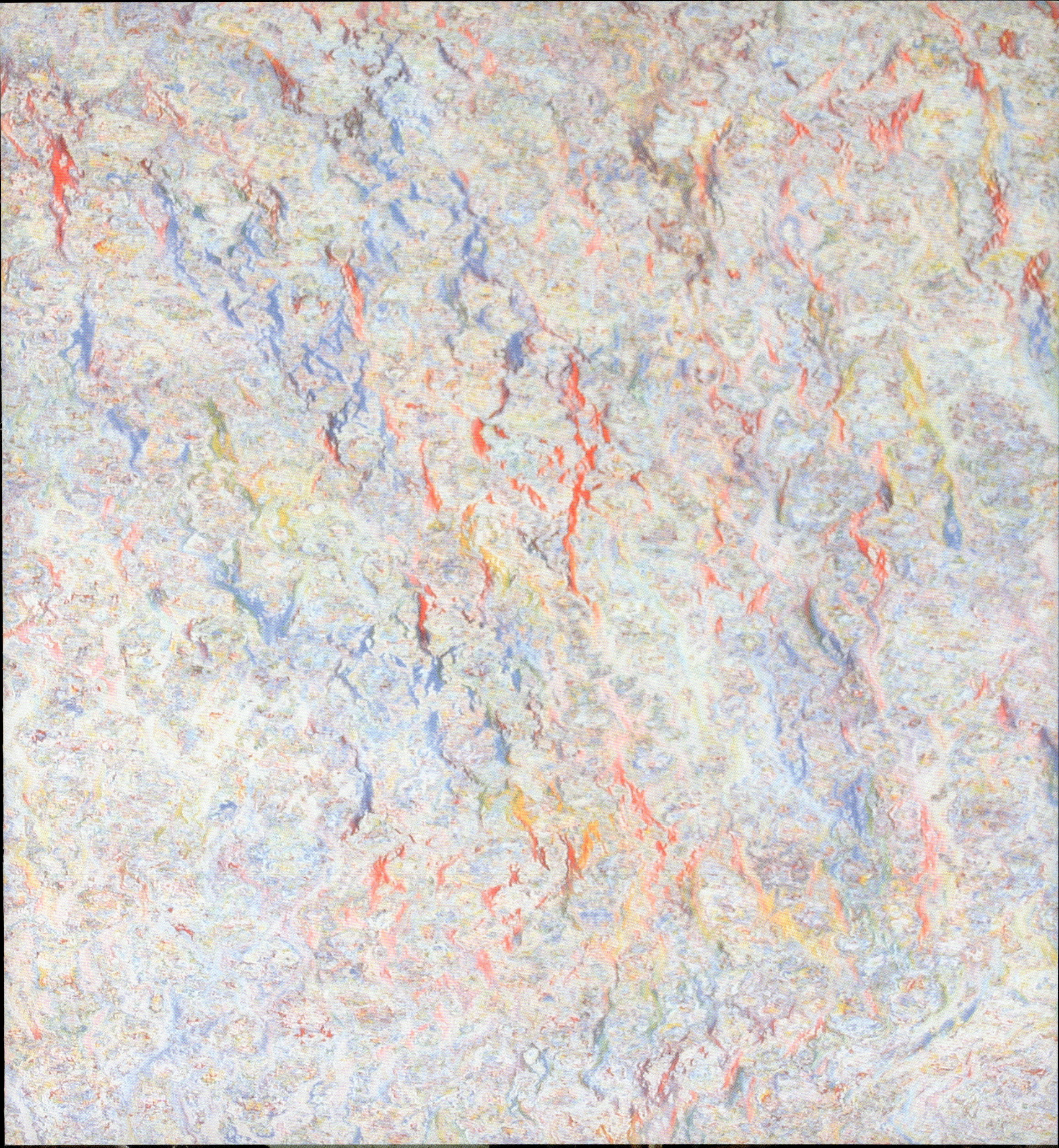
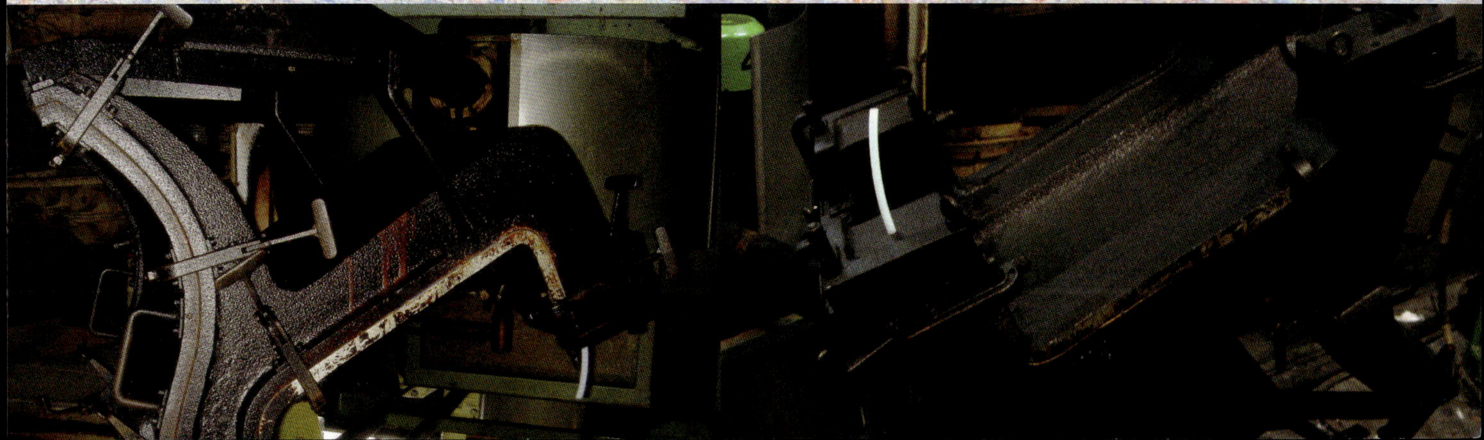

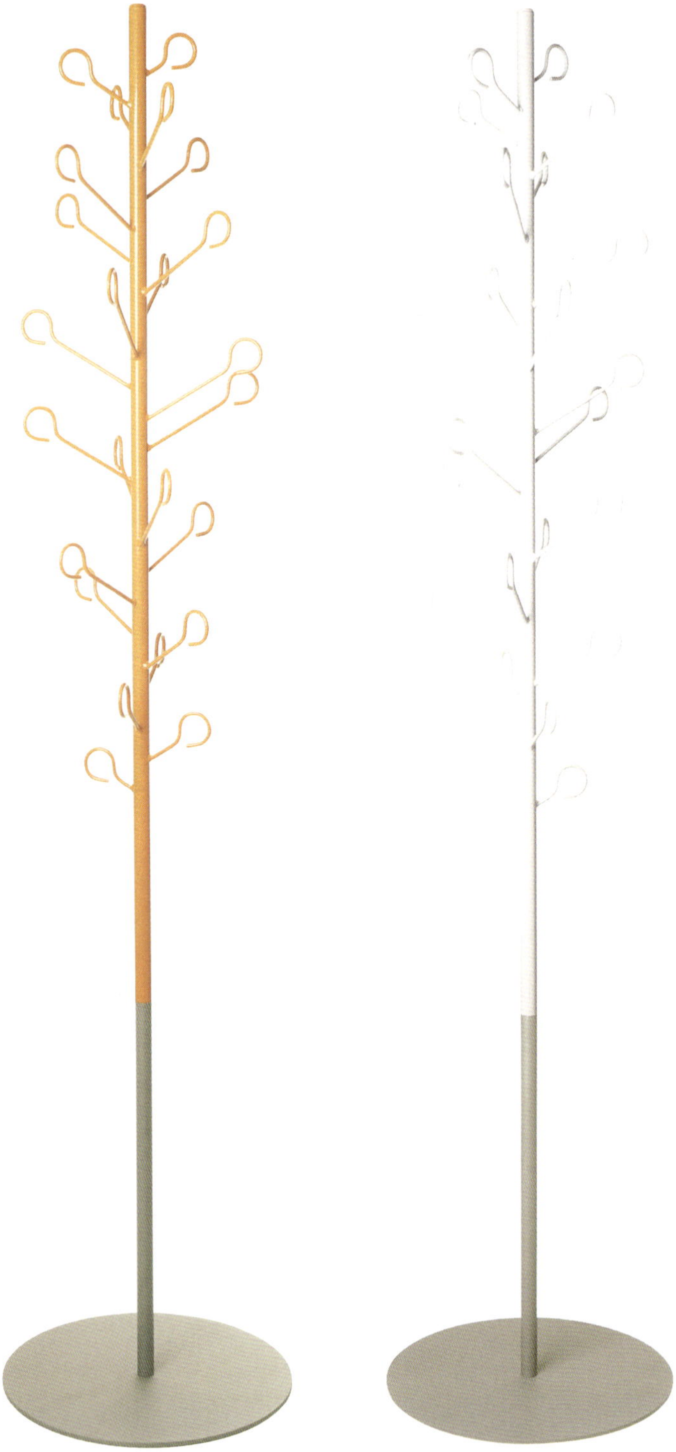

Coattrunk — 2000

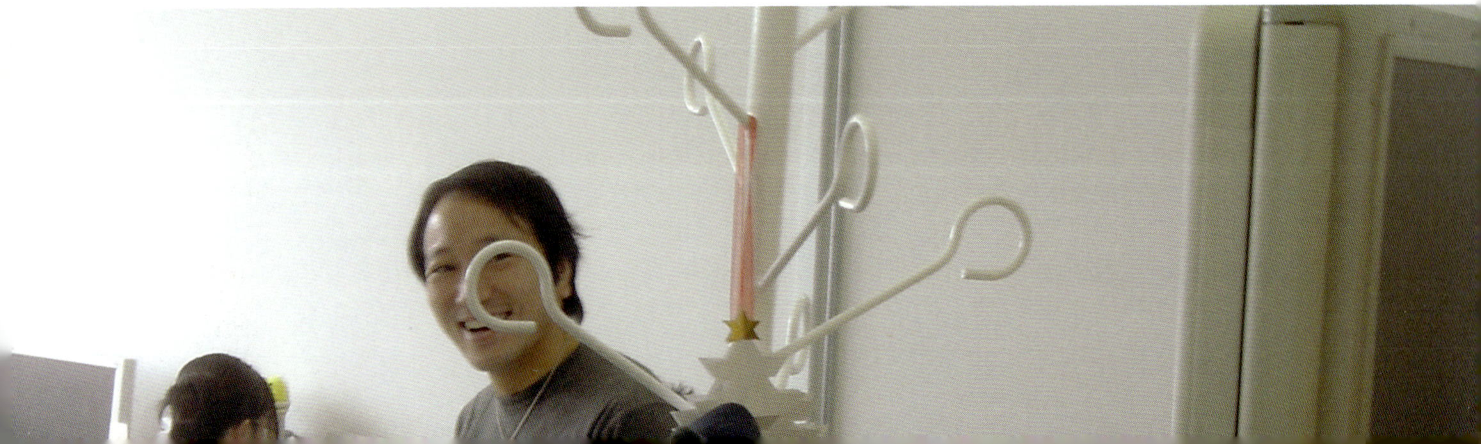

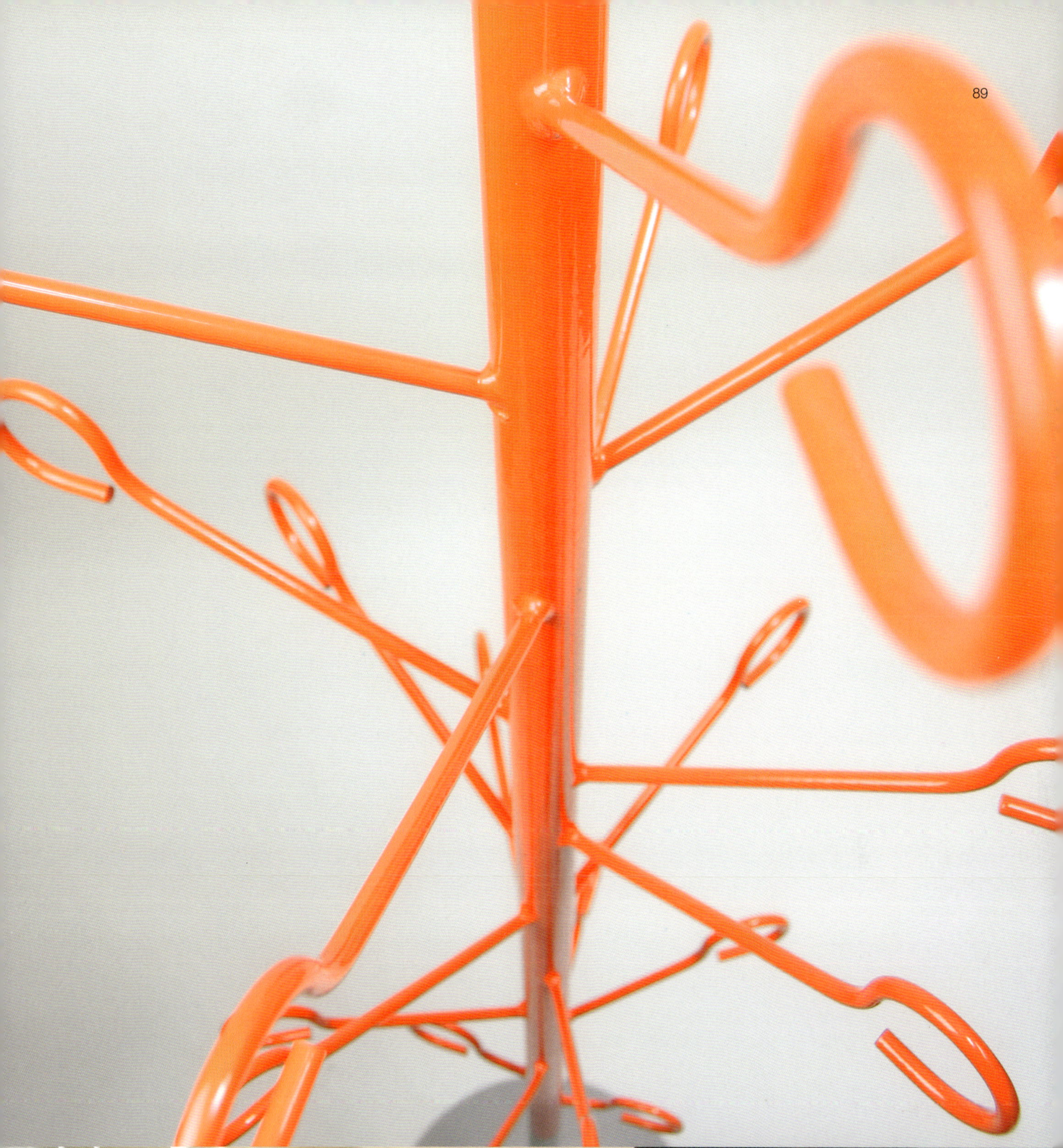

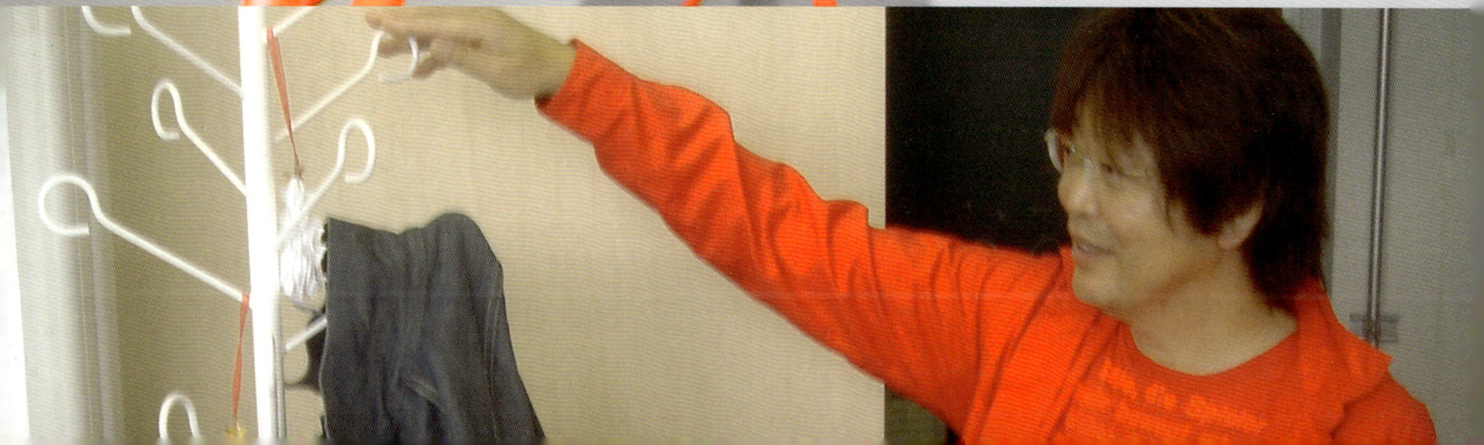

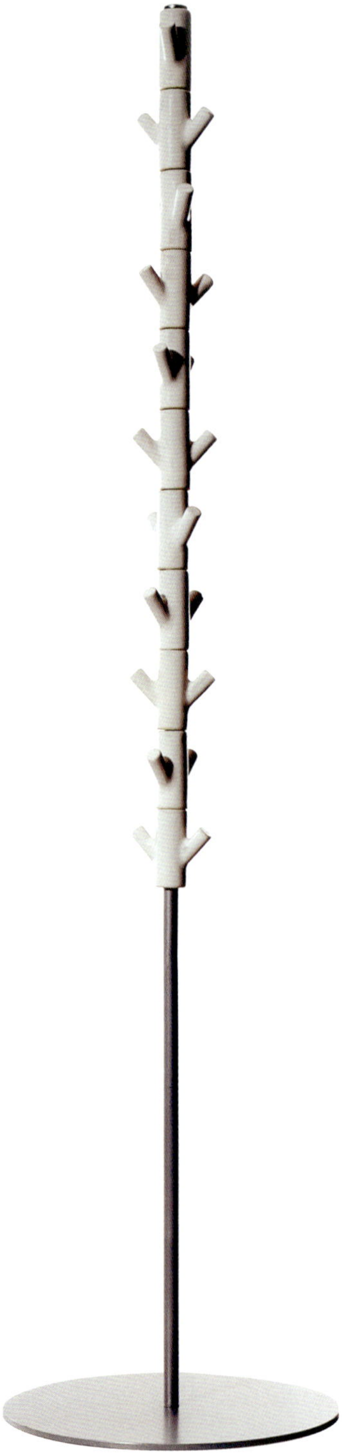

Bone — 2001

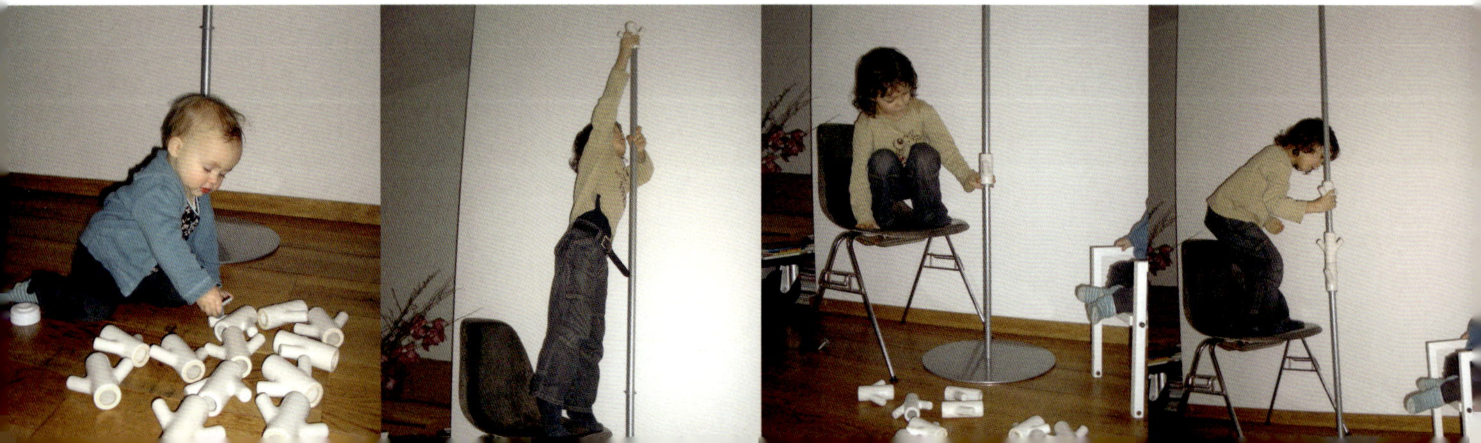

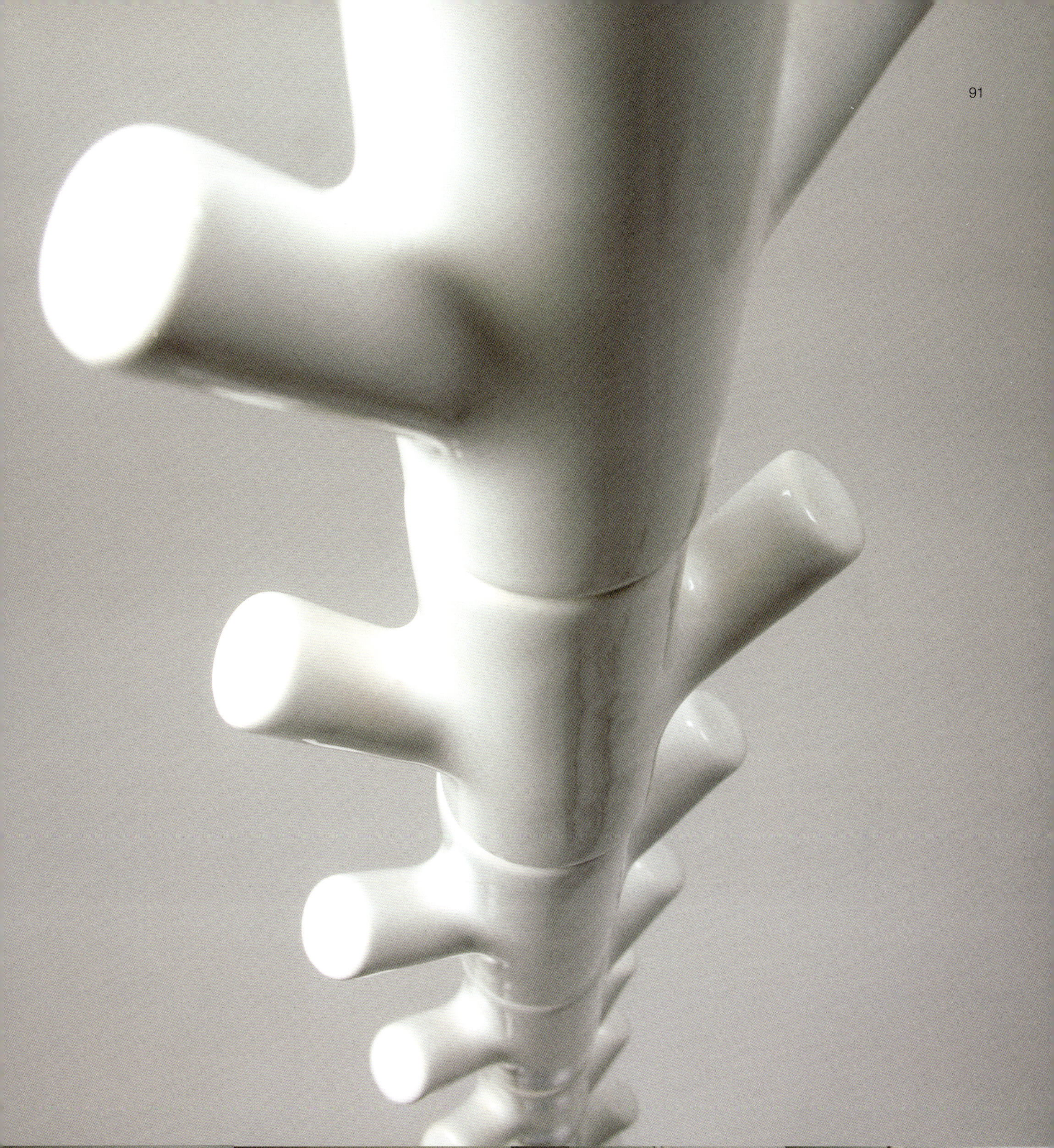

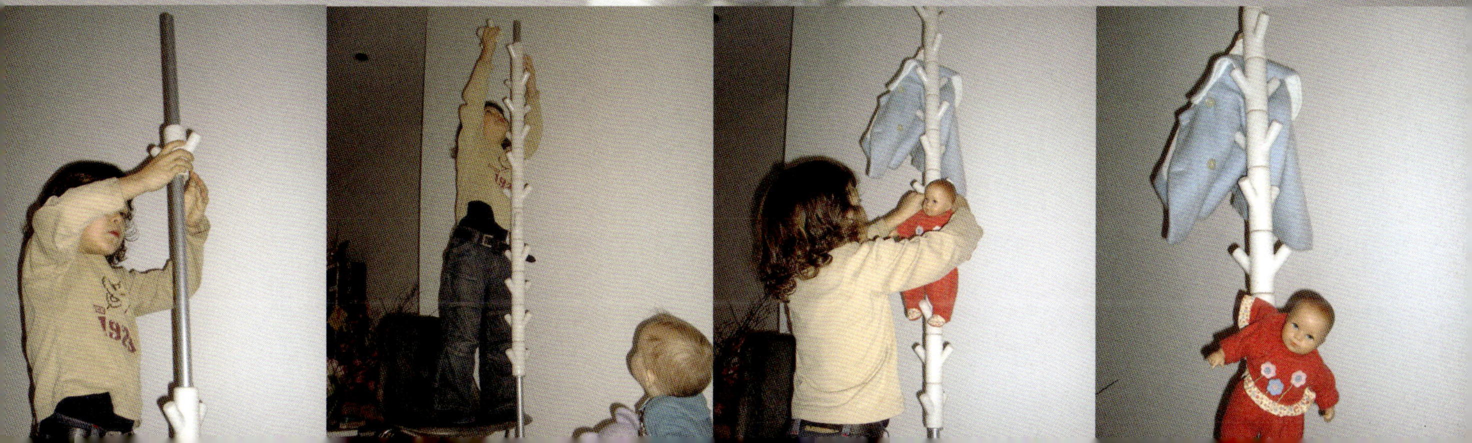

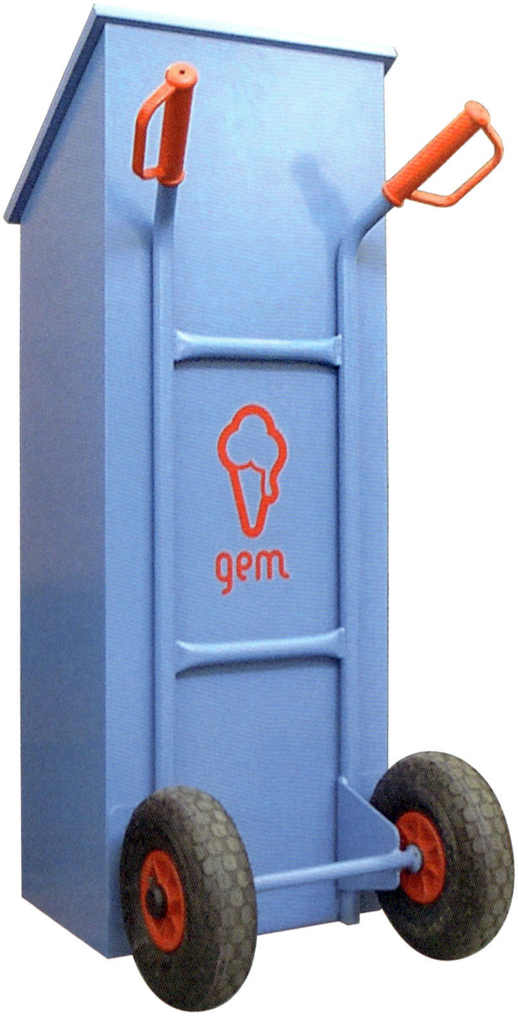

Roll it away — 2004

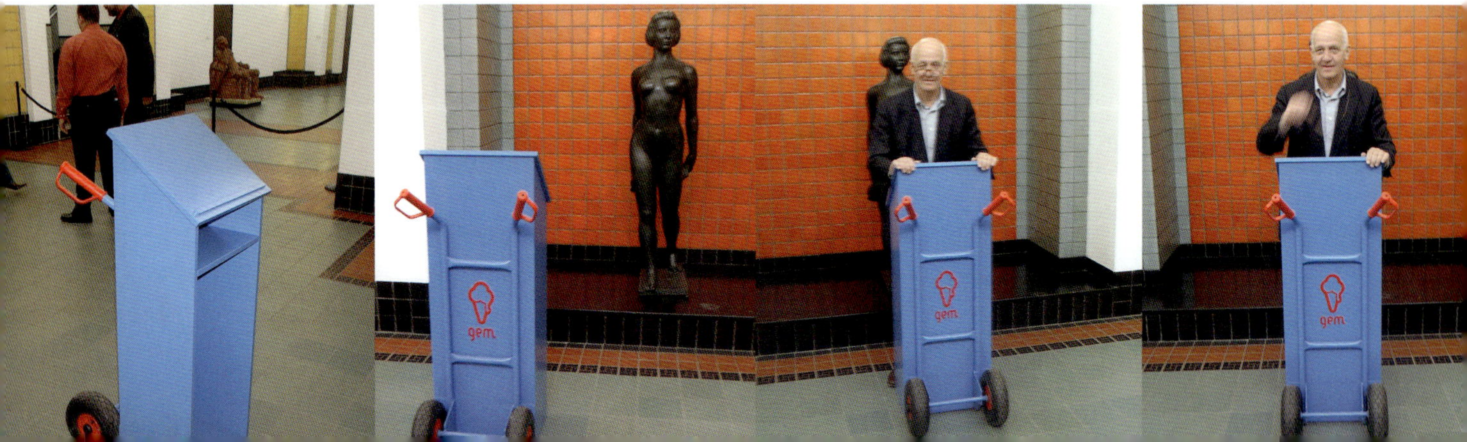

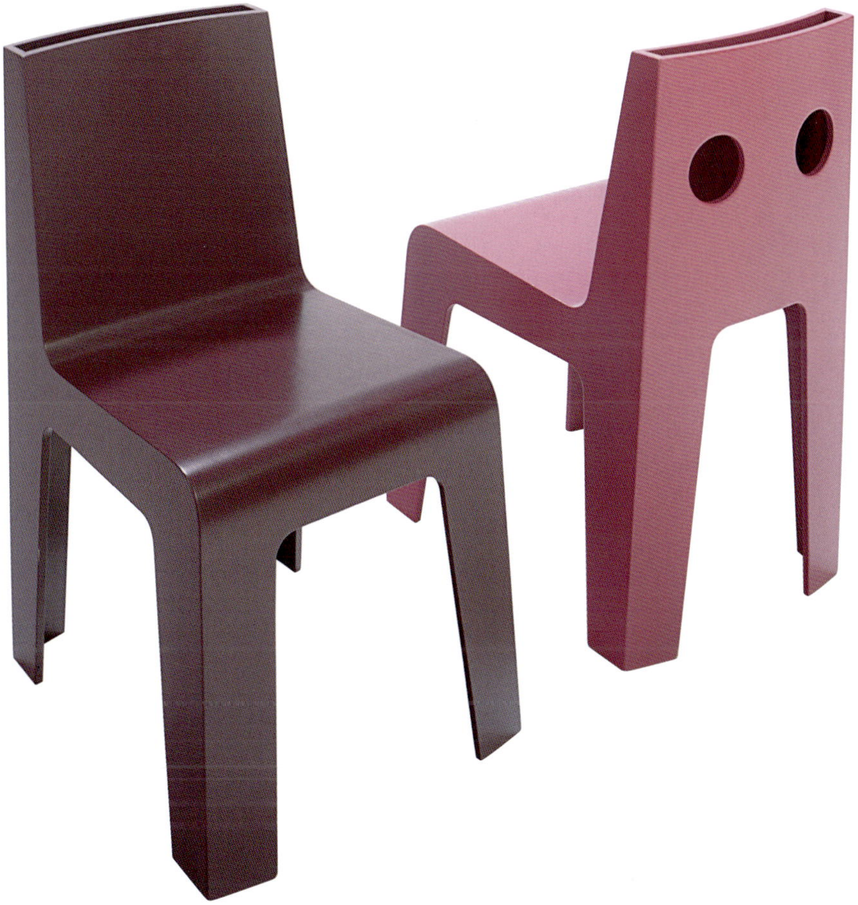

Abel children chair — 2001

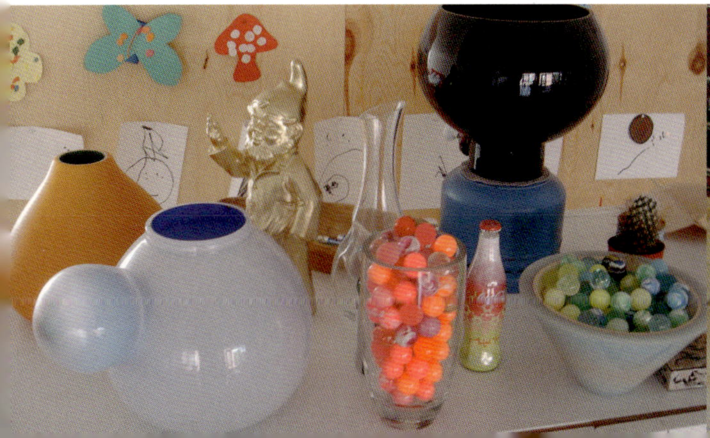

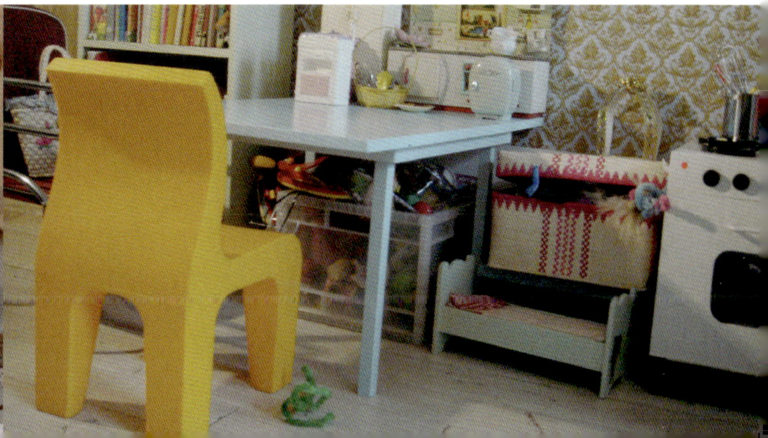

Brigitte Fitoussi

IN USE
A new look at uses

I met Richard Hutten for the first time in Tokyo in October 2003. Clearly adulated in Japan, this thirty-year-old Dutch designer had come for the opening of the exhibition 'Toys for Life'. A retrospective that the Axis Building in Ropponggi had dedicated to him as a part of Tokyo Designers' Week. He had designed the scenography himself, dividing the space in two parts with a long diagonal of seats and small pieces of furniture in between which the floor had been covered with a profusion of scattered light plastic yellow balls. Play… is a recurrent theme with Hutten whose vocabulary of forms through its simplicity bears childlike aspects. The father of two young boys, Abel and Boris whose existence he proudly reveals in his biography, he remains decidedly under the influence. The concept is neither naïve nor simplistic, on the contrary, it inspires a real breath of fresh air into the world of design. The spirit is very inventive, genuinely motivated by a desire to experiment and discover, as children often do. I therefore discovered in Tokyo his colourful, playful world, and was surprised by the simplicity and expressivity of each one of his designs. Hardly any words, a picture, a feeling, as in 'Haïkus', the short Japanese poems that instantly find their just form.

I was also impressed by his humour and his way of slightly confusing the issue as regards to a certain established order dictating notions such as use or functionality. Hutten's objects

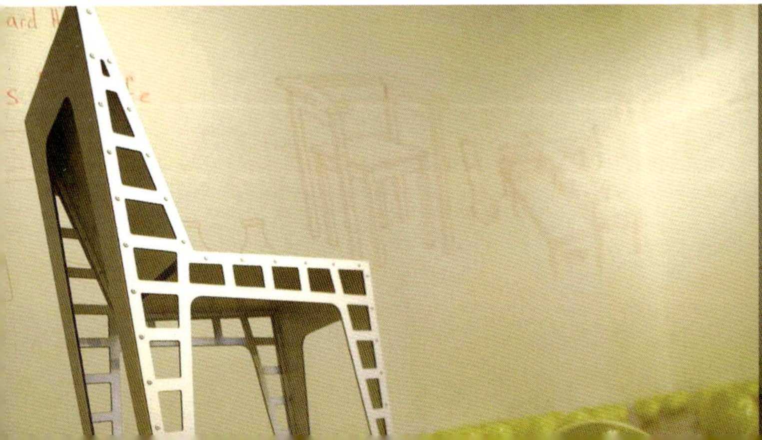

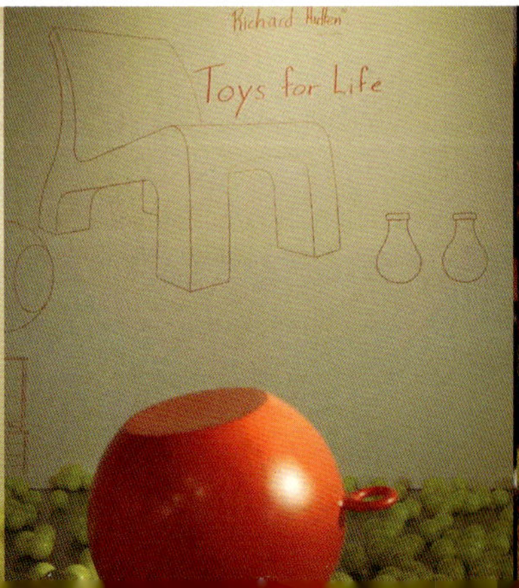

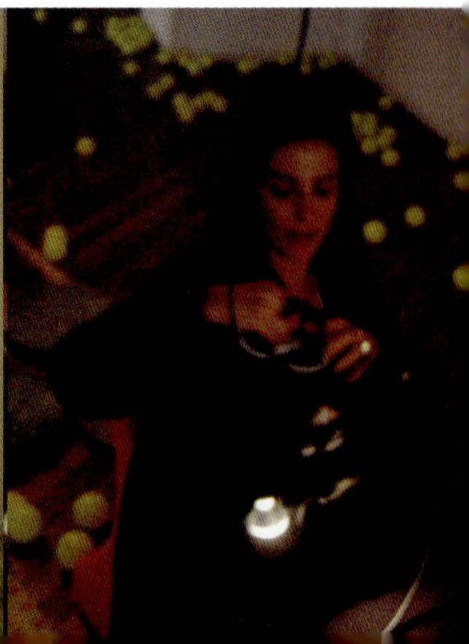

attract attention because they are amusing and because their reinterpretation of archetype is relevant. It would seem that each time, they represent an excuse to confront styles and periods, tradition and avant-garde, to explore matter, reveal a use and tell a new story around an ordinary object… An object which, under his control, always ends up distinguishing itself, through a small inexplicable something else (which he explains), this supplement of soul which defines talent. It is, for example, the case of the 'Sexy Relaxy' chair (my favourite), a quasi monobloc seat, inspired by Sharon Stone in 'Basic Instinct', of which the slit in the middle, marking the two legs, evokes even before one sits on it a very suggestive posture. Without ostentation yet efficient. Very ironic. Or 'Thing', a simple light-coloured wooden bench with a very sparing design but which instantly expresses a positive art of quirkiness. According to how Hutten positions, to the limits of the absurd, backrests or armrests the very expressive bench-object transforms itself and suggests here too different attitudes and postures. It becomes humanised and inspires a new concept in seating. A form of theatricality.

Another one of his leading designs, the plastic Dumbo mug: with its two oversized handles, fluorescent colours and sensual, satin-smooth material, appeals to the eye and the touch. Toy object, emotional object and useful object, all in one, in a few subtle words he revisits the child's metal tumbler… A form of poetry.

In his large studio, situated in a Rotterdam inner suburb where he lives and works, Richard Hutten gives shape to his thoughts through a computer as well as through a multitude of prototypes he produces with his team. He checks out. He also manufactures some of his furniture, often in wood, which he produces in small series, under his own name. The Richard Hutten collection. Reflection upstream of the design process is solitary. He often thinks like an architect by concentrating on the structure of things and on the context these things will be a part of. Shape is not his first

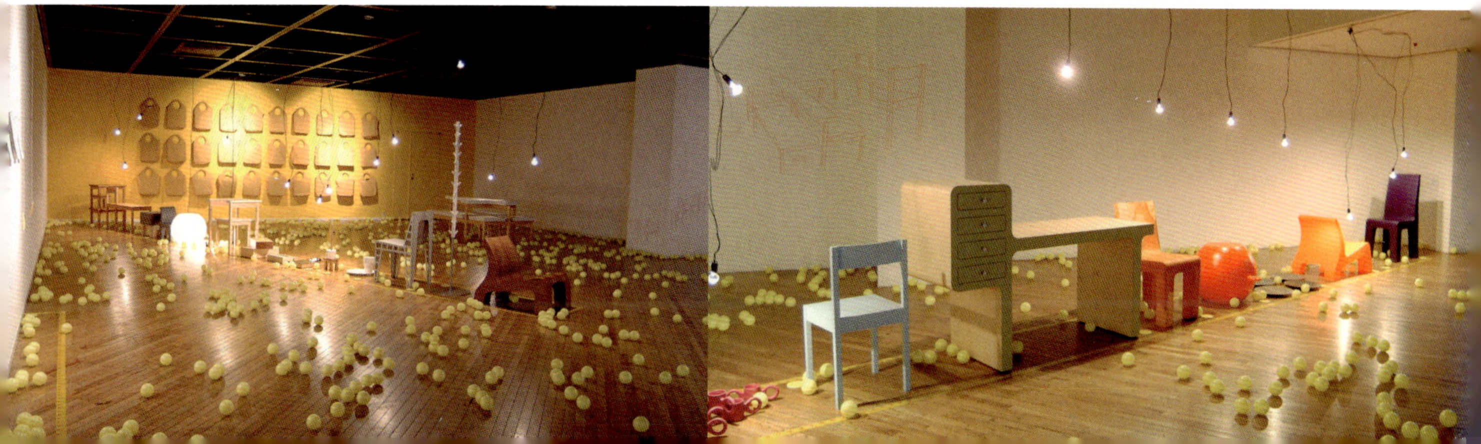

concern, it is the idea or the concept that prevails. Often austere, rarely sophisticated, more often obvious or exacerbated, essentially the concept seeks to be meaningful. It stems from a whole mental process rather than from an aesthetic artistic gesture. By focusing fully on the object's use, by anticipating it, Hutten reveals his conceptual approach, contrasted by his intention to combine the artificial and the natural, acid colours and those of raw material, design and no design, the artistic and the functional, the industrial and the artisanal, the obvious and the subtle… A different world, therefore paradoxical, which proceeds from a reasoned and rational creative process, thought out to the limits of a radicalisation of the idea, but which always leaves 'a door open' to emotion, memory, surprise… A form of freedom.

In fact, I had become acquainted with the designer's work in snatches, through collective exhibitions at the furniture show in Milan and the Dutch Design exhibitions, but mostly through Droog Design with which he has been involved since its creation in 1993, and of which he remains still today one of the emblematic figures. The famous Dutch collective helped him at the beginning – after he graduated from the Design Academy of Eindhoven where he studied industrial design from 1986 to 1991 – to become well known internationally. It also helped to establish the reputation of other designers of his generation (who have become friends) such as Marcel Wanders, Jurgen Bey and Hella Jongerius.

To be perfectly honest, this encounter in Tokyo with Richard Hutten was a little more decisive for me. It enabled me to become better acquainted with the creative spirit of the Dutch, I who so far had my attention firmly turned towards Italian design. Of course, there is no question of promoting a specifically Dutch style given that design, having become an international language, cannot be summed up as a territorial identity. However, it is undeniable that contemporary local culture, in its way of integrating modernity and vernacular, makes a strong impression on the new

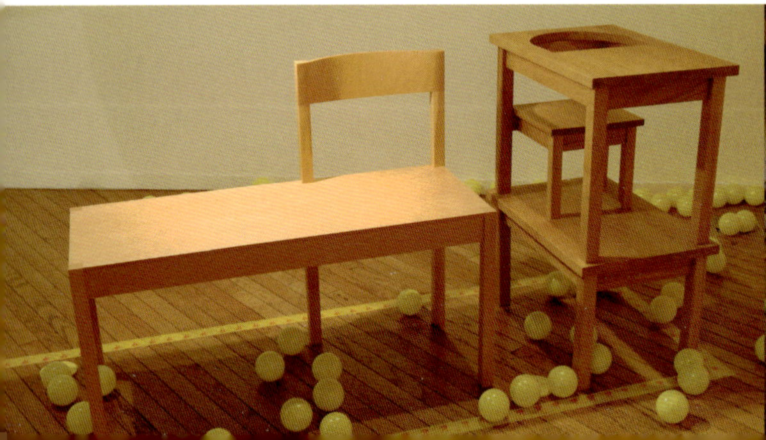

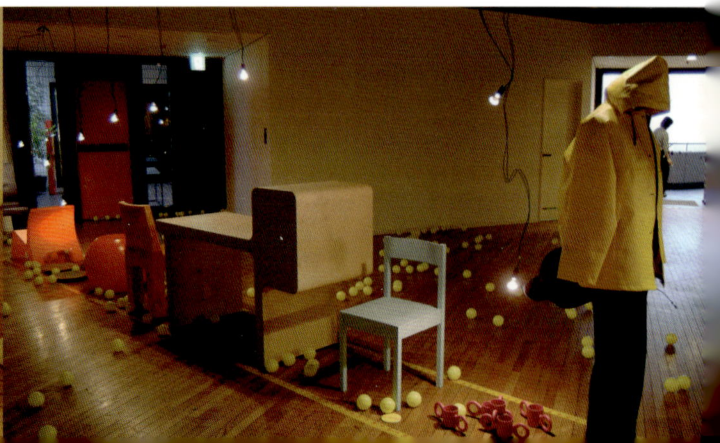

generations of designers more than ever conscious of the perverse effects of globalisation. It reinforces an identity, singularises it. Richard Hutten is no exception to the rule.

Dutch inventiveness, of which he is the new herald, is not necessarily revolutionary, as in the modernist 'tabula rasa', but is to be found in the diversions, recuperations, citations of existing objects, materials or practises, in the desire to propose a new poetry for daily life. Innovation results from a personal reinterpretation, in a singular diversion of pre-existing points of view. One observes a cultural attitude shared by many contemporary Dutch designers which operates through elective affinities and which, as with Richard Hutten, consists of taking another look at ordinary things, to reveal in them another form of beauty.

Hutten's eye intelligently questions use which becomes the connecting thread of his work. 'In use' marks a point of view. Contextually it seeks to debunk the myth of the object of design, to take it out of the glossy pages of fashion magazines and the confines of museums to place it once again in a real context that is more intimate, more diverse, less refined… in one word 'real life'. Purposefully 'indiscrete', the designer – in search of answers to the intentions he nourishes for his projects – seeks here to establish ties, to provoke a dialogue between the user and the acquired object, to insist on the circulation of the objects he has created and the way they are used. By that very fact he underlines beyond an artistic or industrial expression (of his own), his intention to make design a part of daily life. A form of humanism.

Brigitte Fitoussi — March 2006

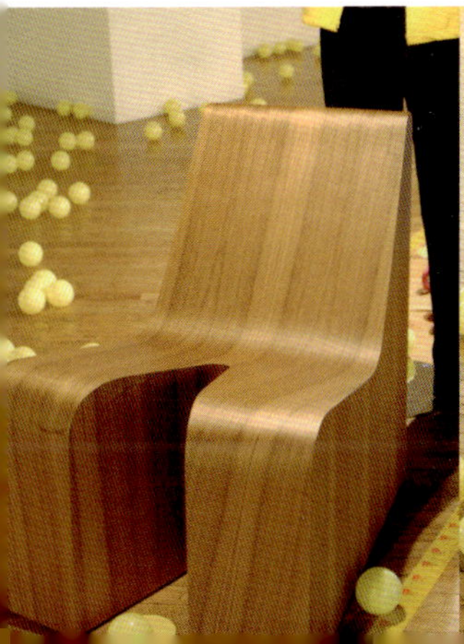
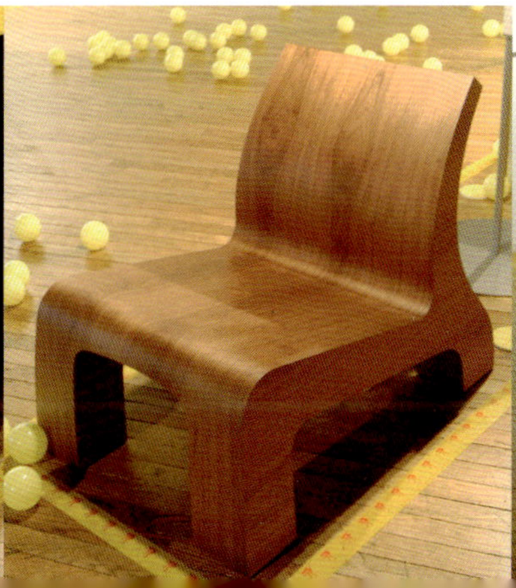
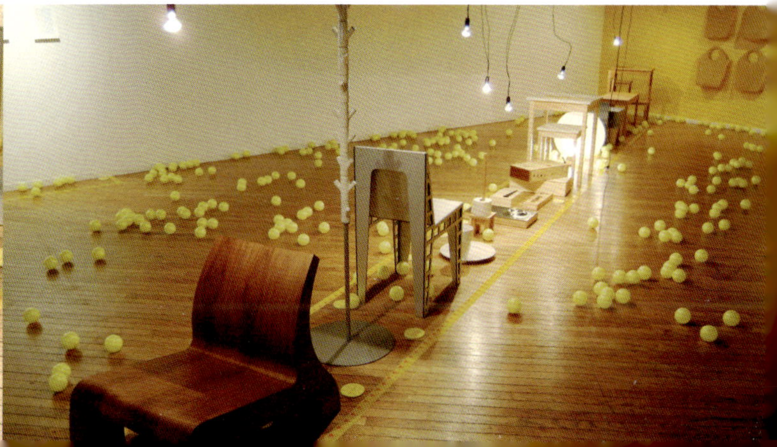

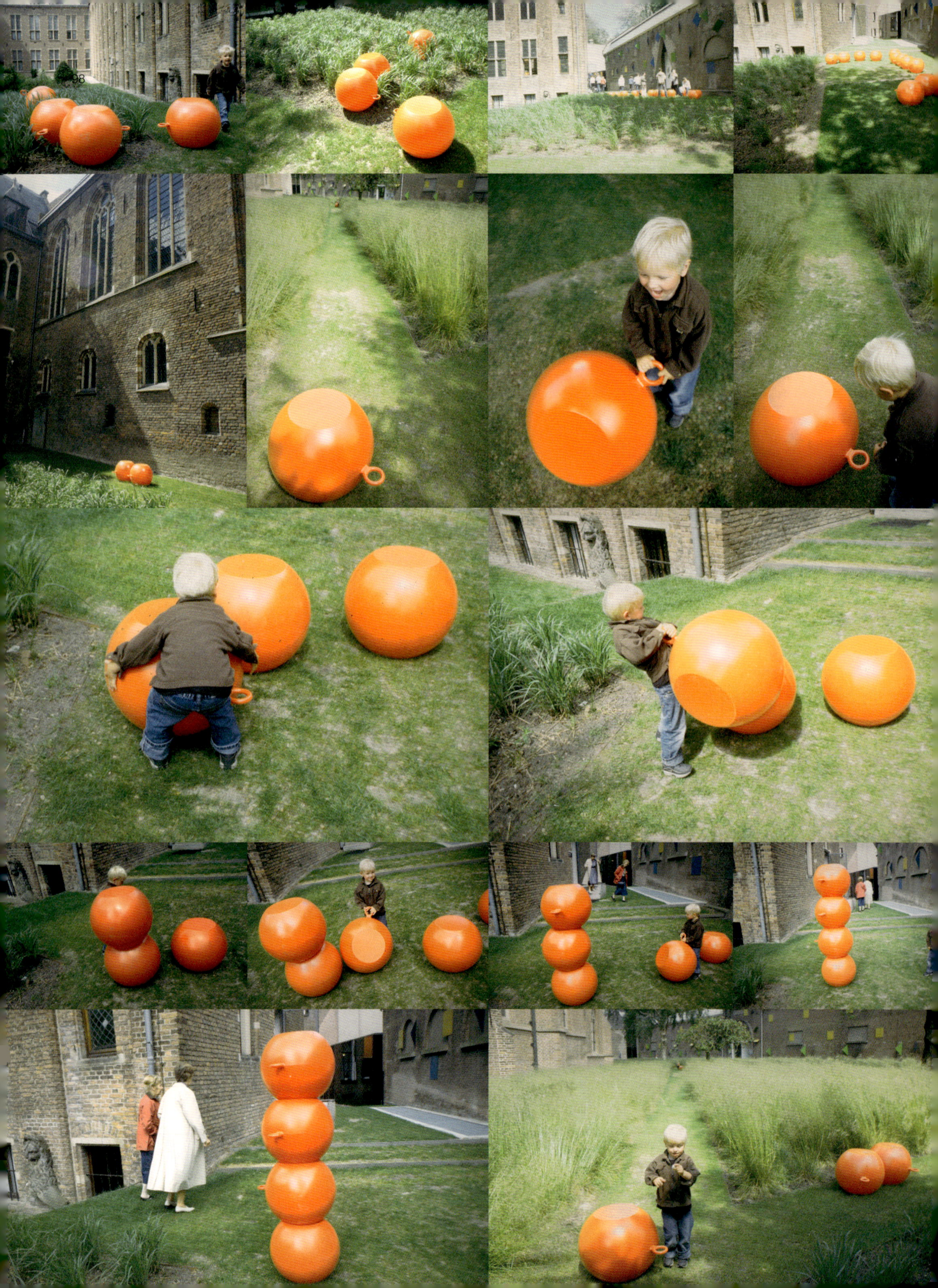

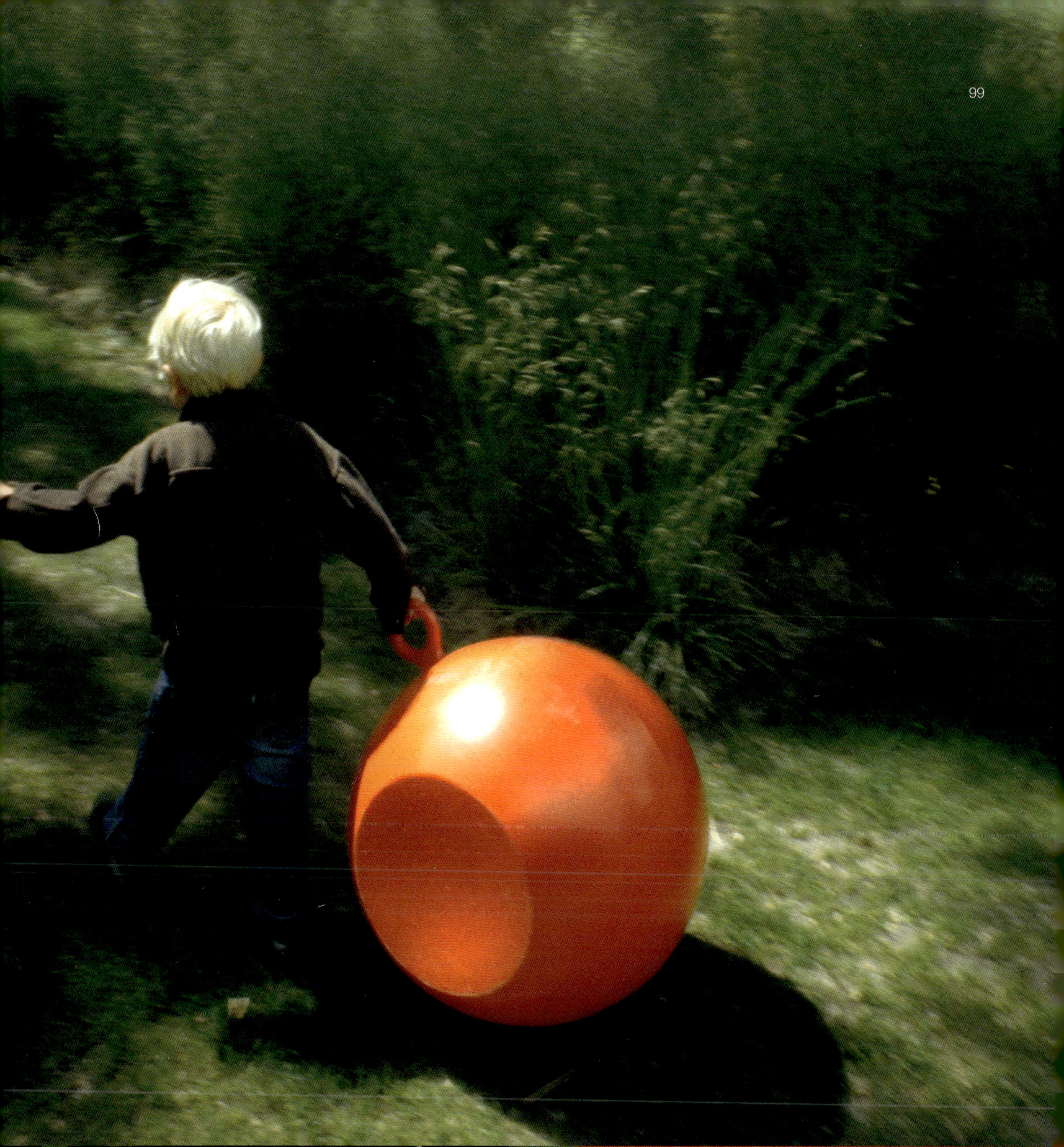

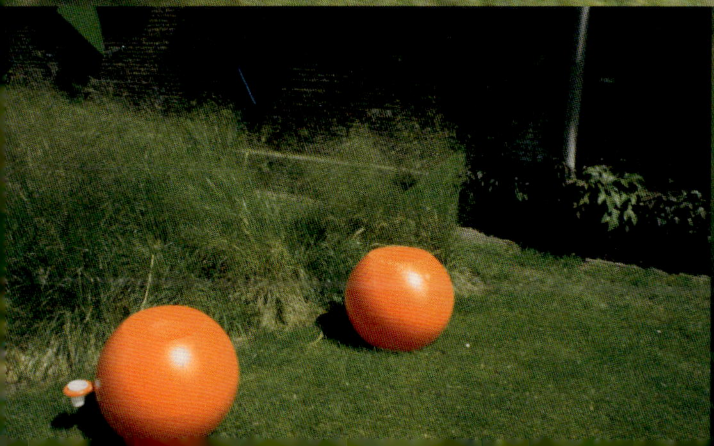

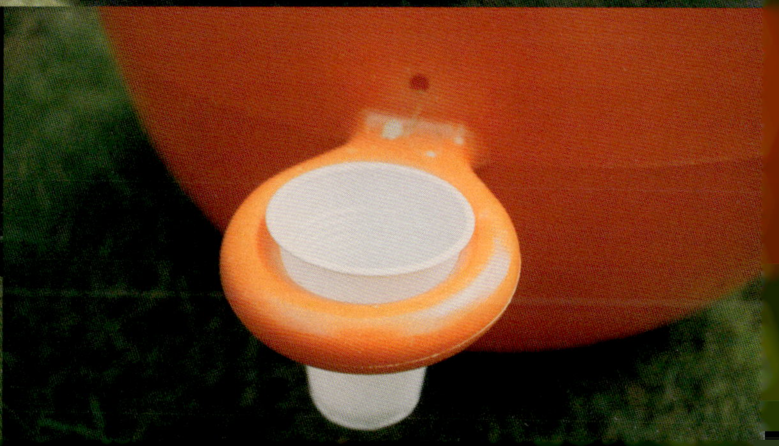

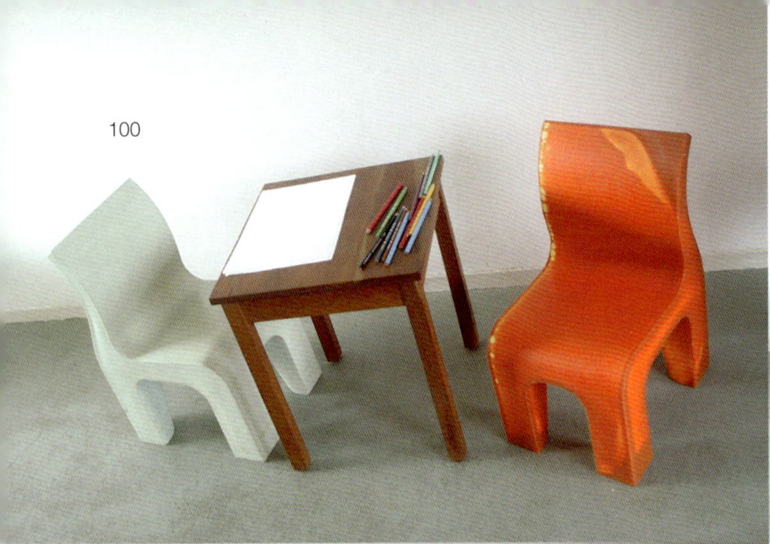
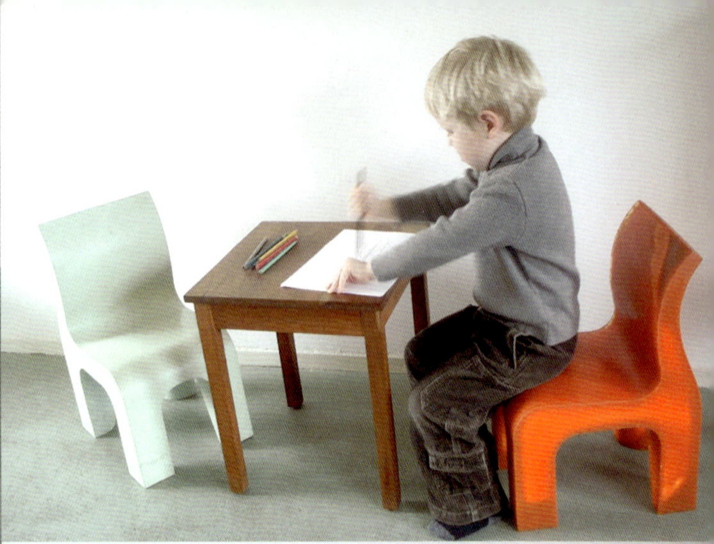
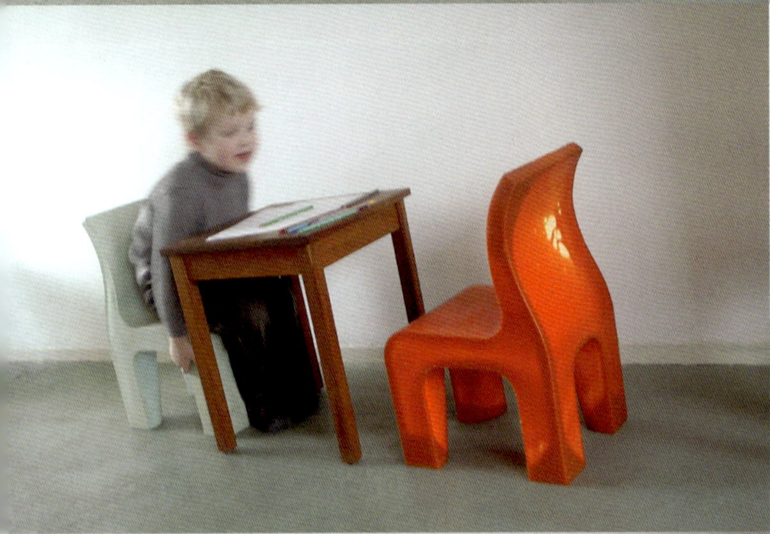
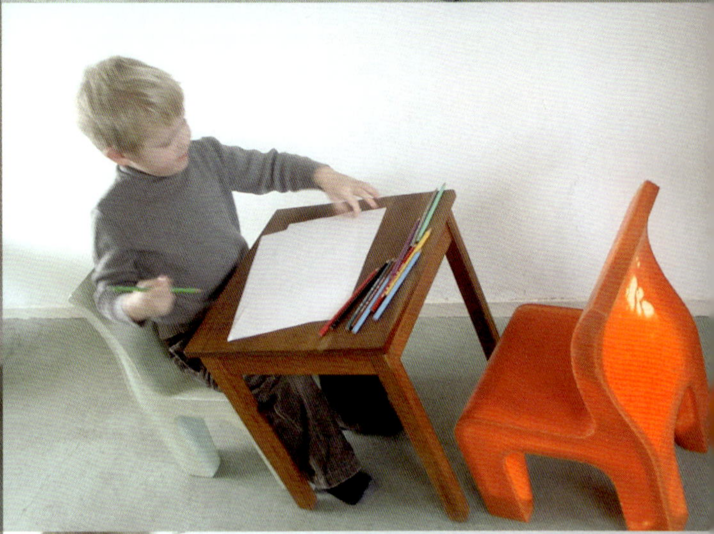
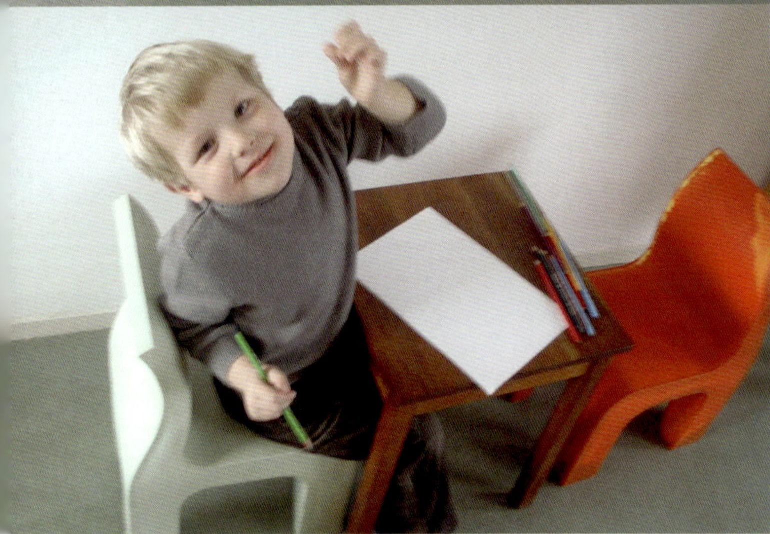
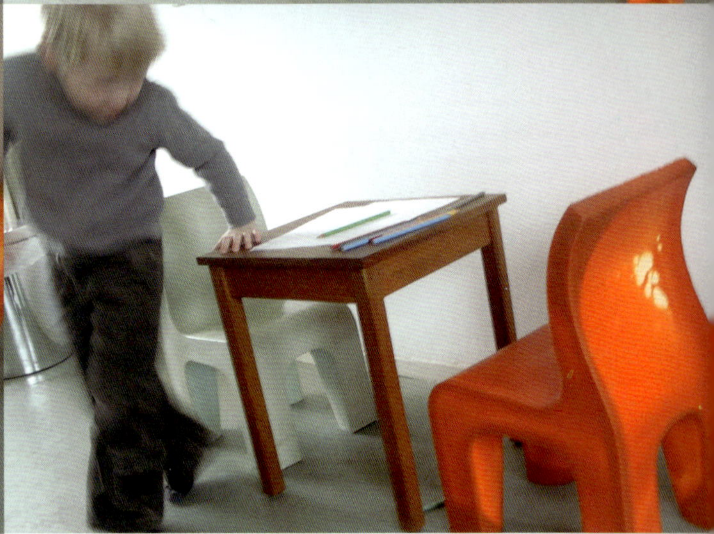
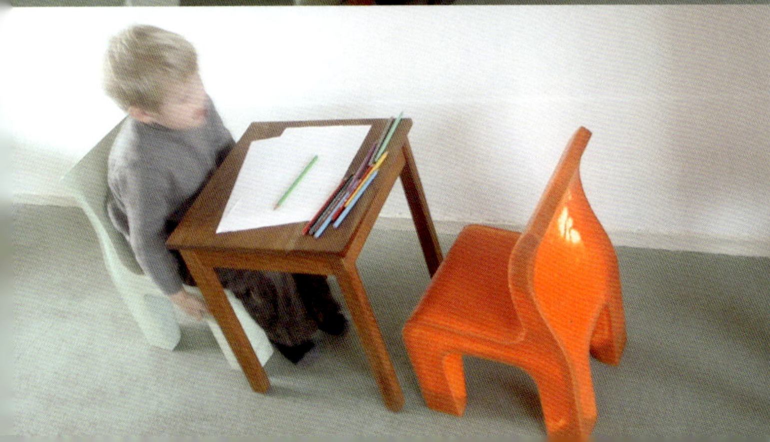
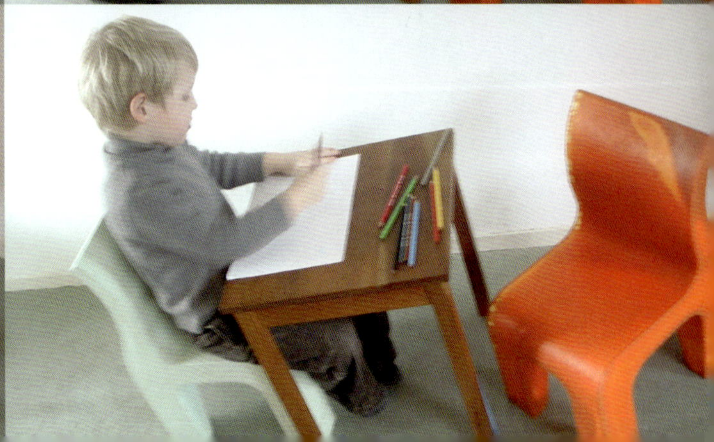

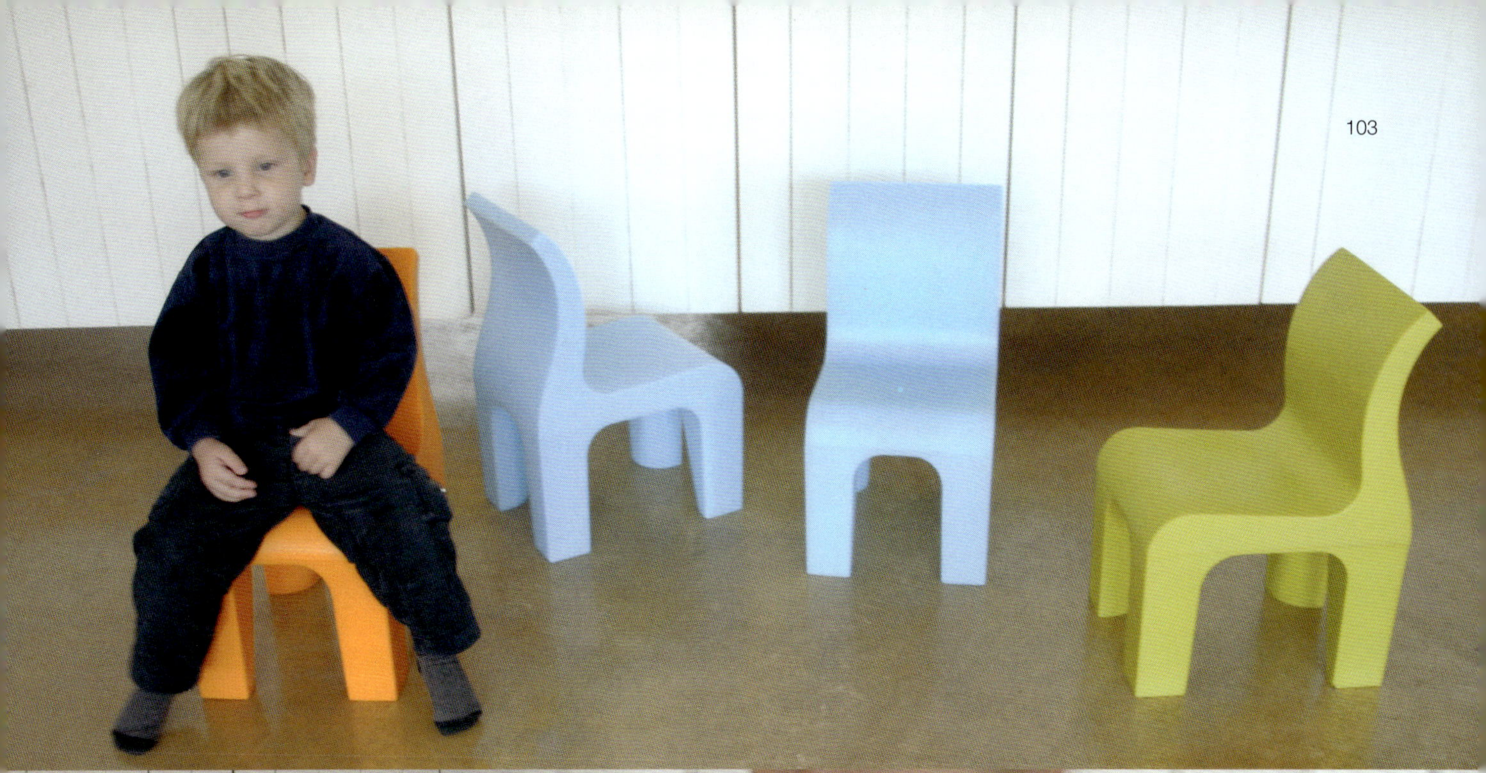

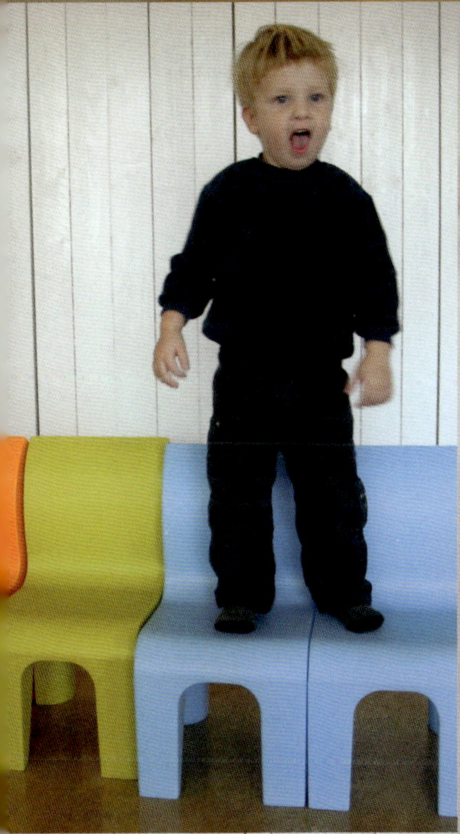

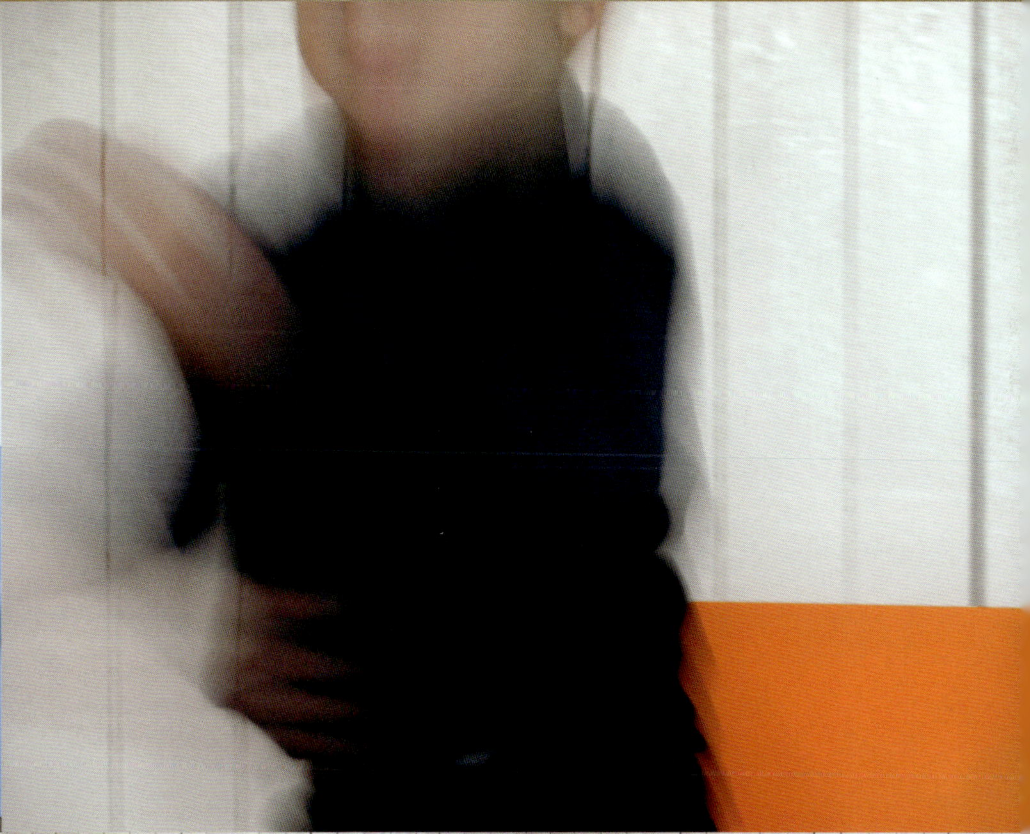

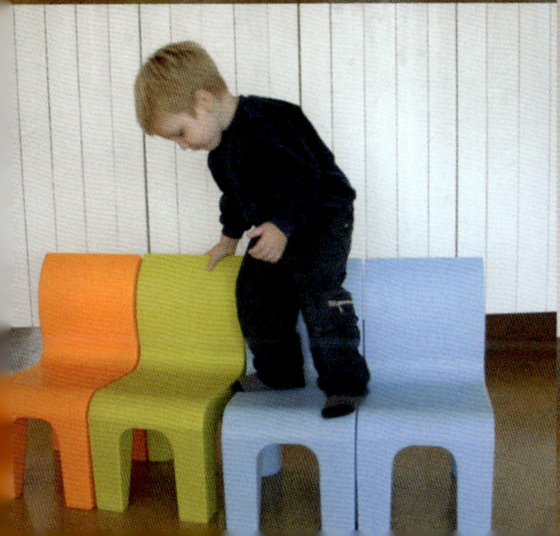

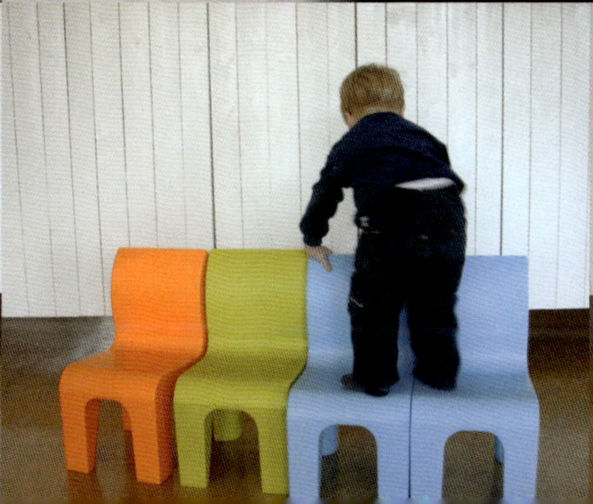

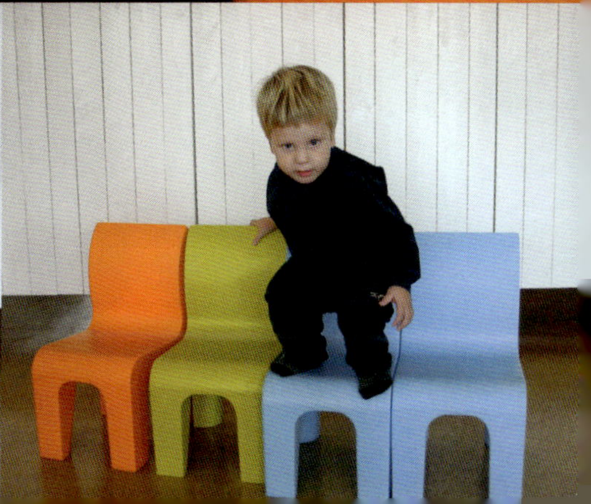

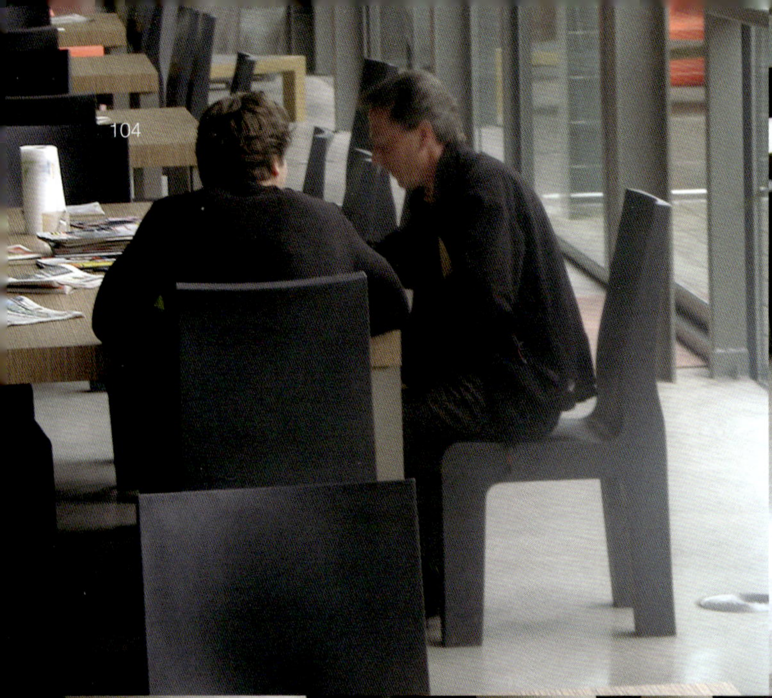

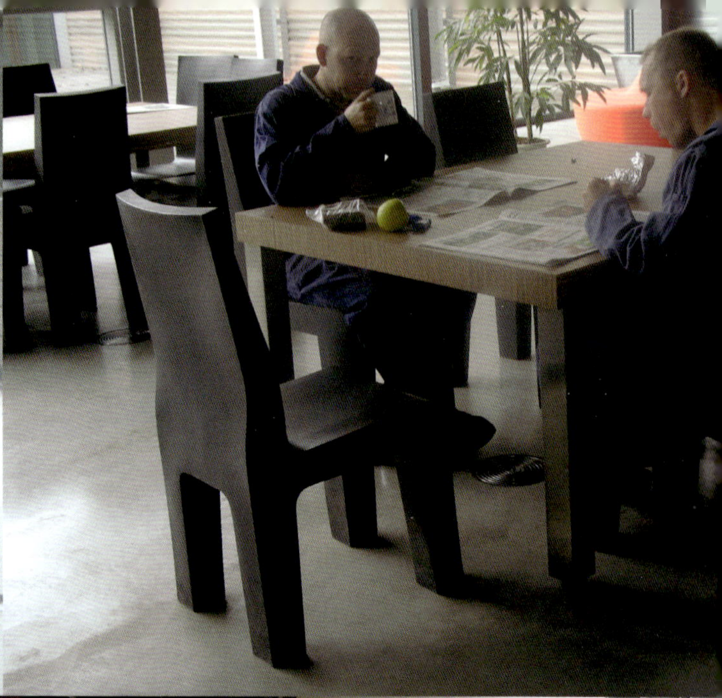

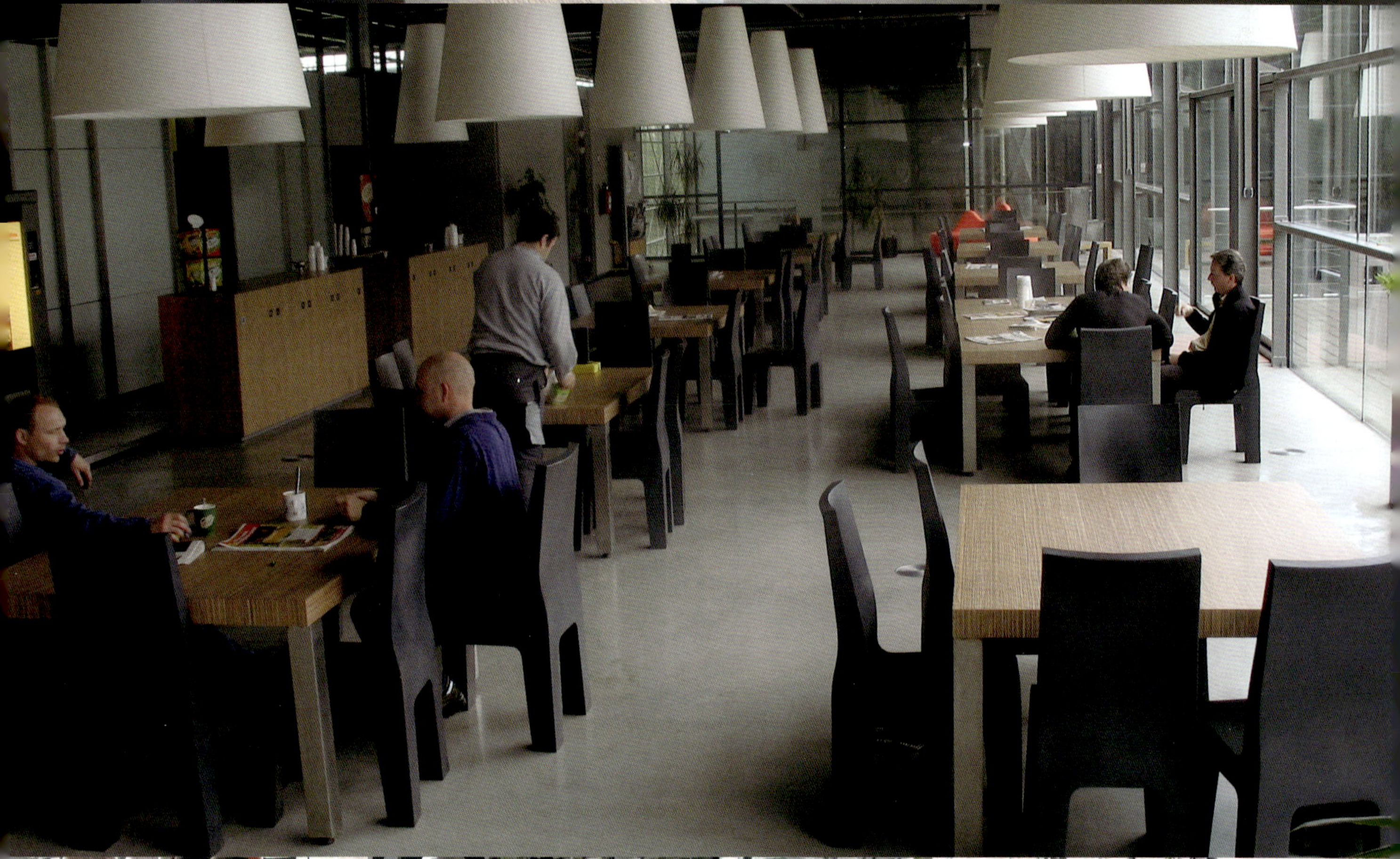

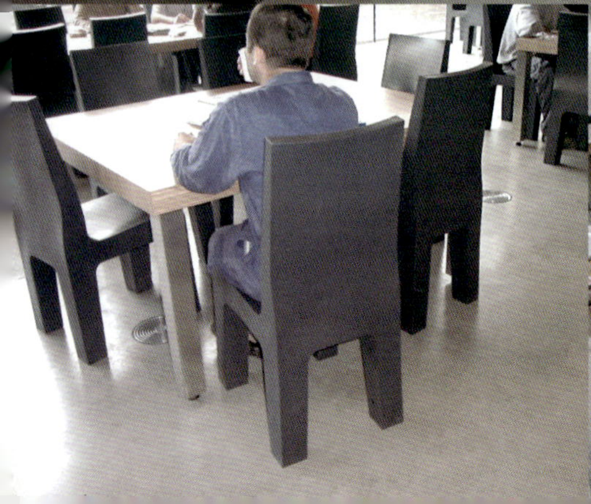

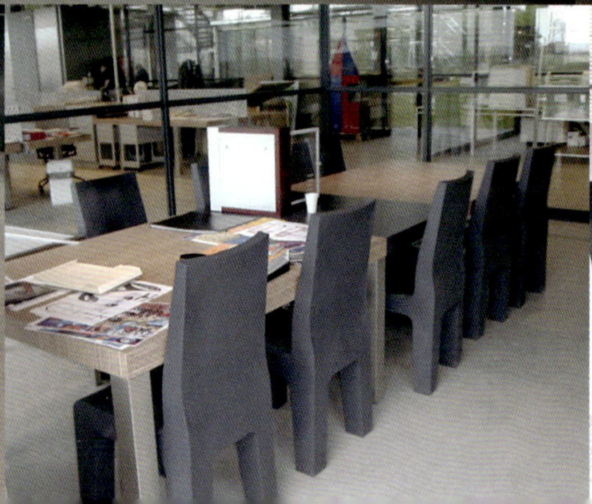

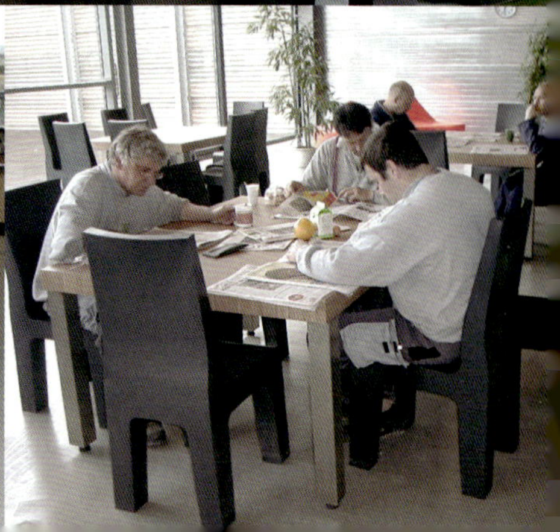

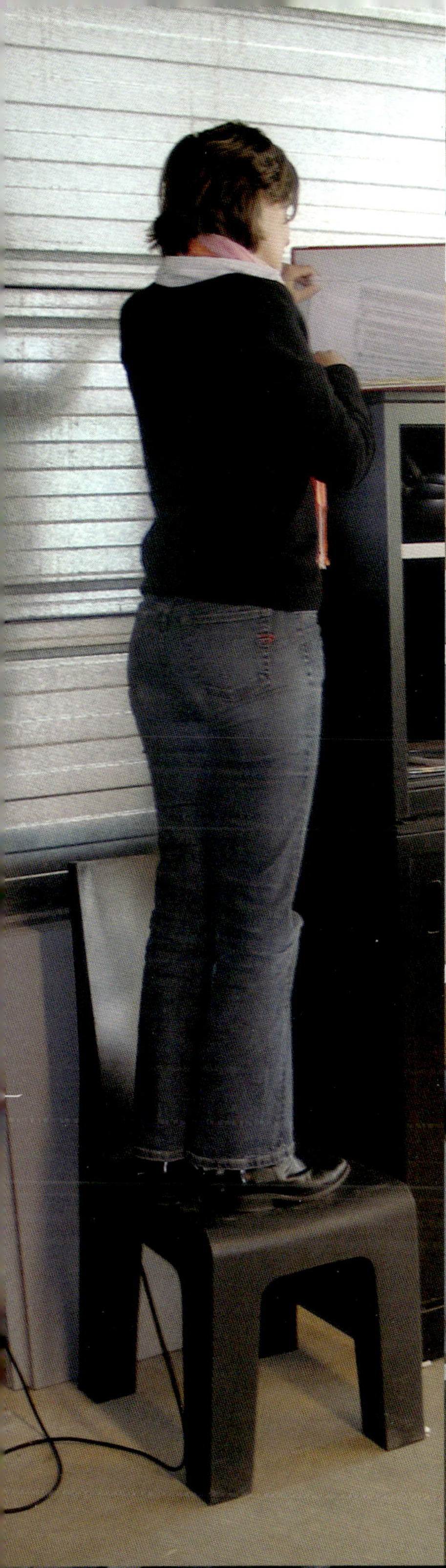

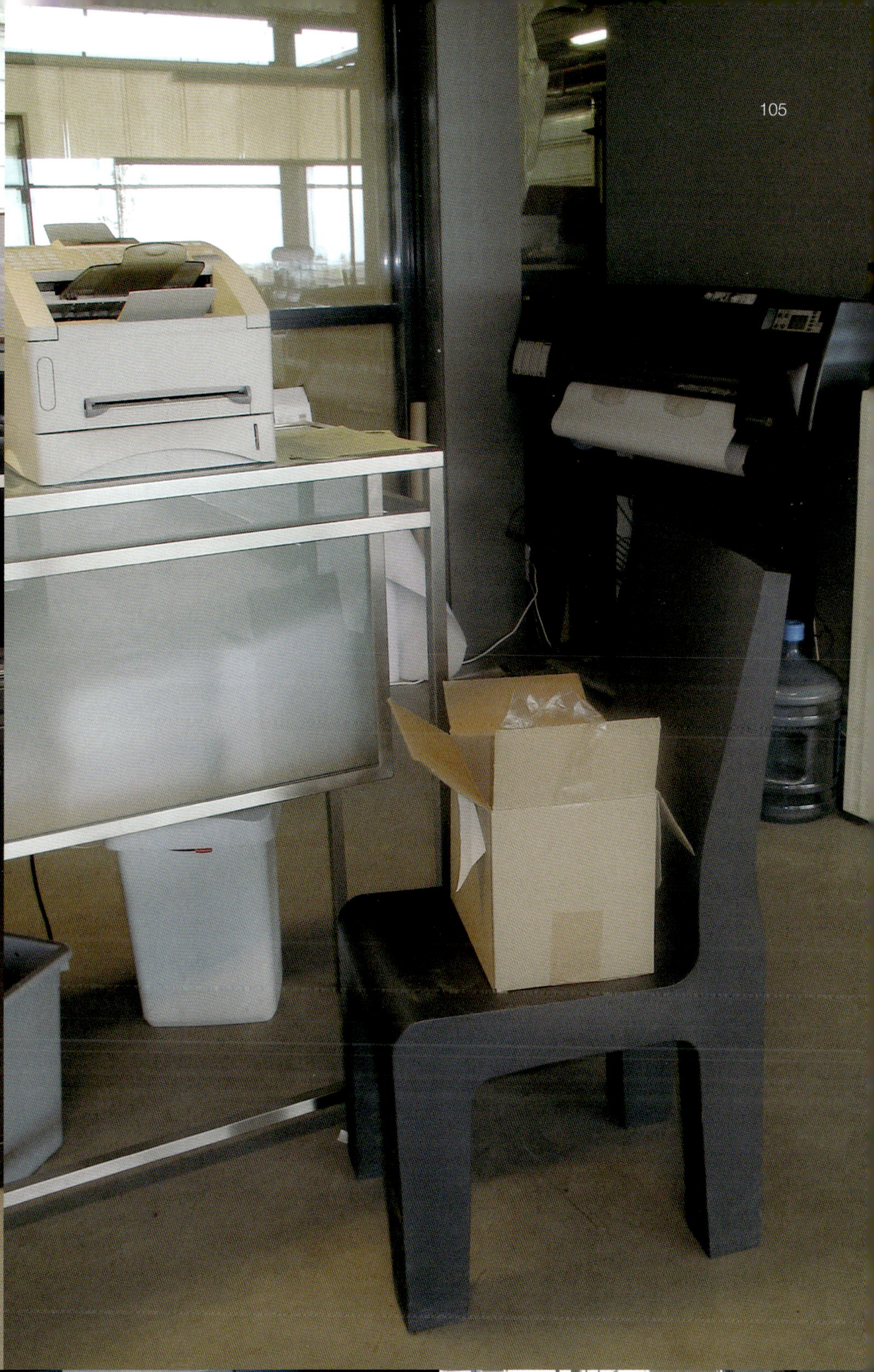

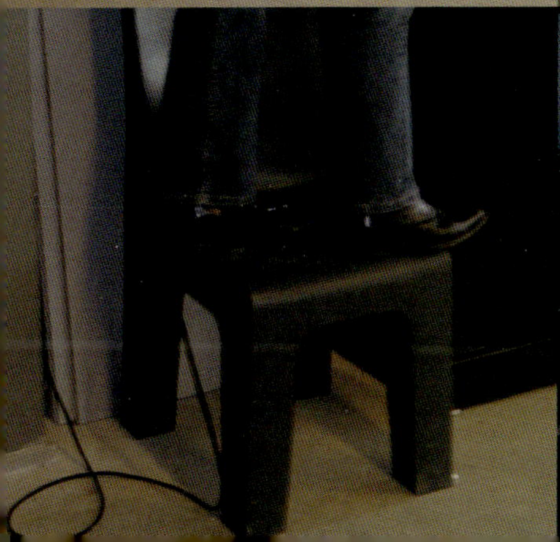

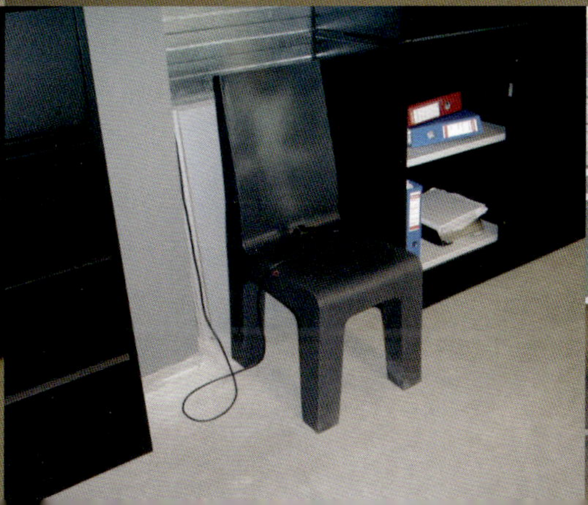

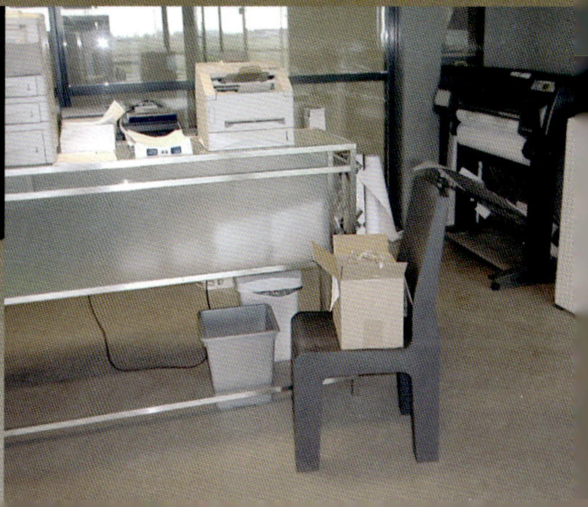

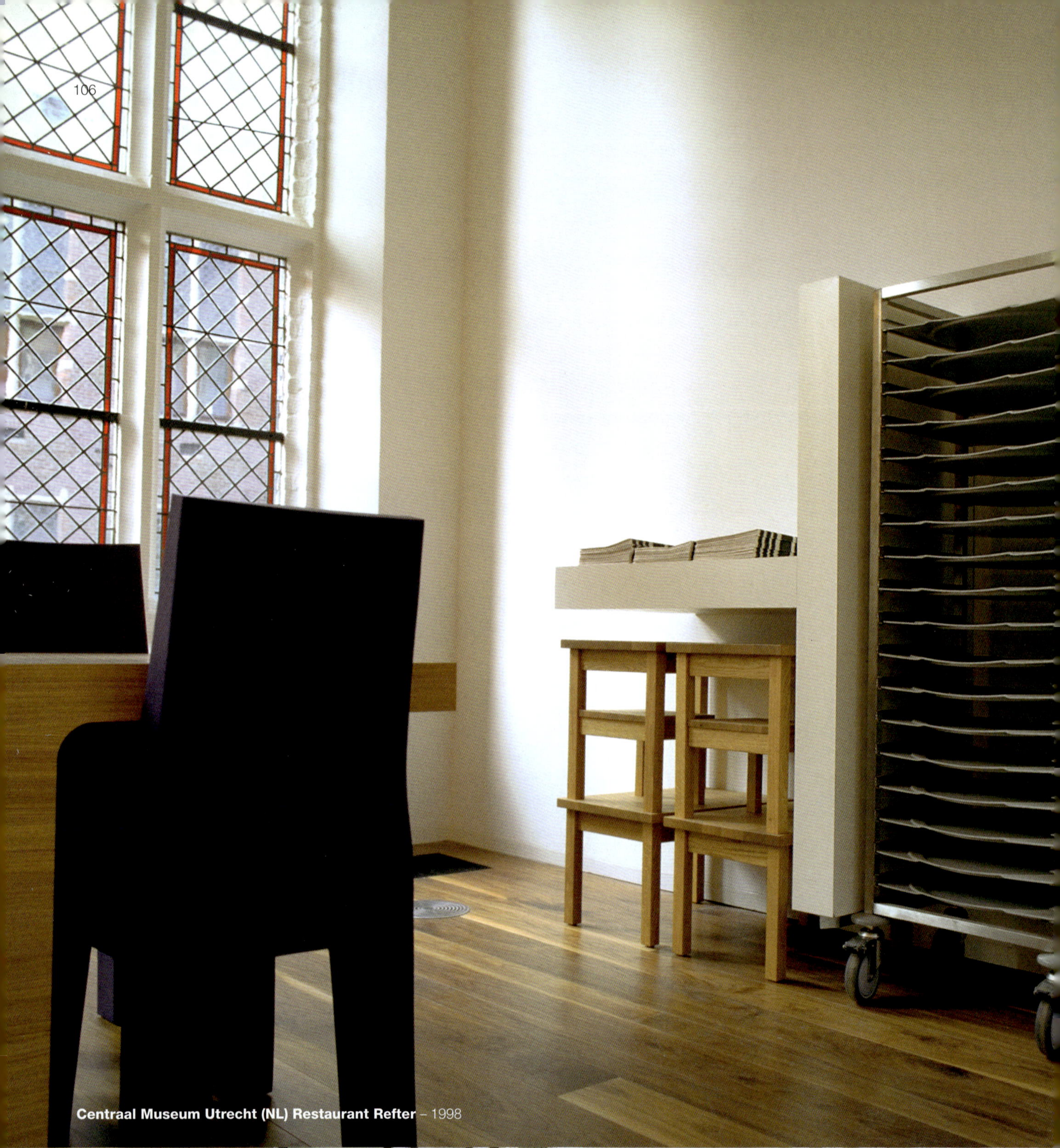

Centraal Museum Utrecht (NL) Restaurant Refter – 1998

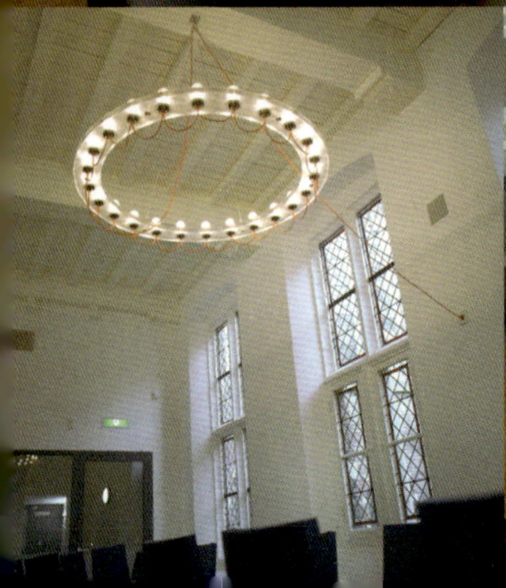

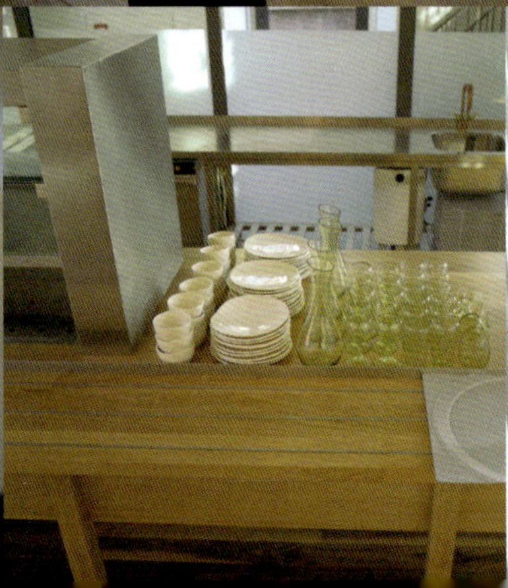

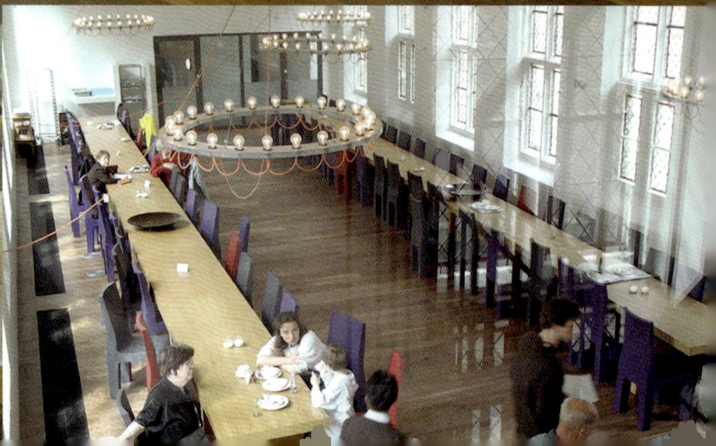

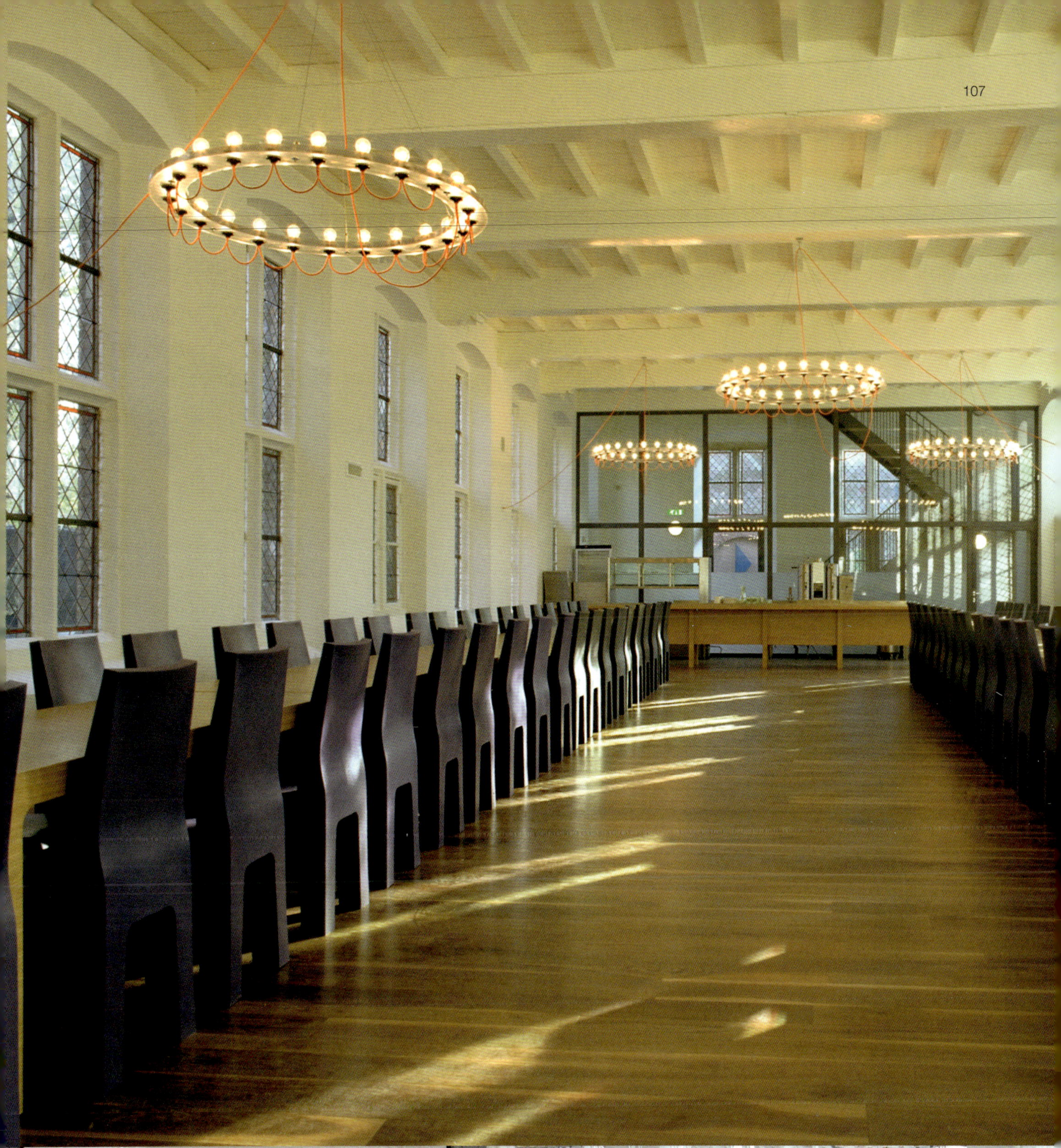

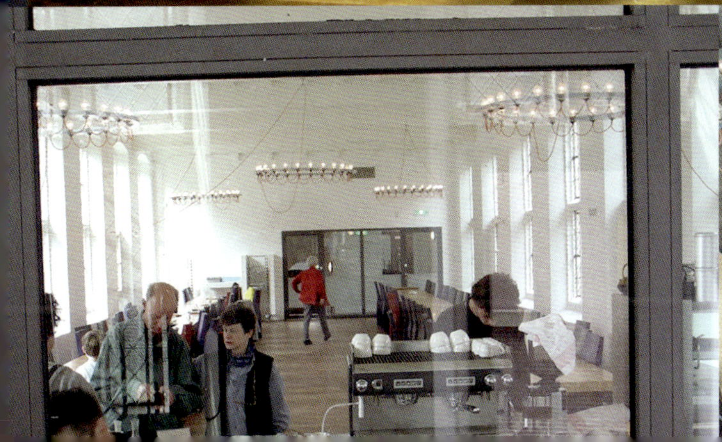

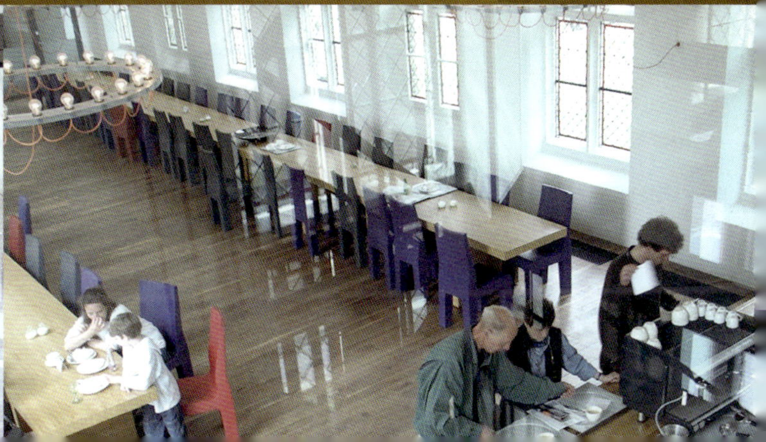

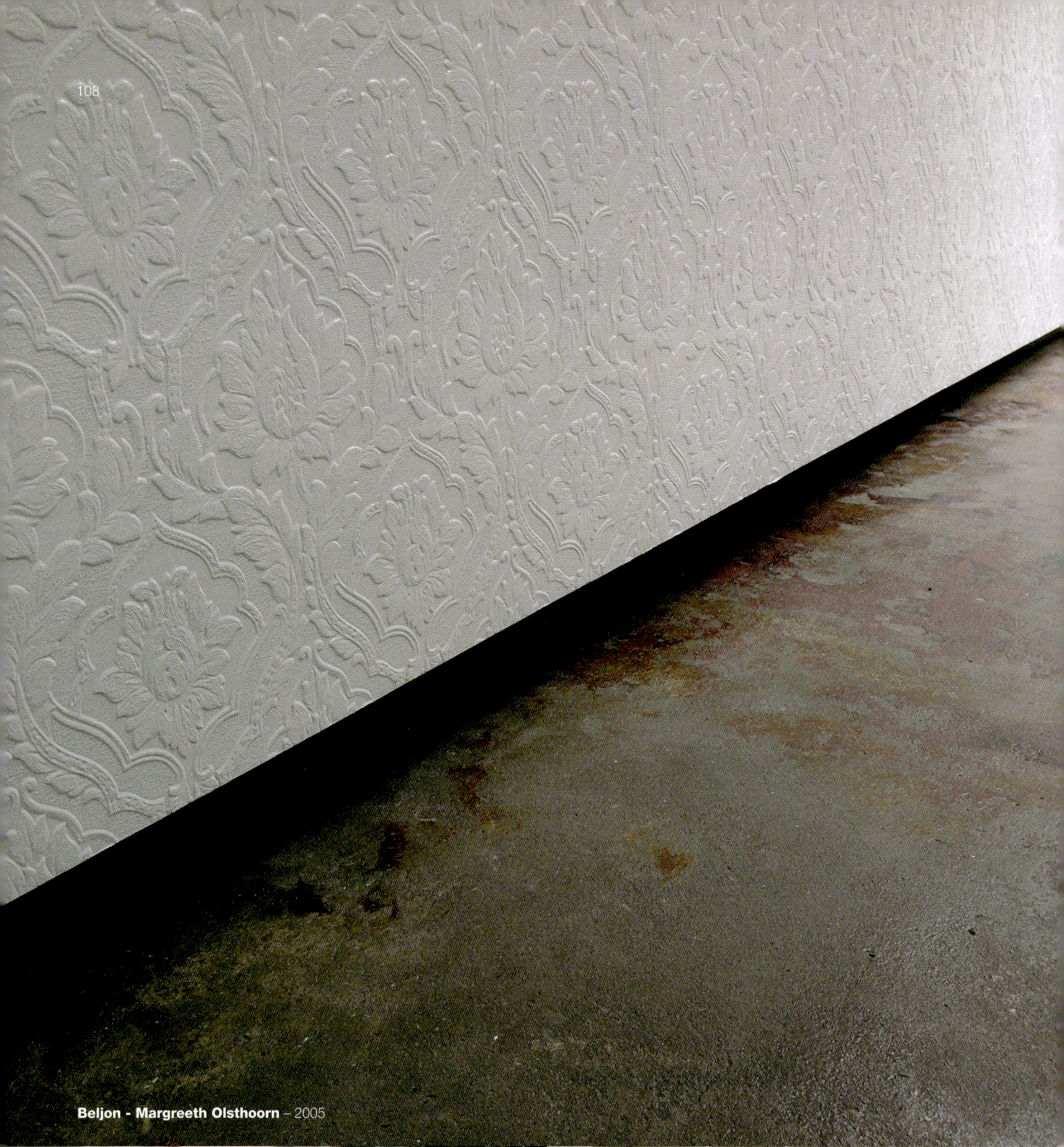

Beljon - Margreeth Olsthoorn – 2005

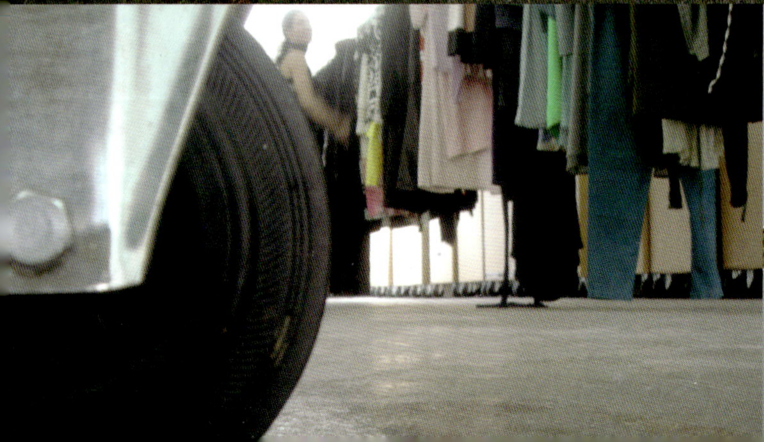

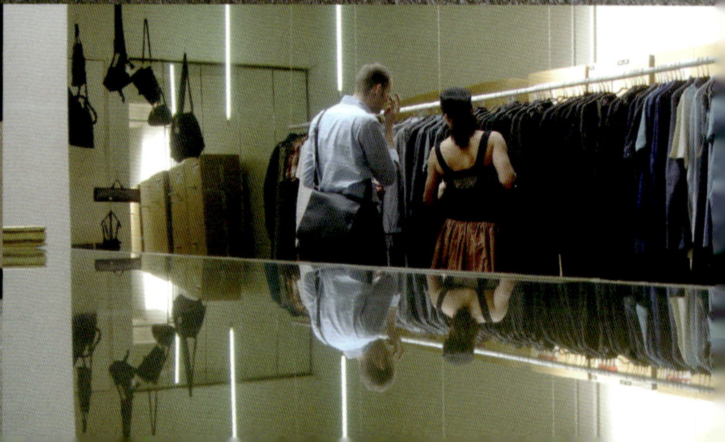

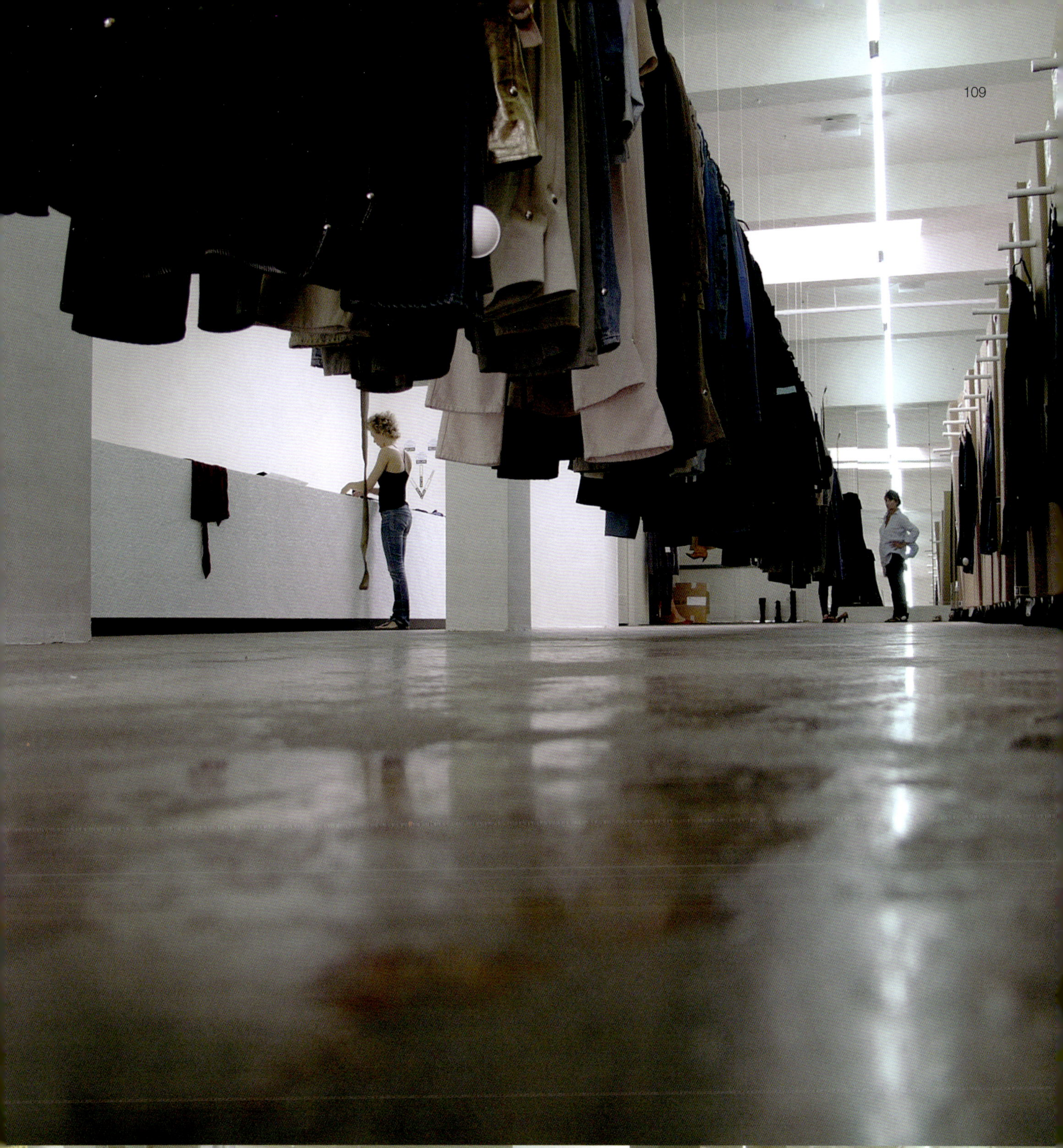

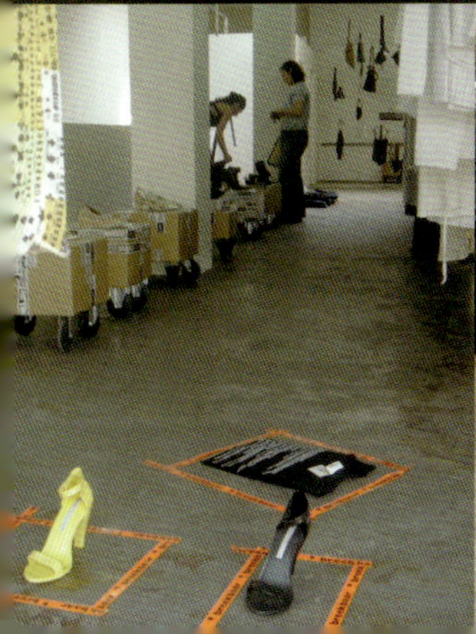

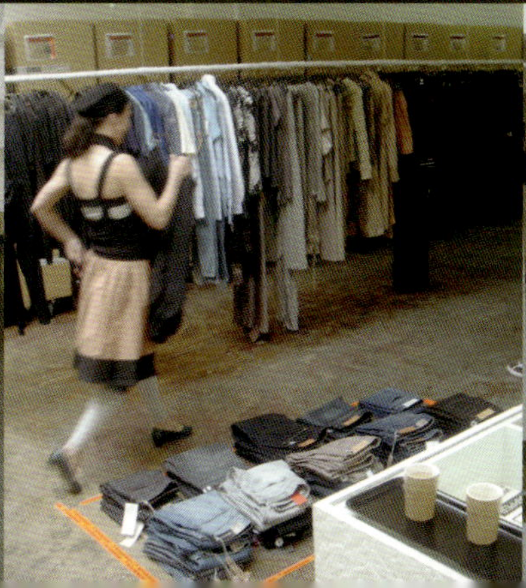

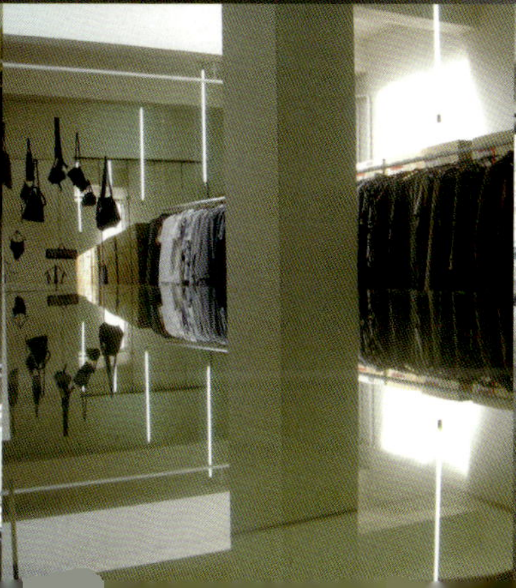

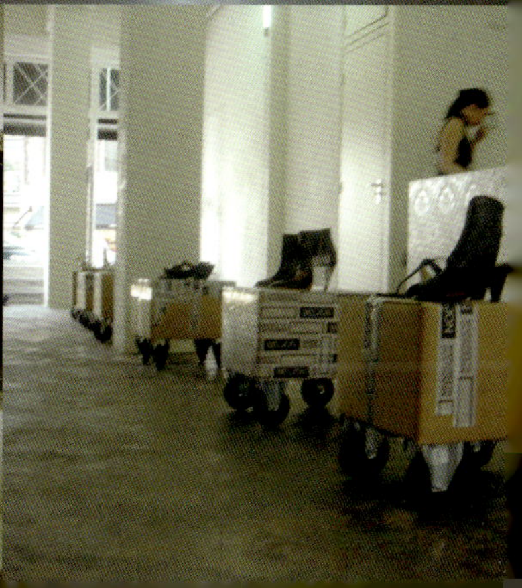

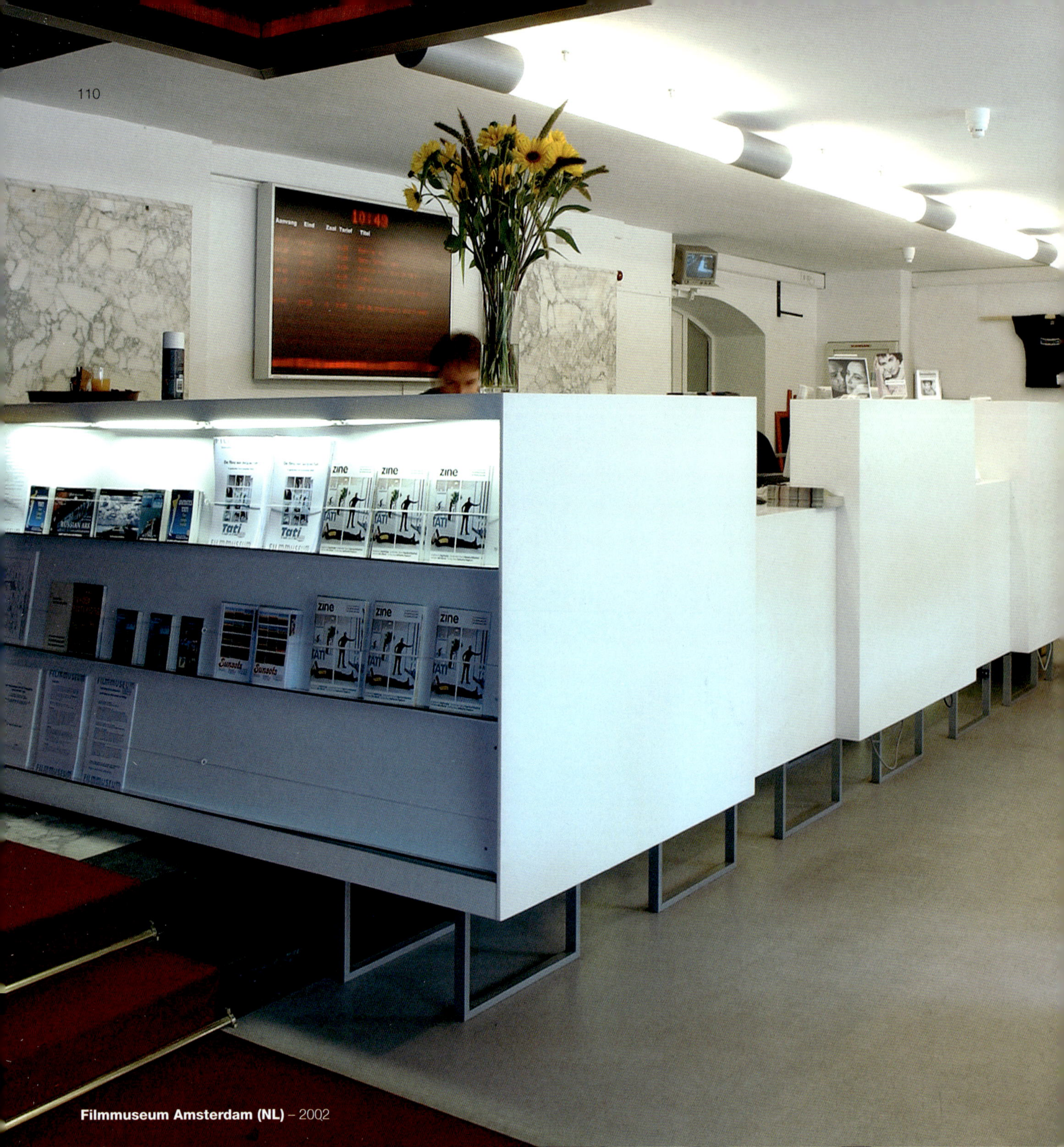

Filmmuseum Amsterdam (NL) – 2002

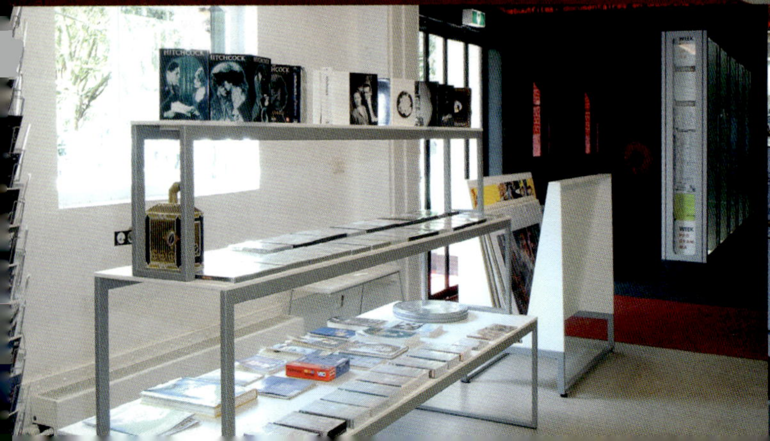

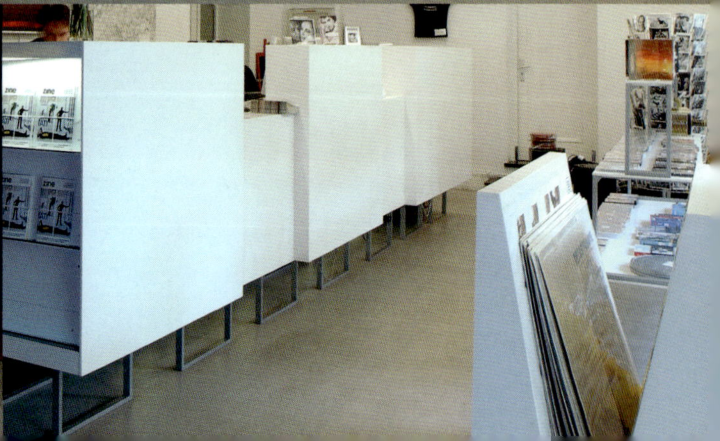

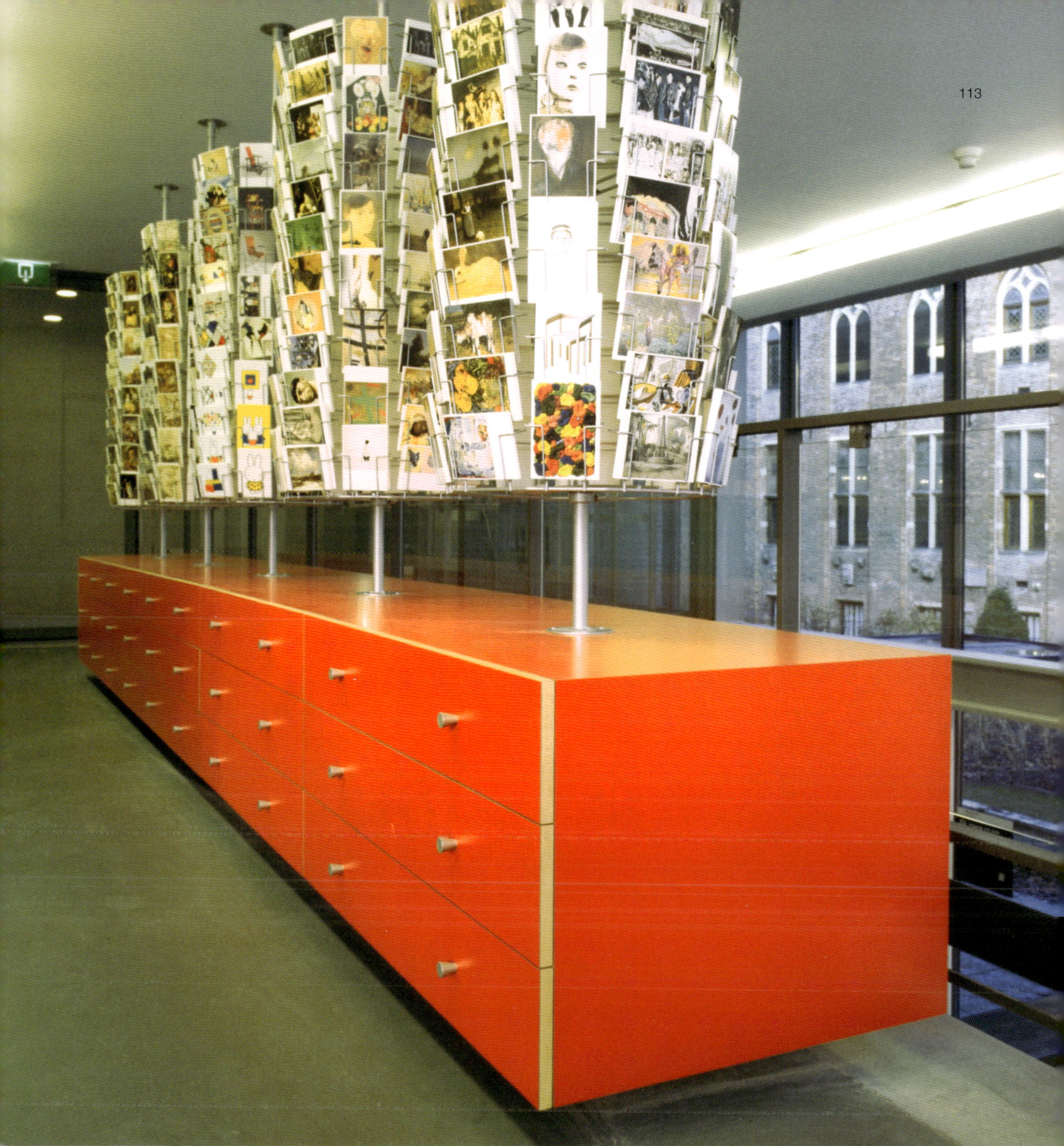

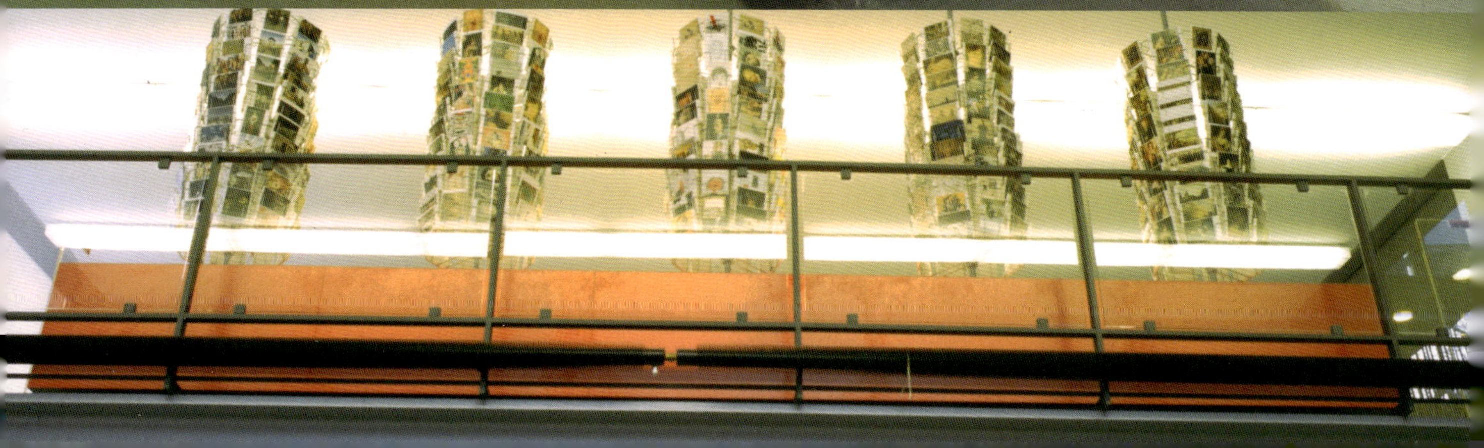

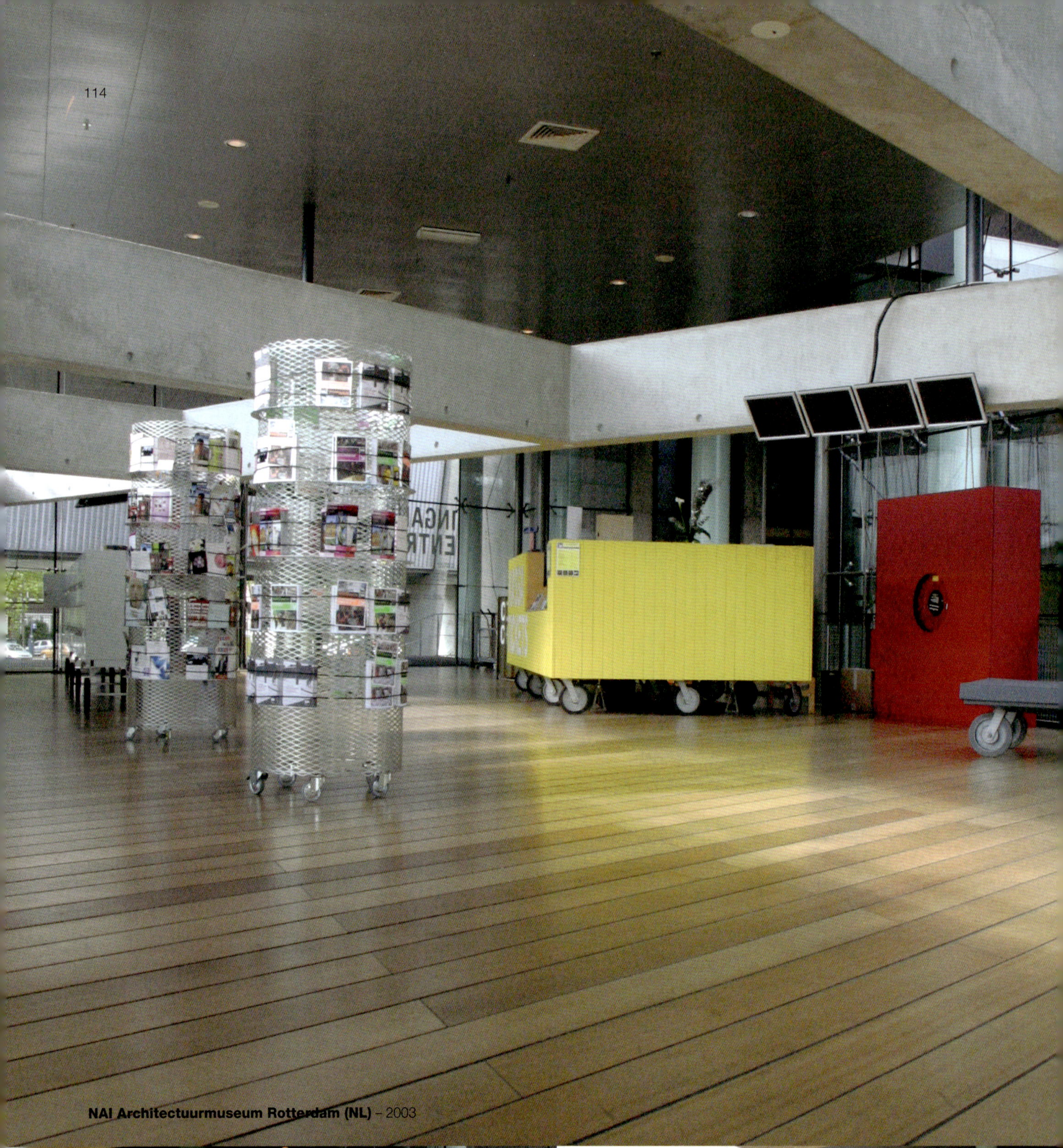

NAI Architectuurmuseum Rotterdam (NL) – 2003

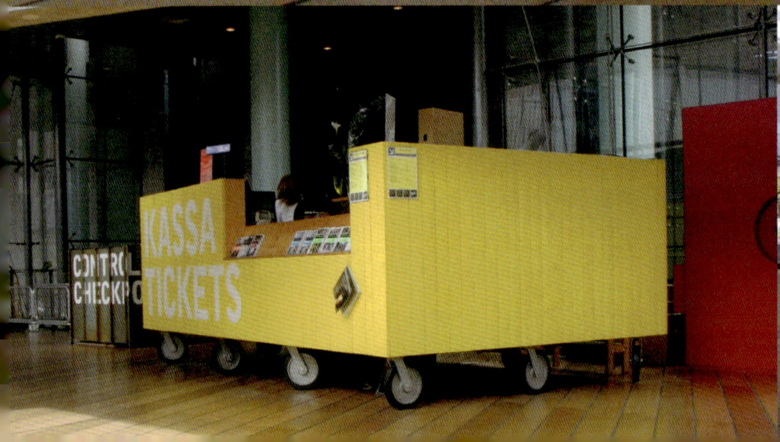

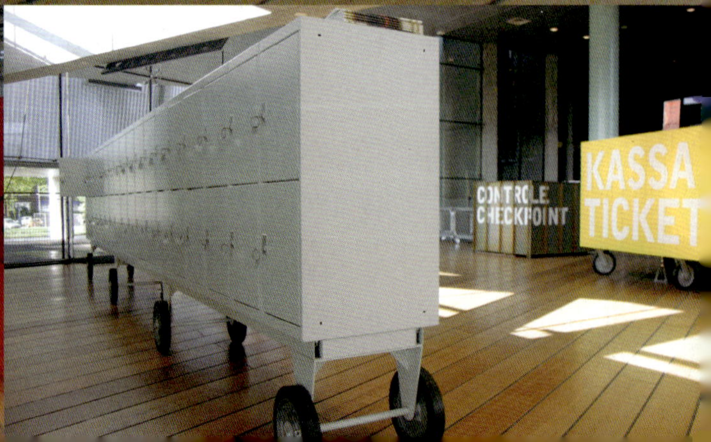

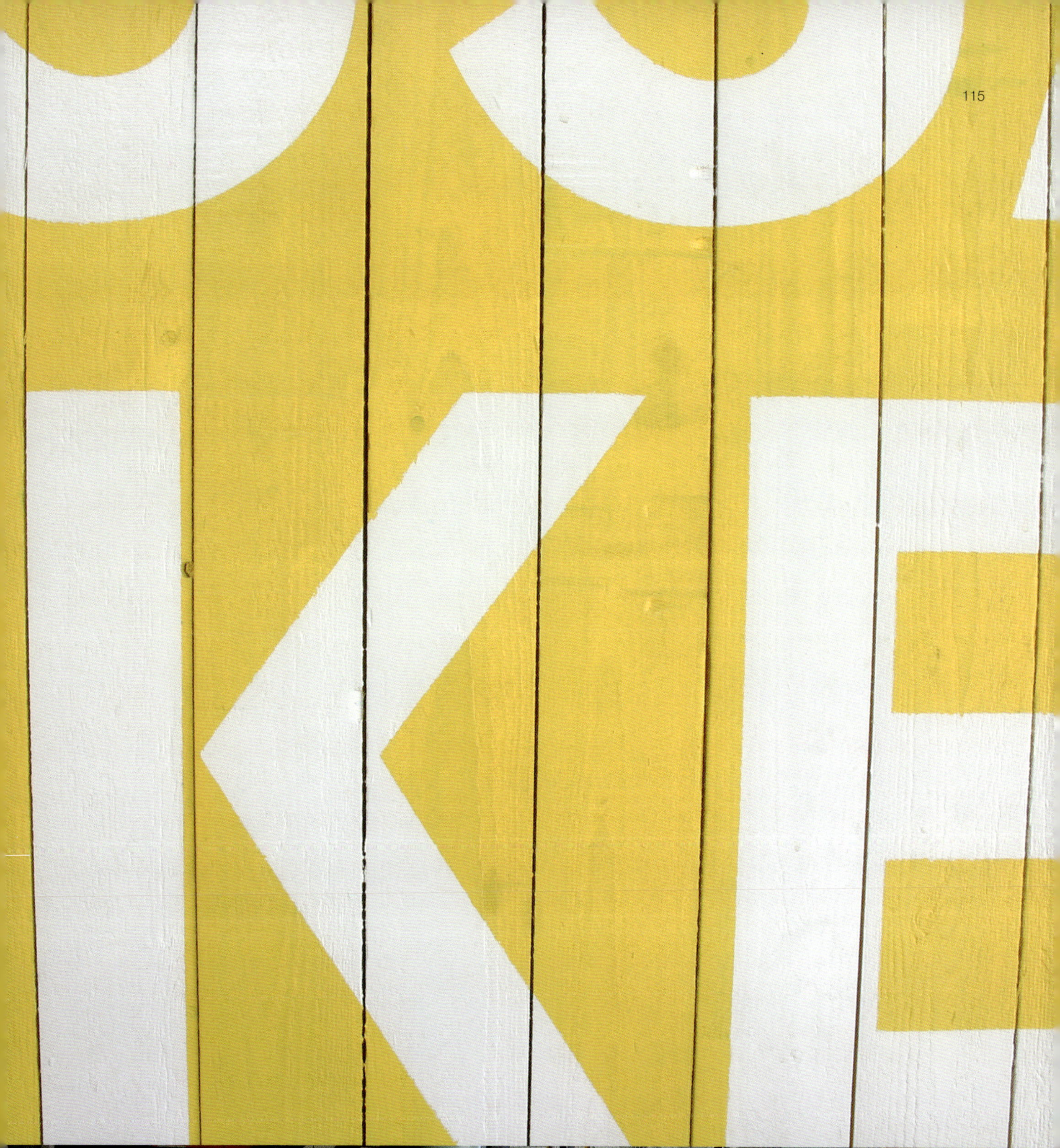

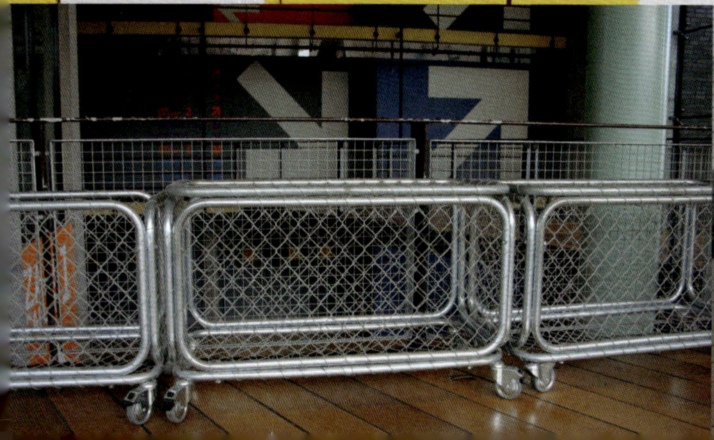

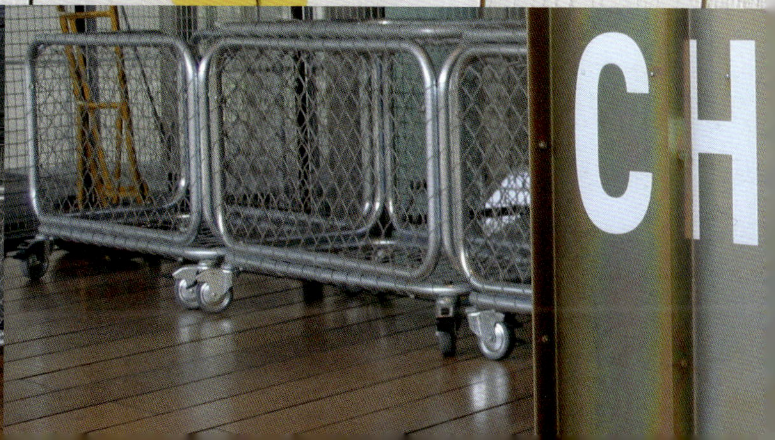
CH

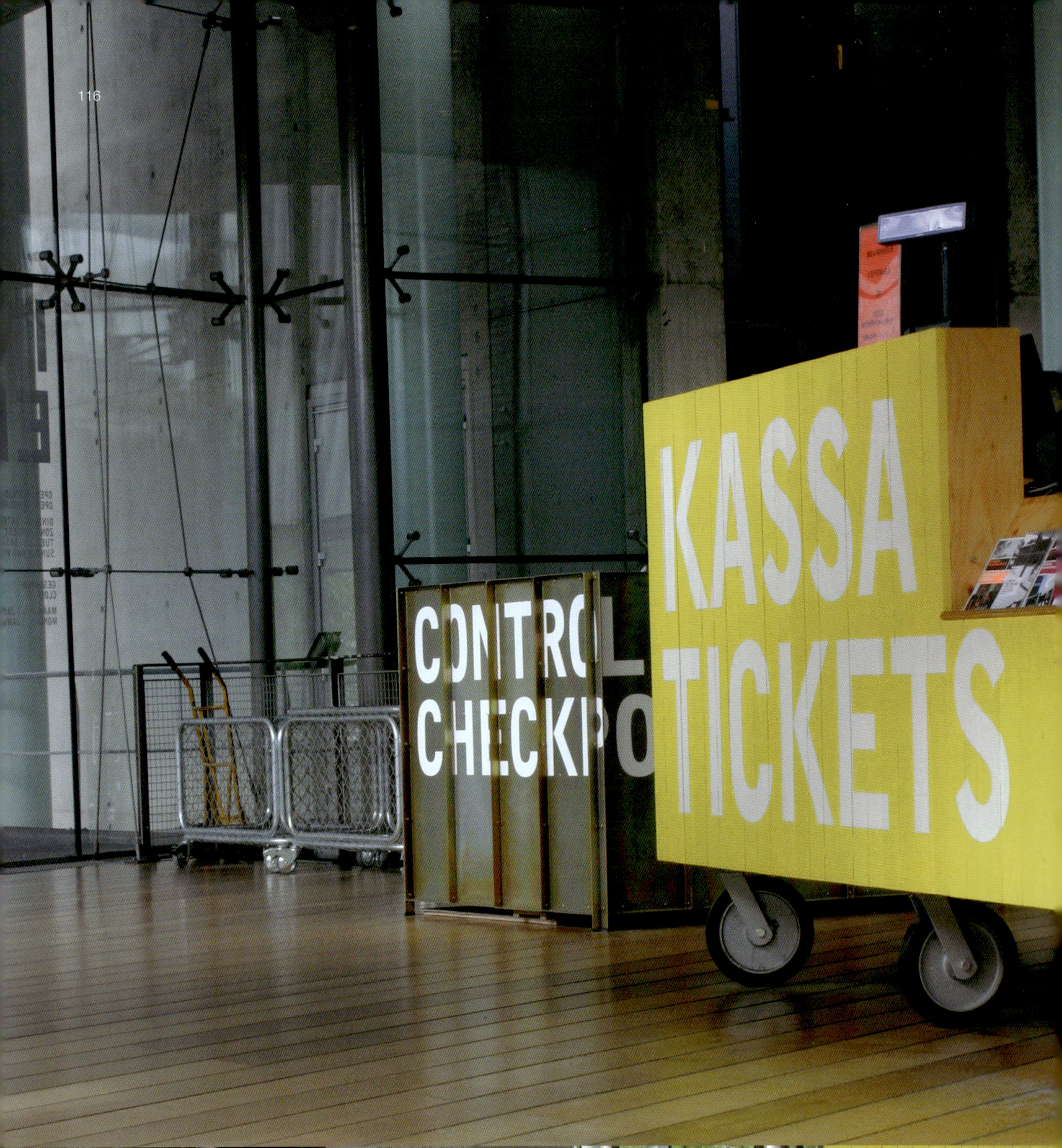
CONTROL
CHECKPO
KASSA
TICKETS

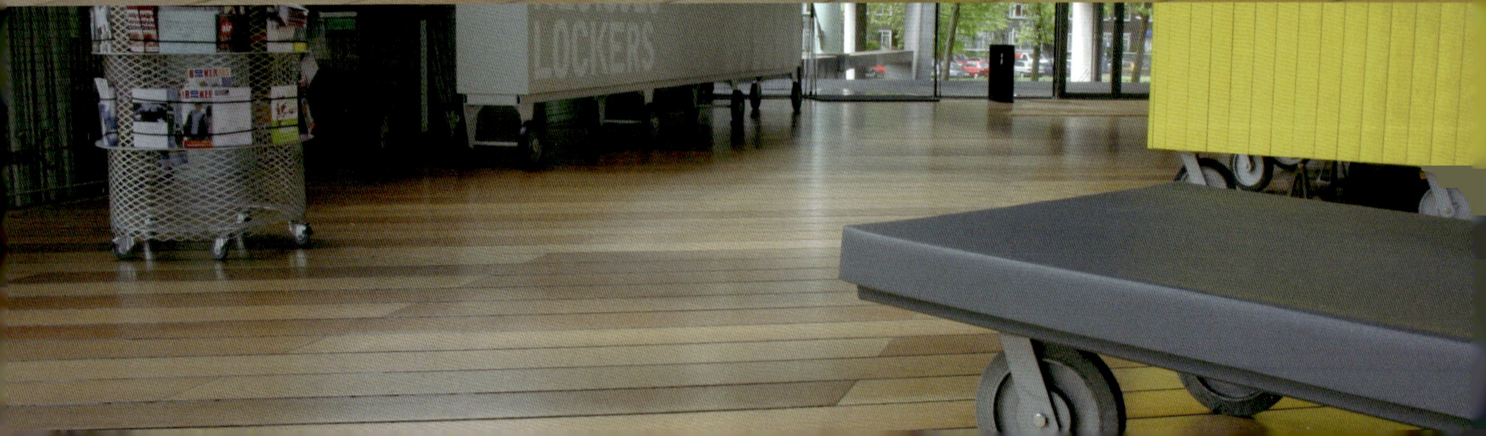
LOCKERS

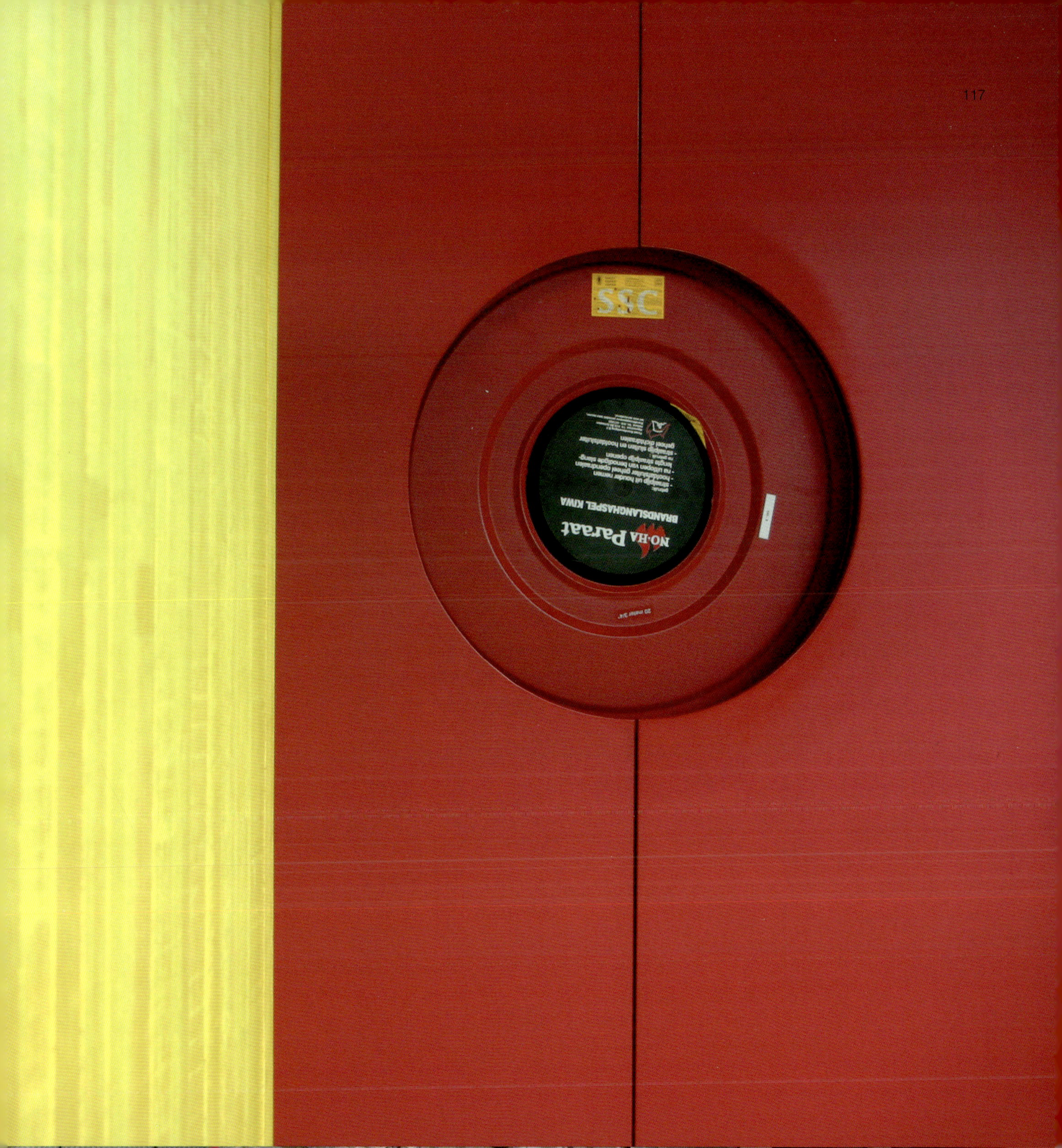
SSC

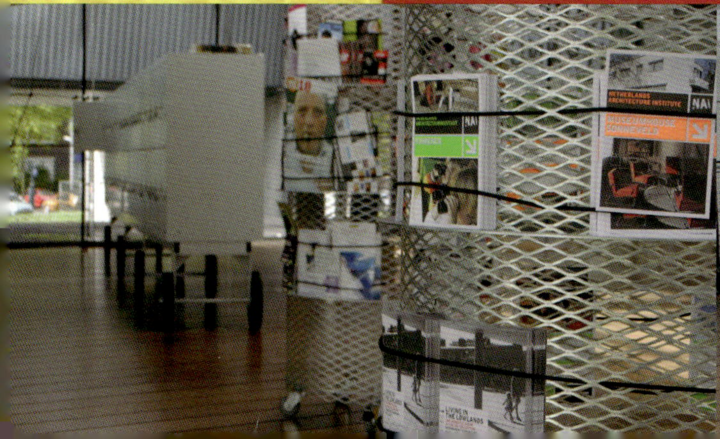

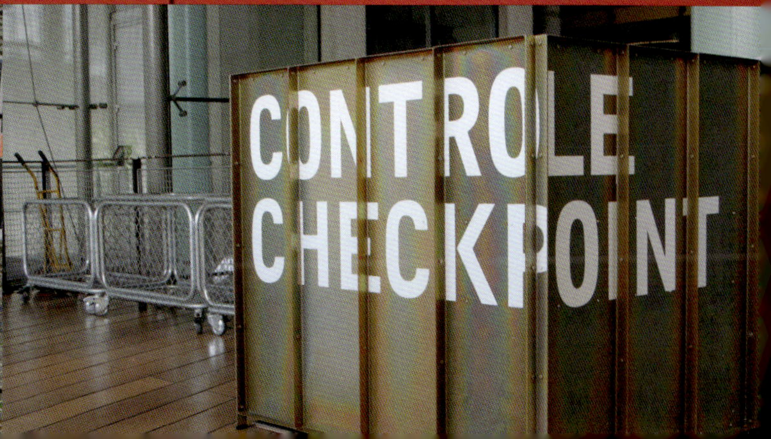
CONTROLE
CHECKPOINT

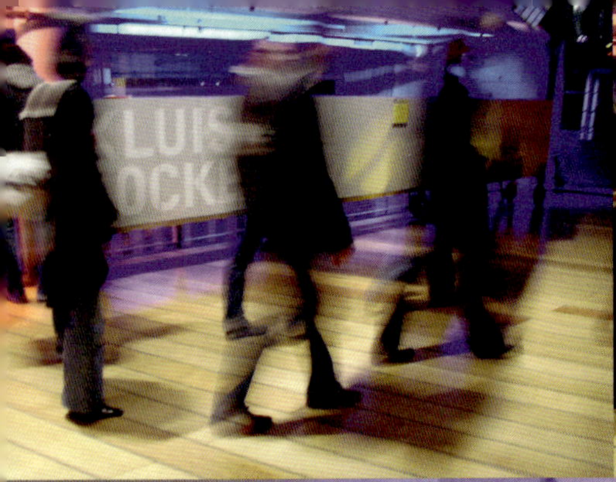
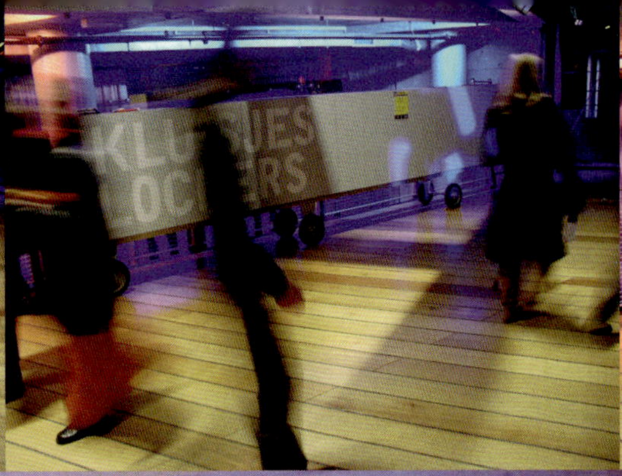
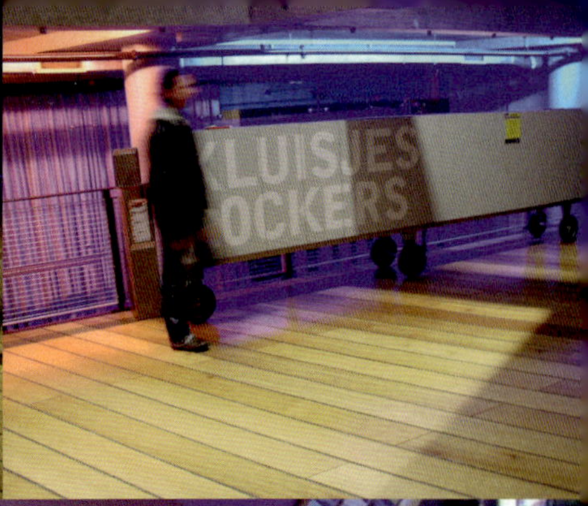
KLUISJES
LOCKERS
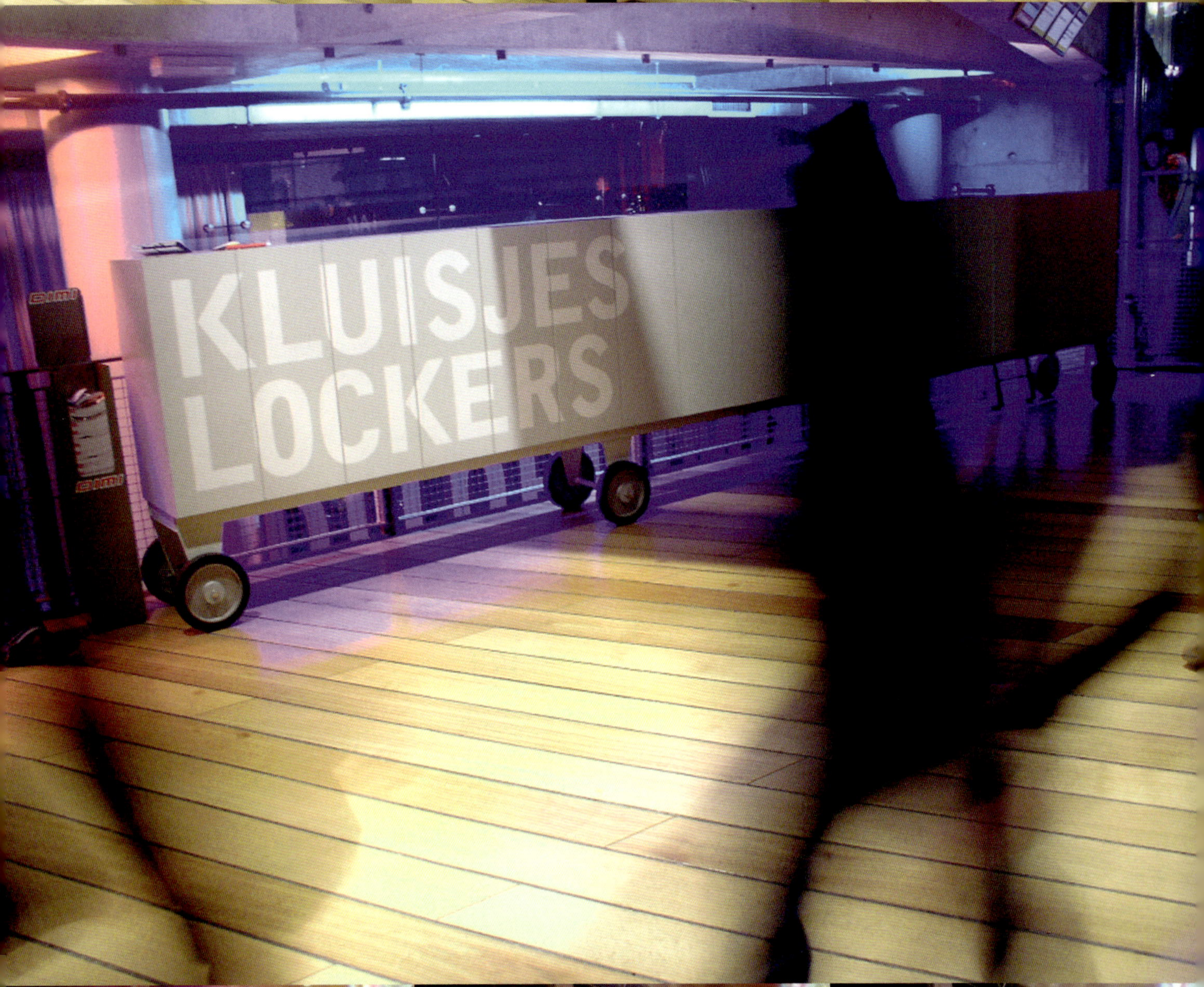
KLUISJES
LOCKERS
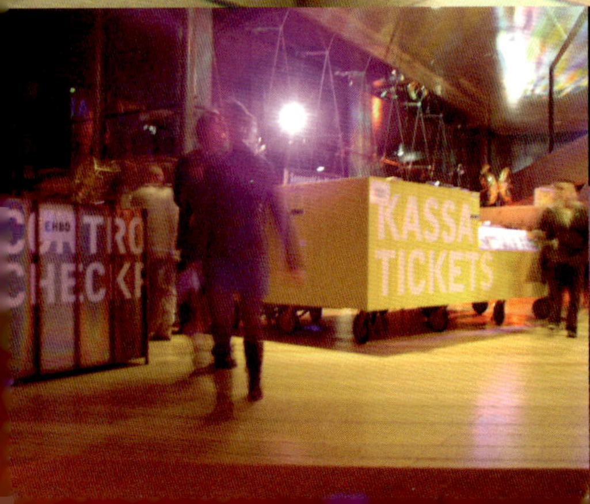
KASSA
TICKETS
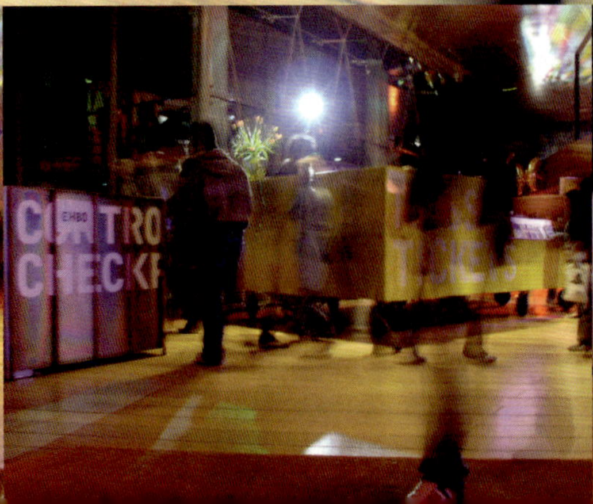
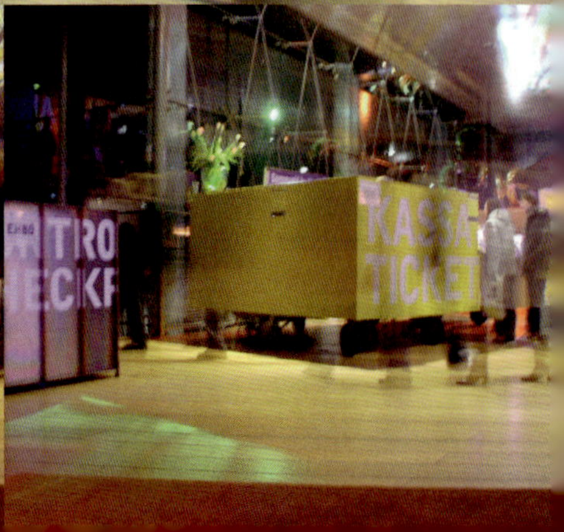

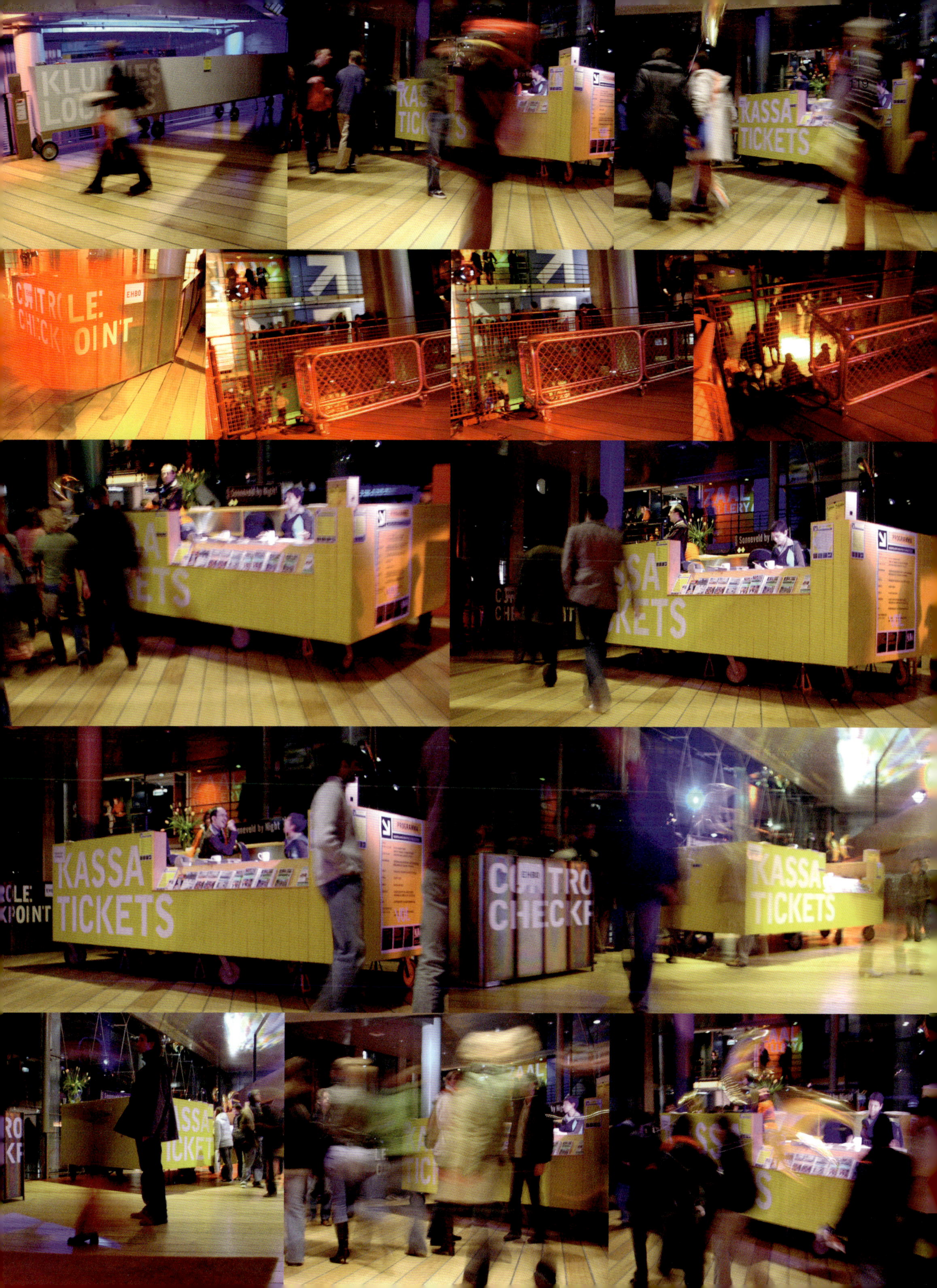
KASSA
TICKETS
EHBO
Sonneveld by Night
PROGRAMMA

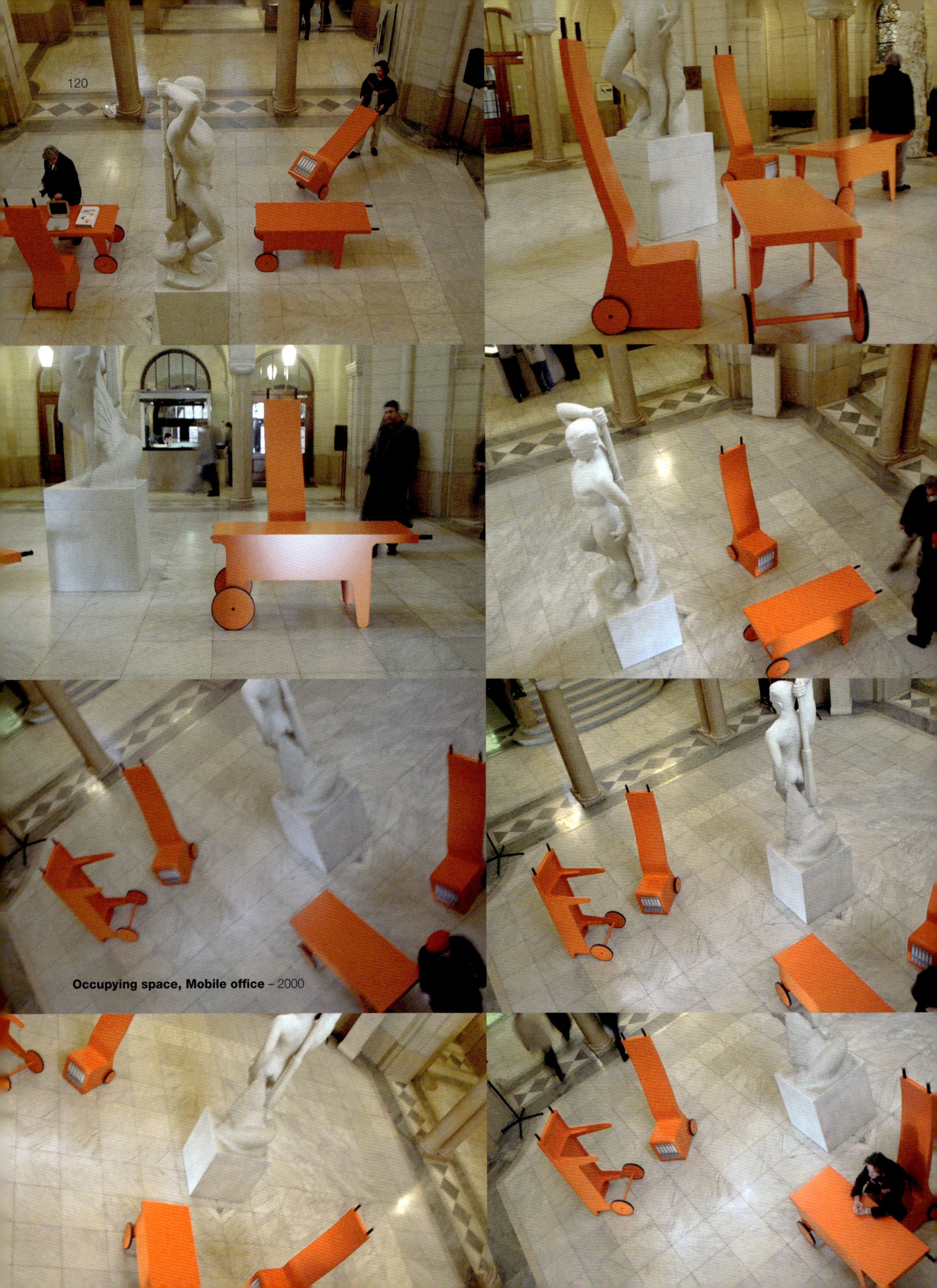

Occupying space, Mobile office – 2000

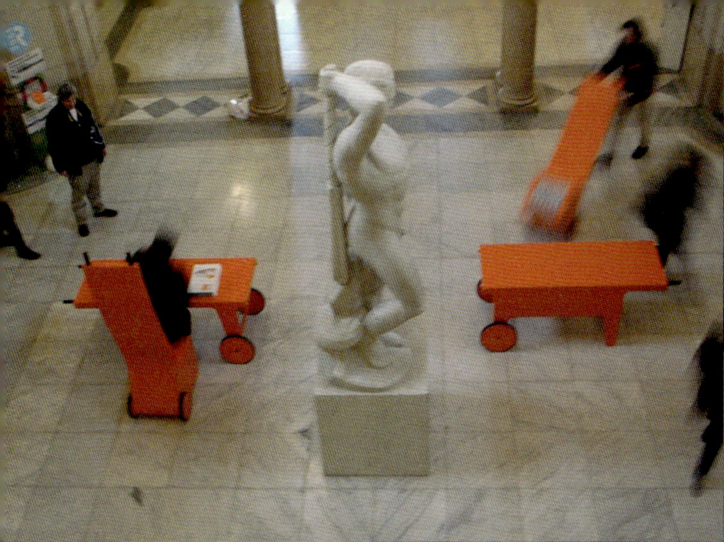

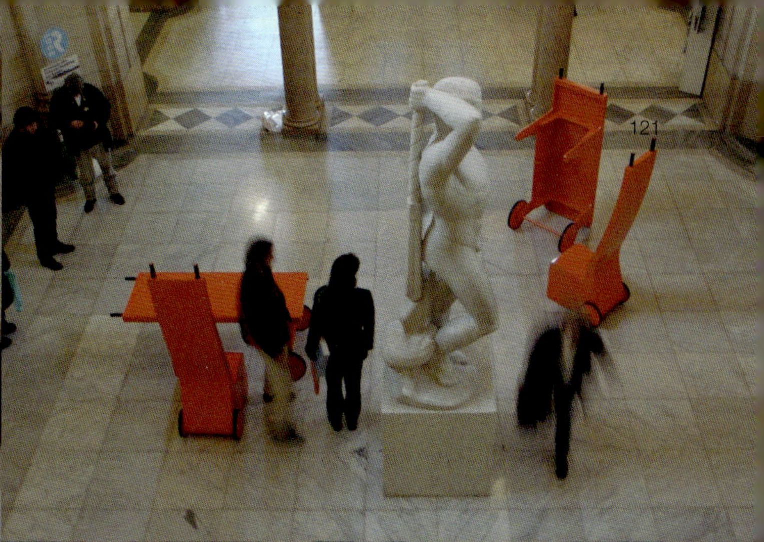

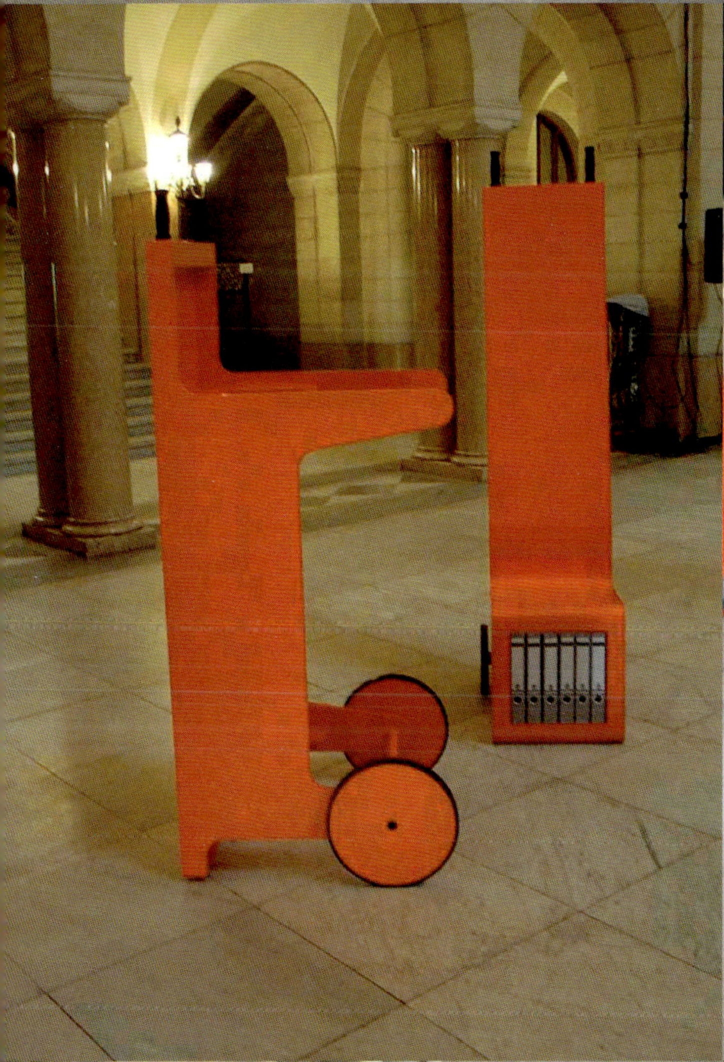

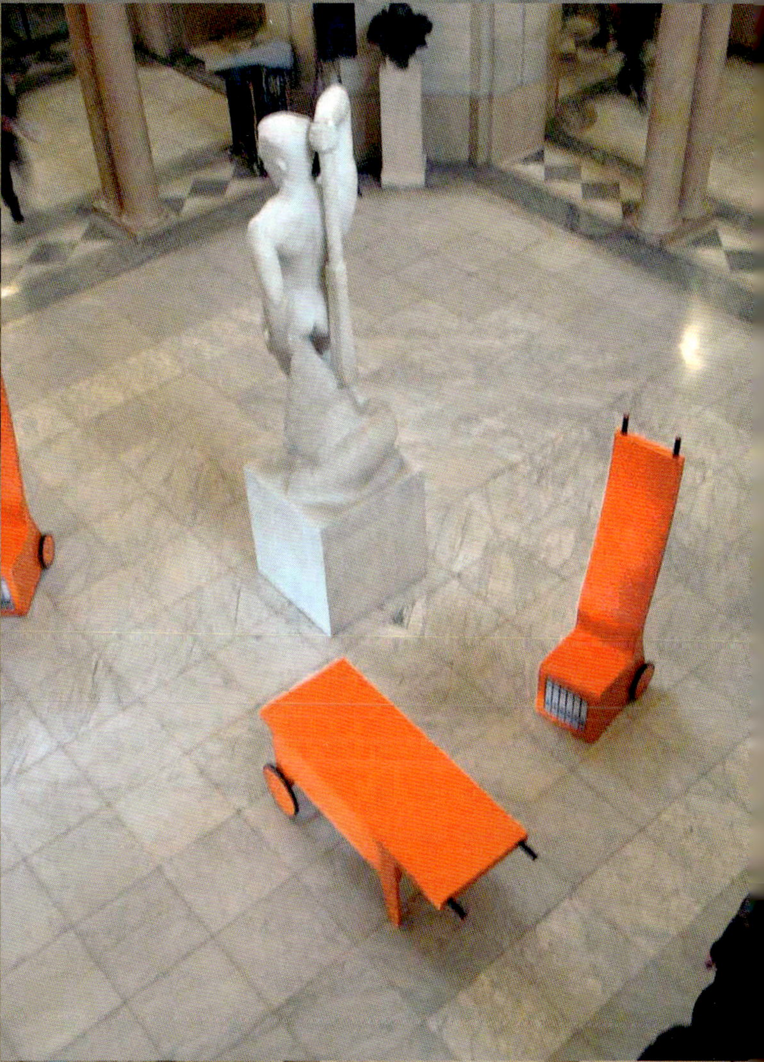

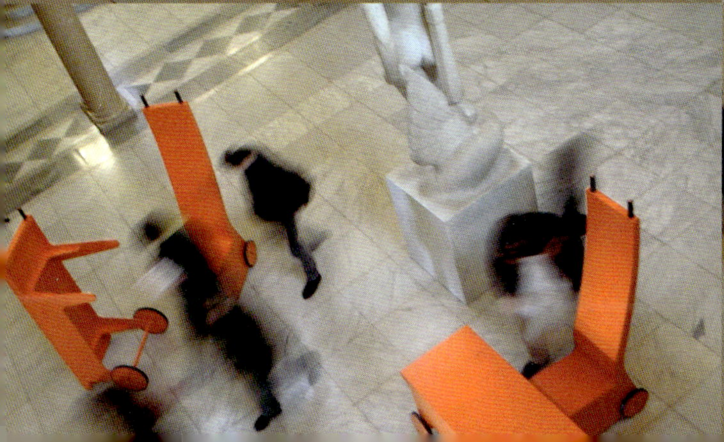

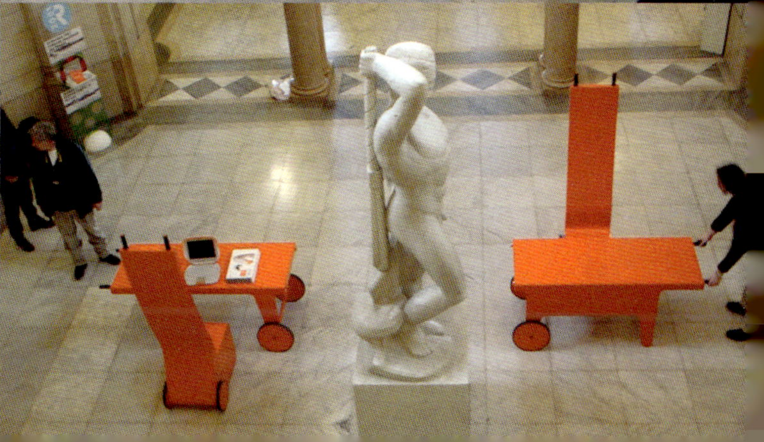

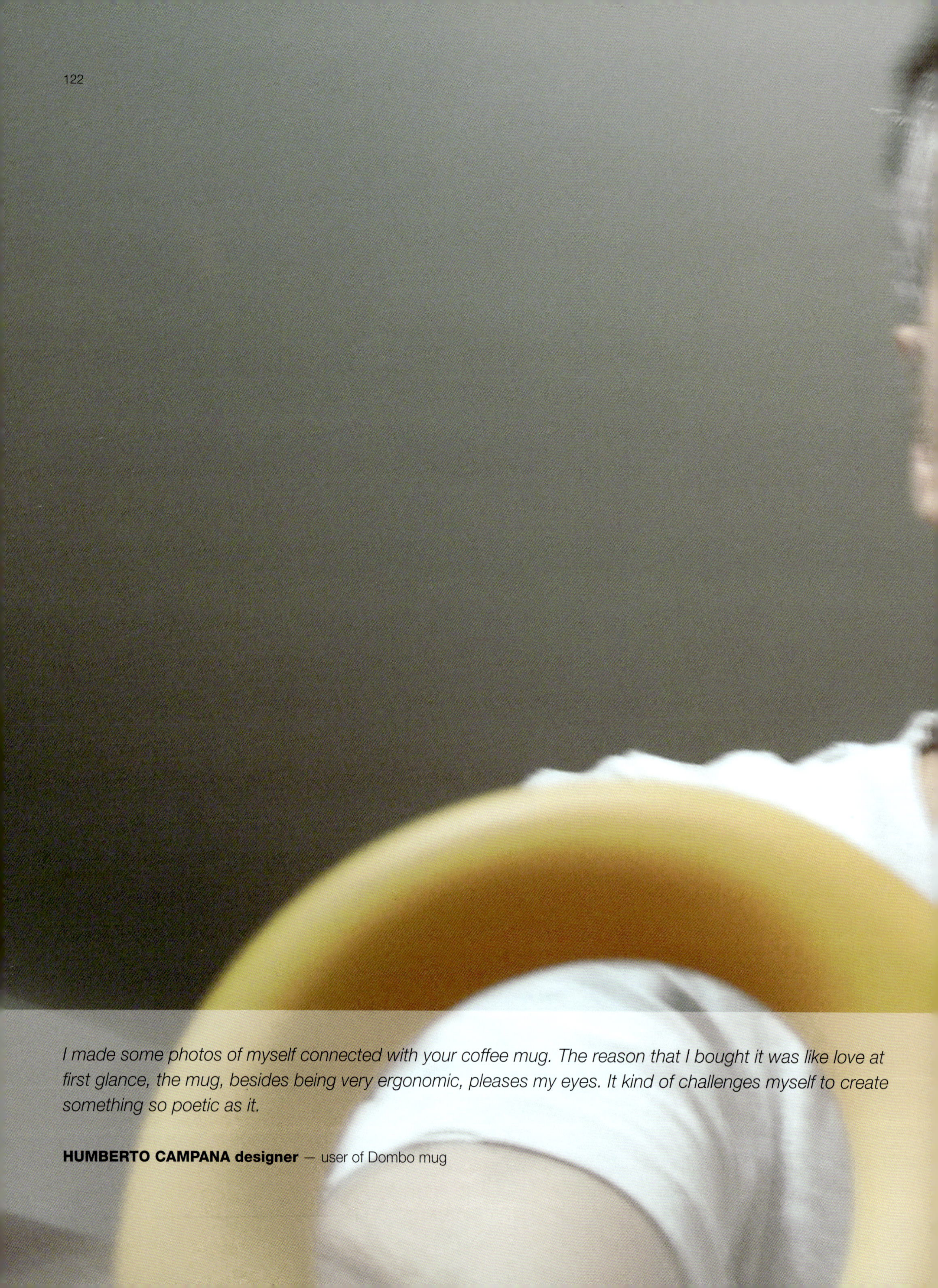

I made some photos of myself connected with your coffee mug. The reason that I bought it was like love at first glance, the mug, besides being very ergonomic, pleases my eyes. It kind of challenges myself to create something so poetic as it.

HUMBERTO CAMPANA designer — user of Dombo mug

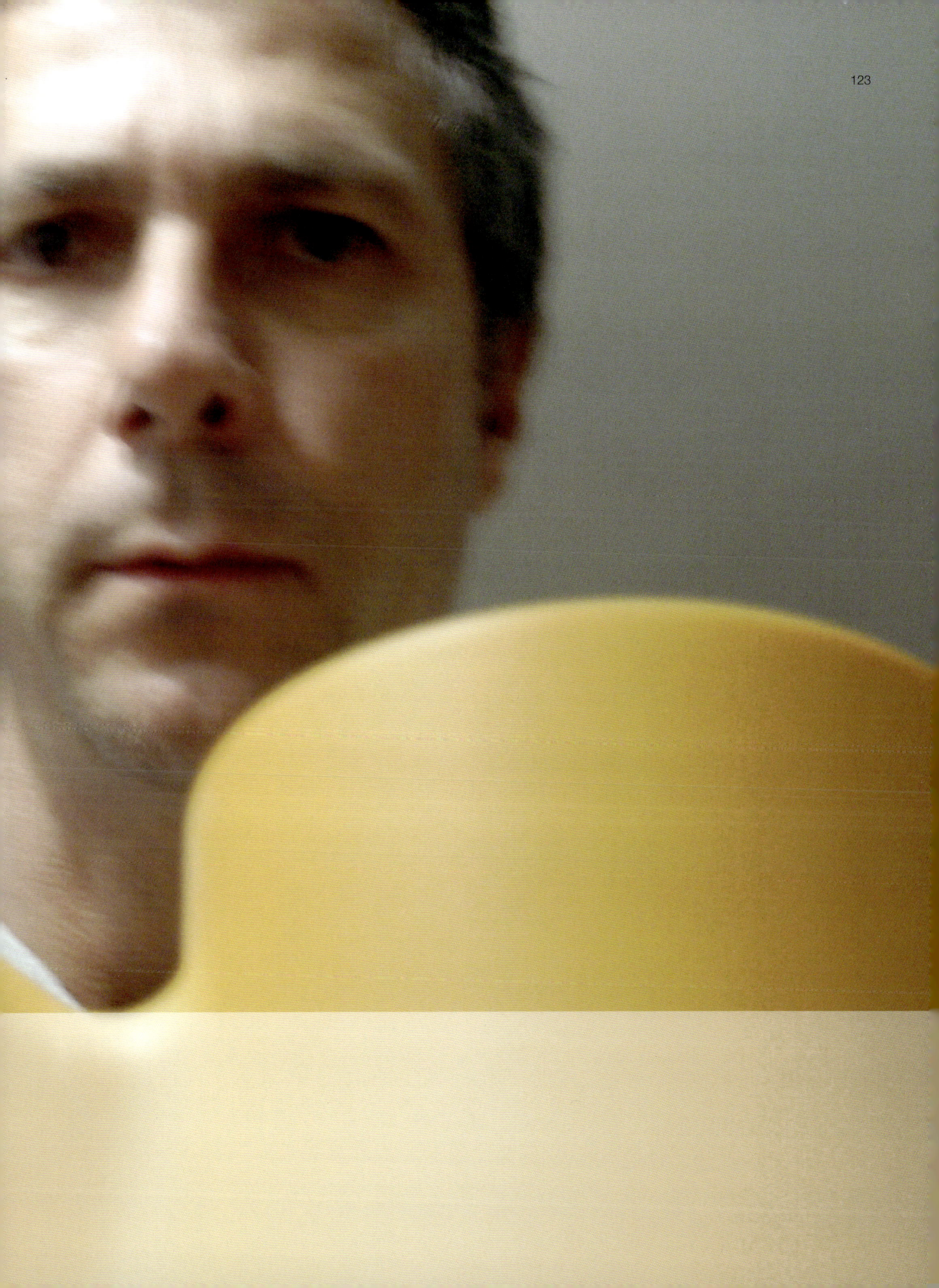

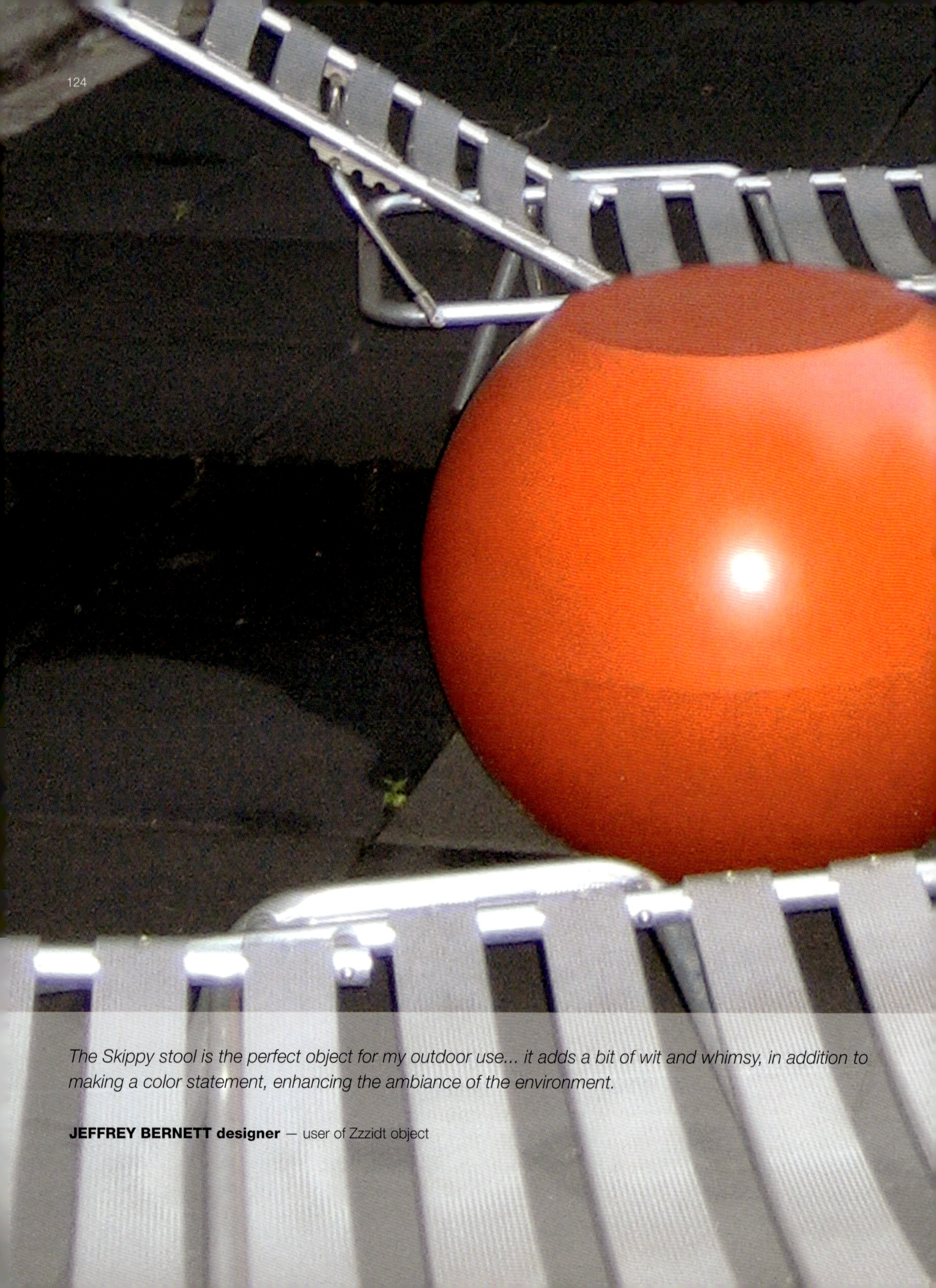

The Skippy stool is the perfect object for my outdoor use… it adds a bit of wit and whimsy, in addition to making a color statement, enhancing the ambiance of the environment.

JEFFREY BERNETT designer — user of Zzzidt object

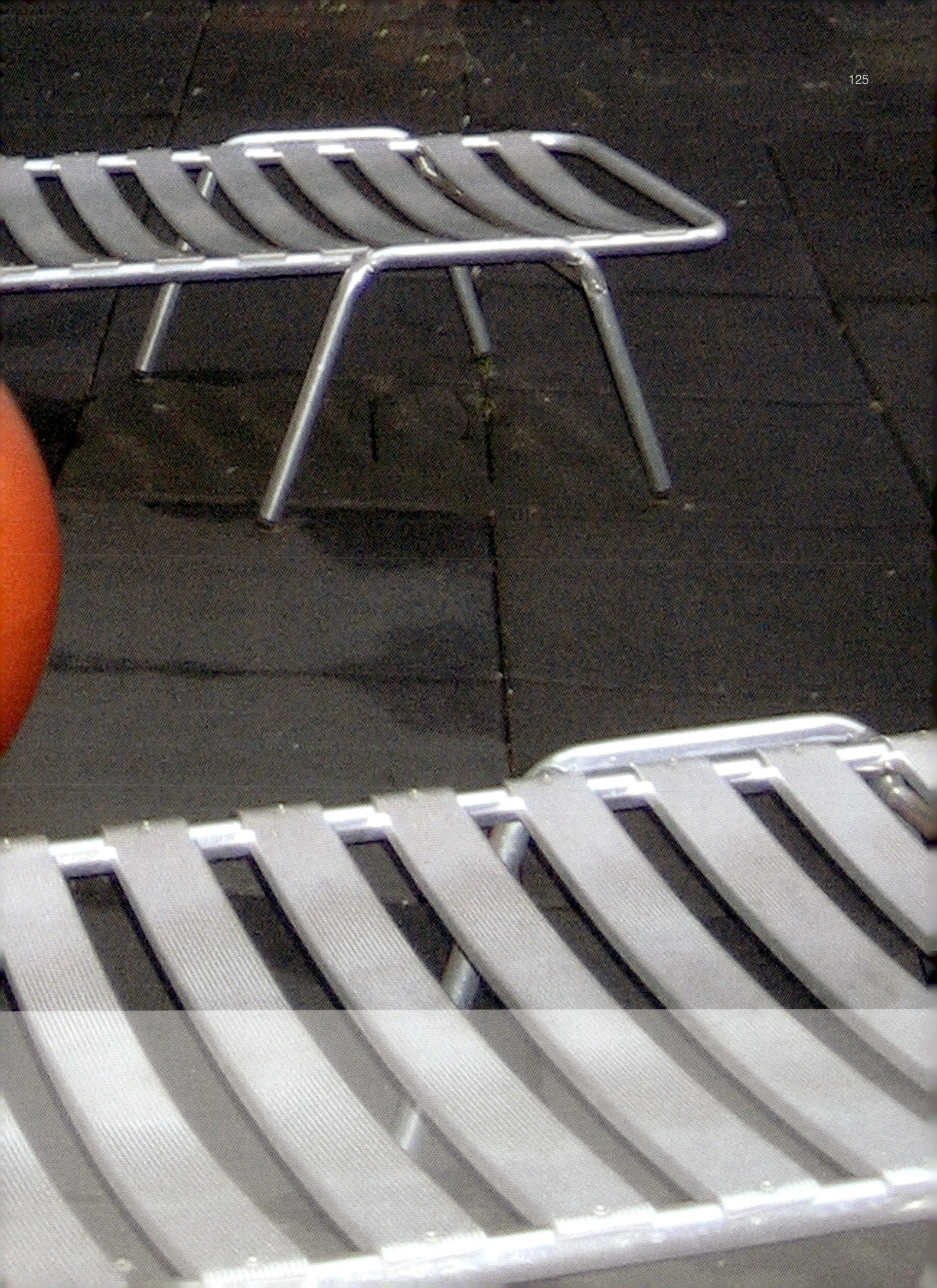

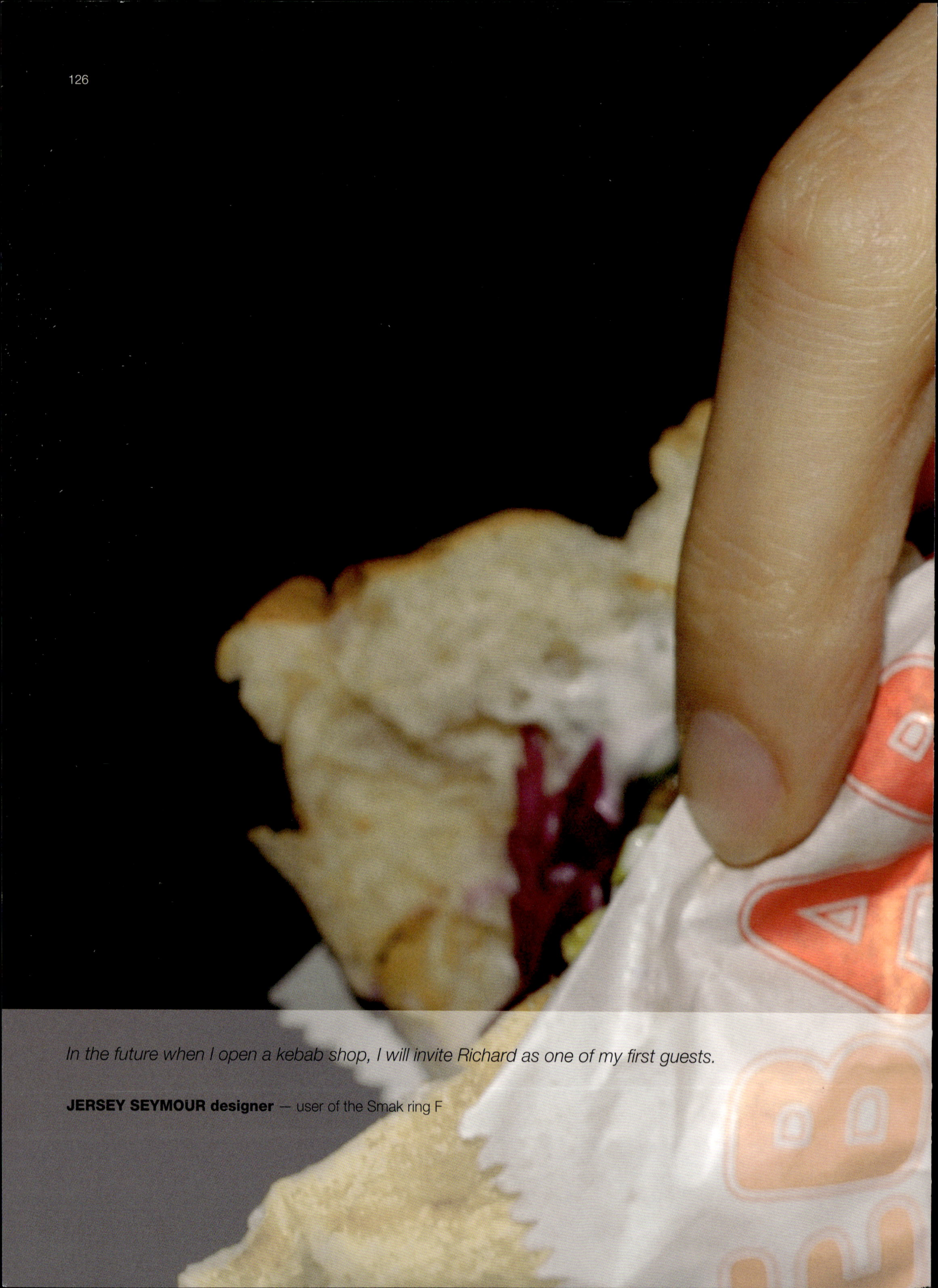

In the future when I open a kebab shop, I will invite Richard as one of my first guests.

JERSEY SEYMOUR designer — user of the Smak ring F

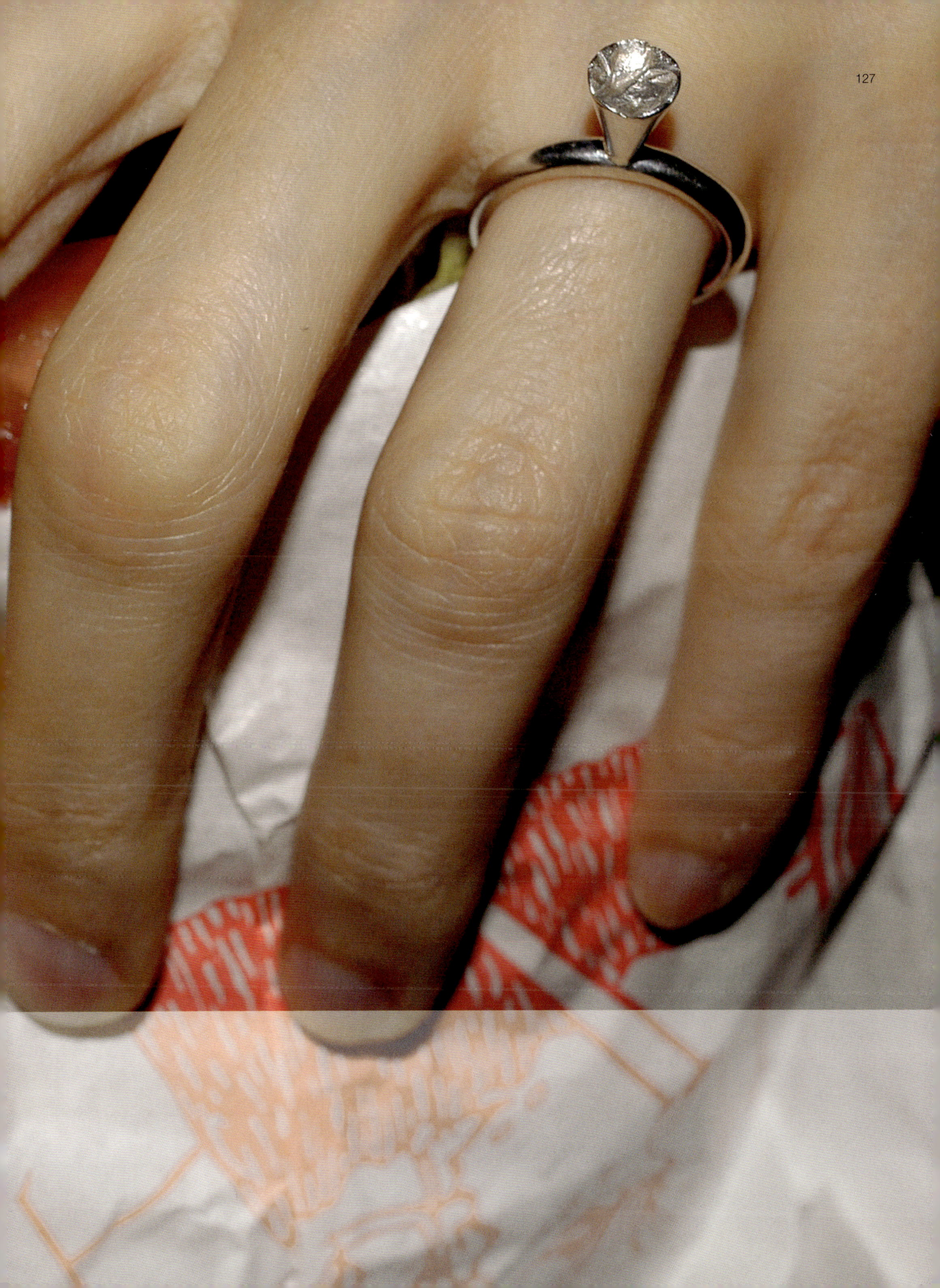

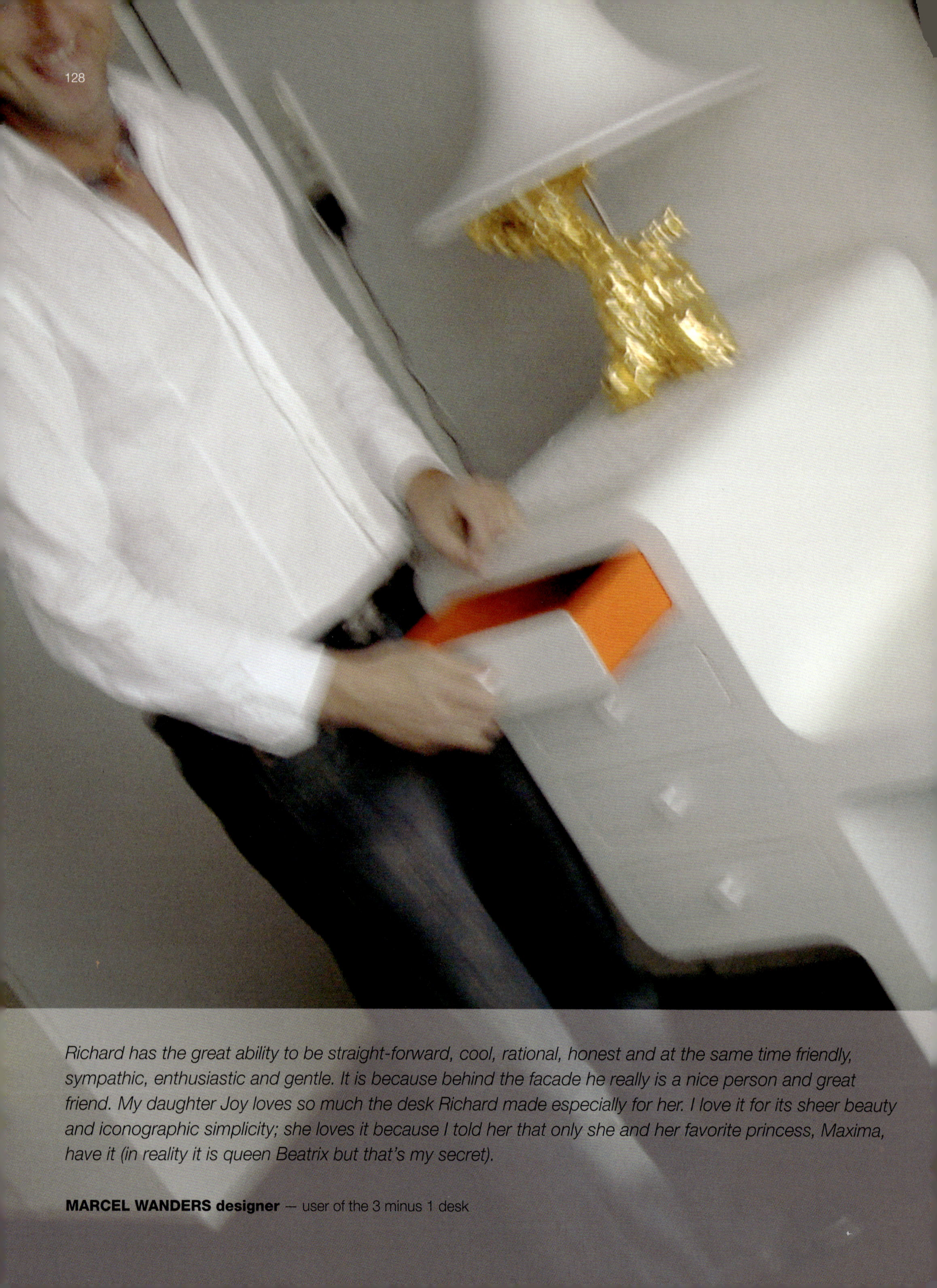

Richard has the great ability to be straight-forward, cool, rational, honest and at the same time friendly, sympathic, enthusiastic and gentle. It is because behind the facade he really is a nice person and great friend. My daughter Joy loves so much the desk Richard made especially for her. I love it for its sheer beauty and iconographic simplicity; she loves it because I told her that only she and her favorite princess, Maxima, have it (in reality it is queen Beatrix but that's my secret).

MARCEL WANDERS designer — user of the 3 minus 1 desk

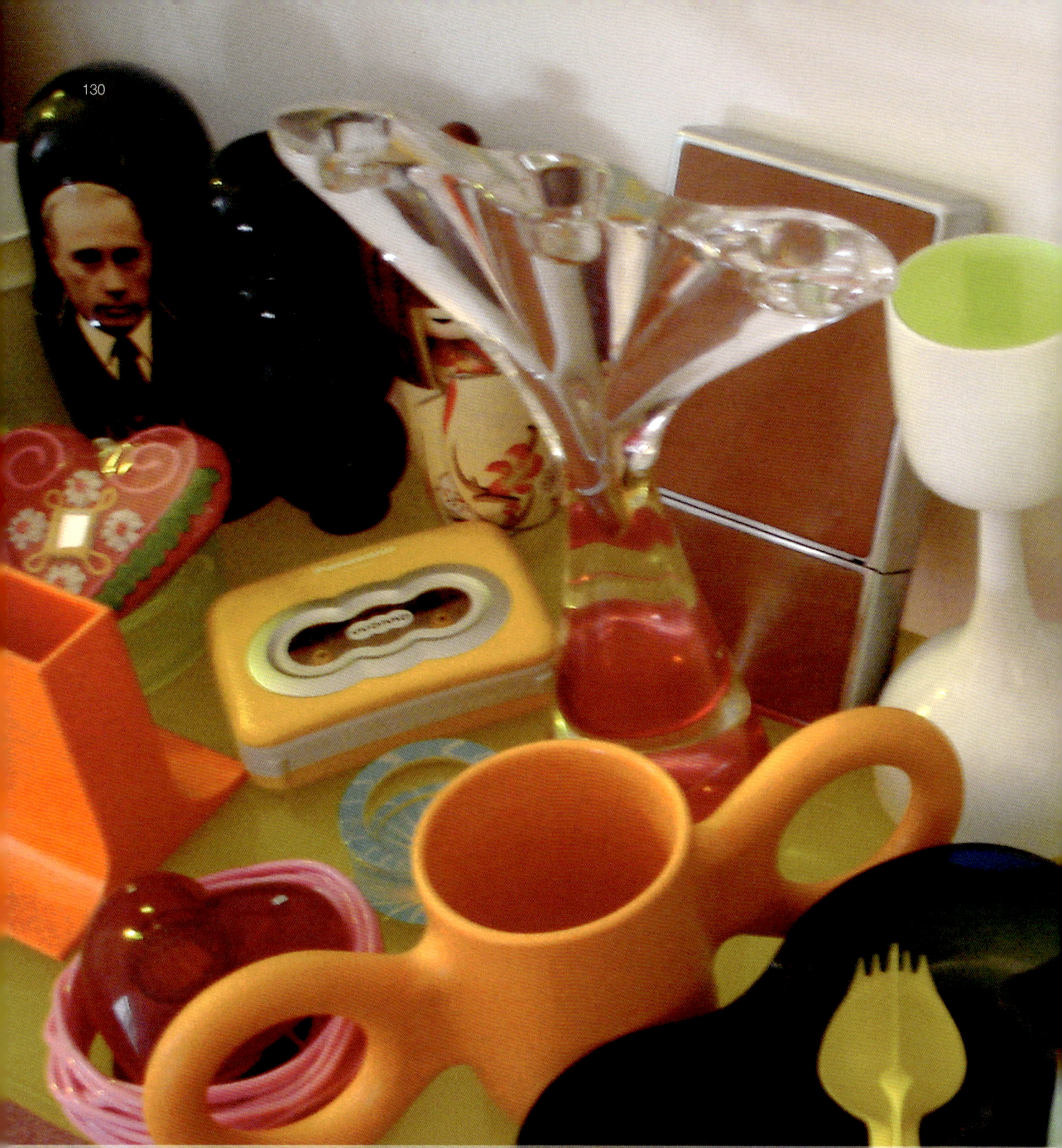

I have always noted a real bifurcation in Dutch design. There is 'hard-core' industrial design that came from the Hague in Holland, a city that shaped brilliant pragmatic utilitarian companies such as Philips. There is the 'cottage industry' of designers doing small one-off batch production. And there is hand-made design, working with old techniques and traditions and – merging them with new technologies or new poetics – more discourse than industry, more craft than machine, more ornate than industrial commodity, more language. His work is set apart from his peers of Droog whose work became steeped in traditions and decorum. Richard is pure and sexy in a Dutch way – frank sexiness – and he demarcates himself with astute poignant object sarcasm. I first noticed Richard's work at the Mondrian hotel where you can see the 'cross table' in the lobby. As a homage to Russian constructivism as poetic rudimentary assemblages (which I see throughout

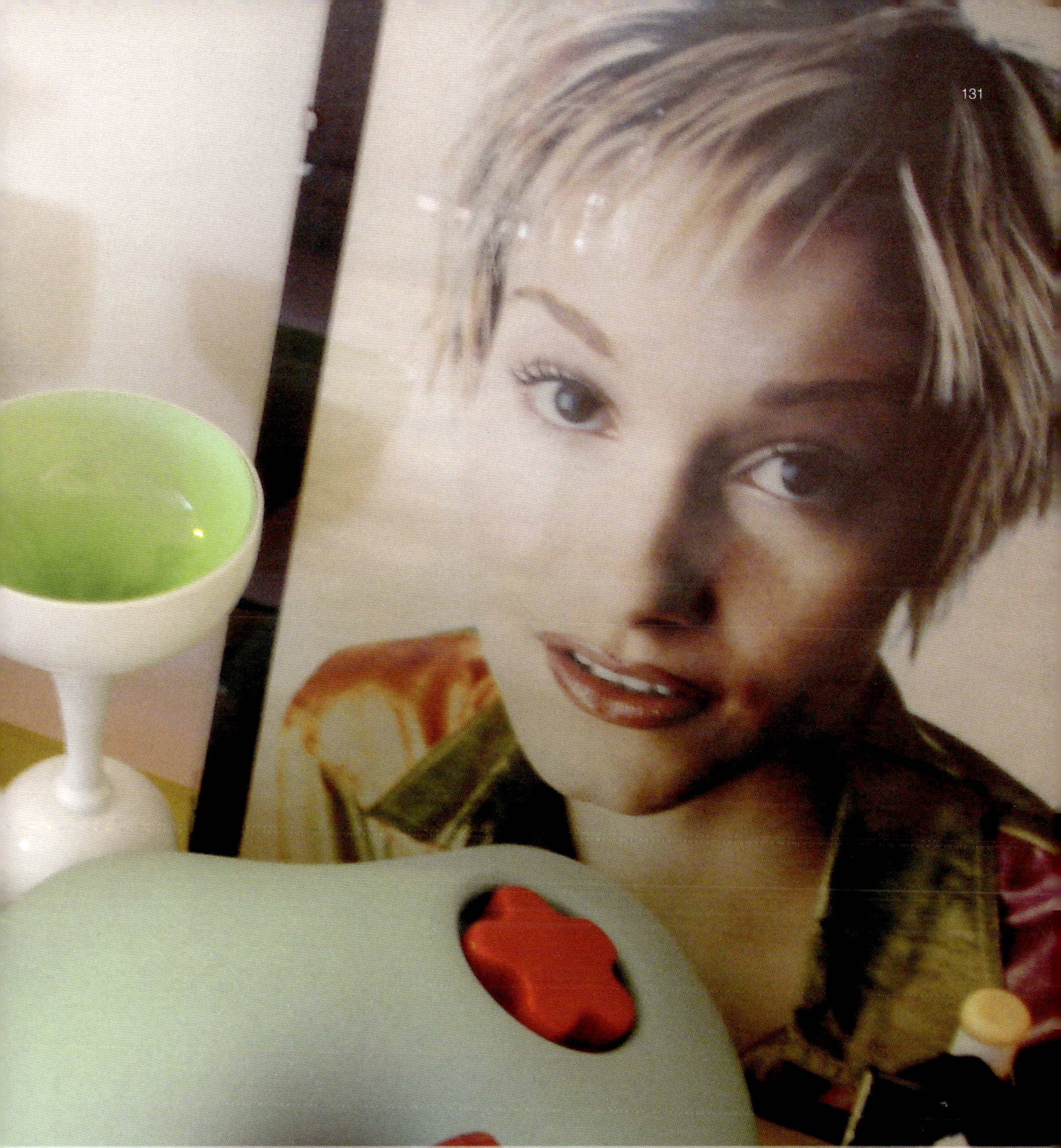

many of his works such as 'table-chair, table-lamp, table-upon-table barstool, lootable) I also see dry humor in a cup with two handles, a restrained cuteness and a beautiful approach, a dialectic of minimalism versus poetics, almost singly creating what I would refer to as Frugal Design – made with modest materials or methods, so simple yet proactive and stately – where the signifiers and motivators of his language tell a narrative, oneliner or statement that sends a message for us to rethink our banalities in life, to question them, to almost spin a quirky twist on our everyday Cartisian world.

KARIM RASHID designer — user of the Dombo mug
Remember it is better to be Rich and happy, than poor and happy.

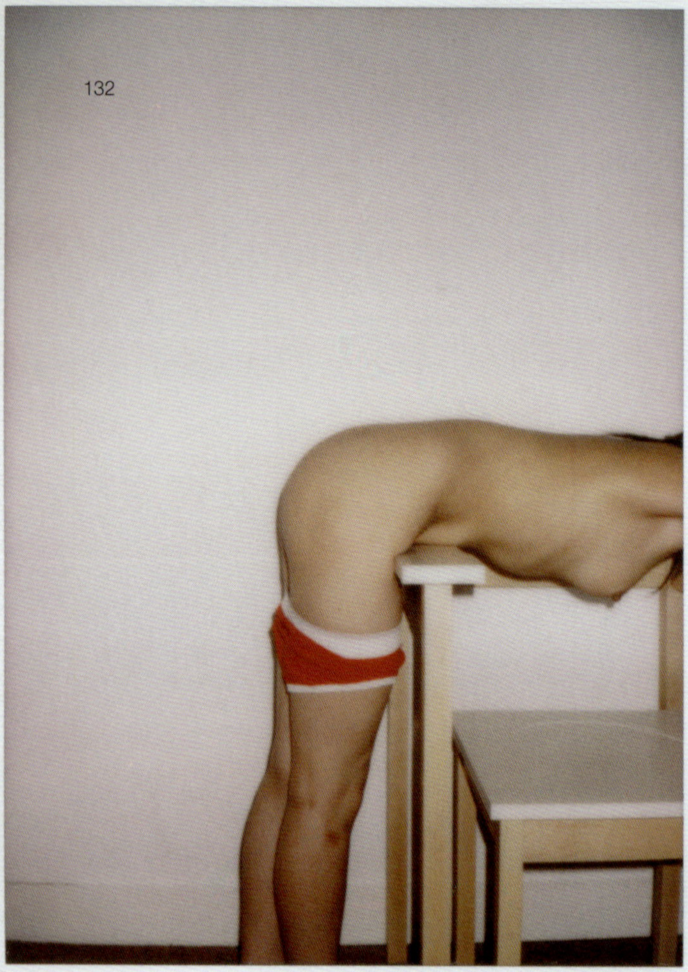

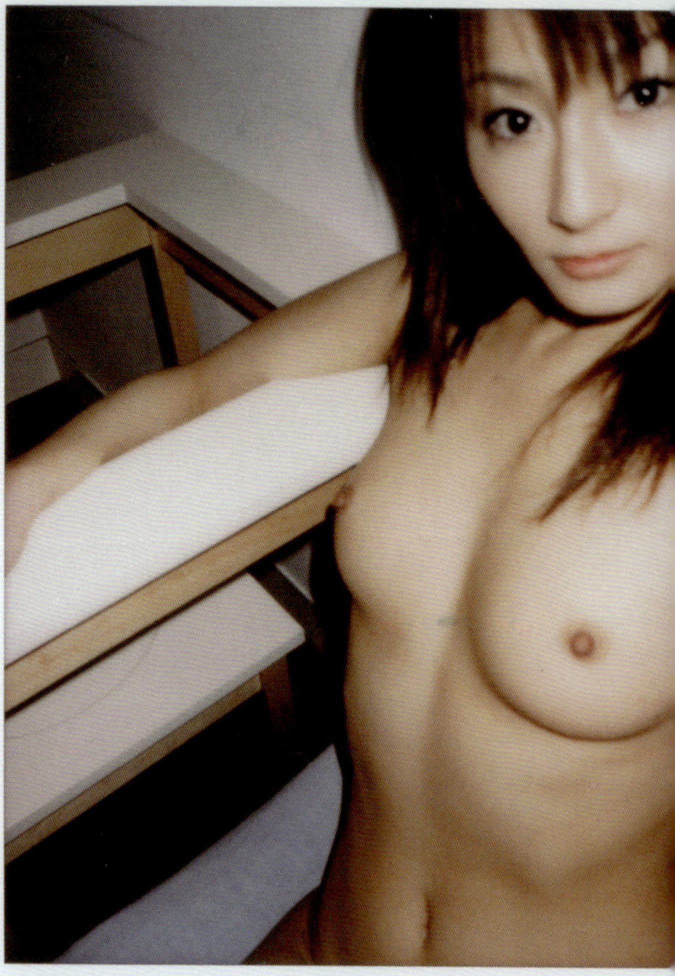

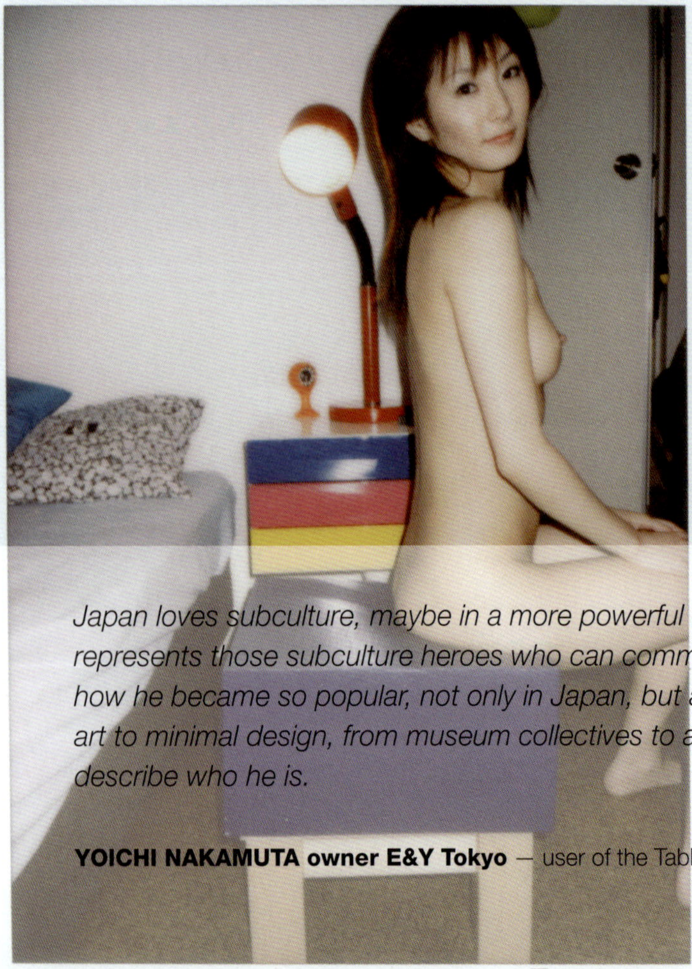

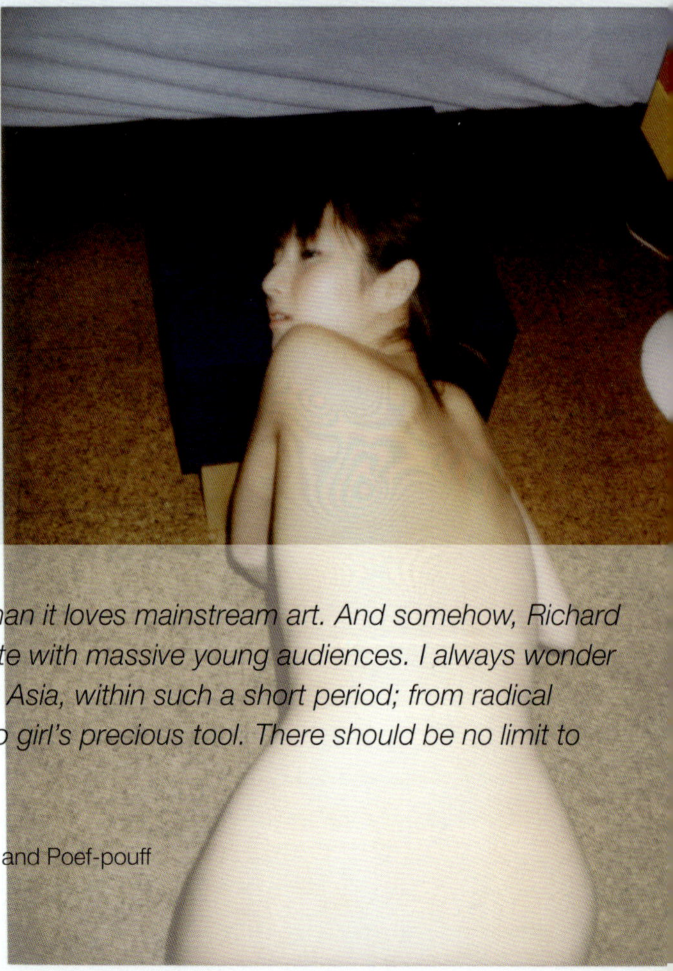

Japan loves subculture, maybe in a more powerful way than it loves mainstream art. And somehow, Richard represents those subculture heroes who can communicate with massive young audiences. I always wonder how he became so popular, not only in Japan, but also in Asia, within such a short period; from radical art to minimal design, from museum collectives to a Tokyo girl's precious tool. There should be no limit to describe who he is.

YOICHI NAKAMUTA owner E&Y Tokyo — user of the Table-chair and Poef-pouff

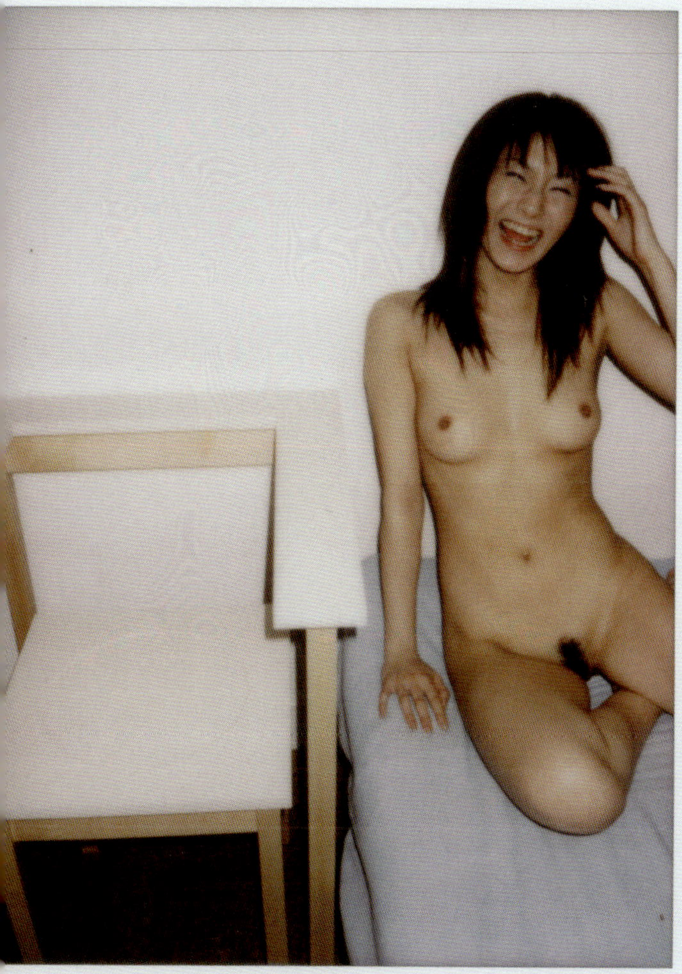

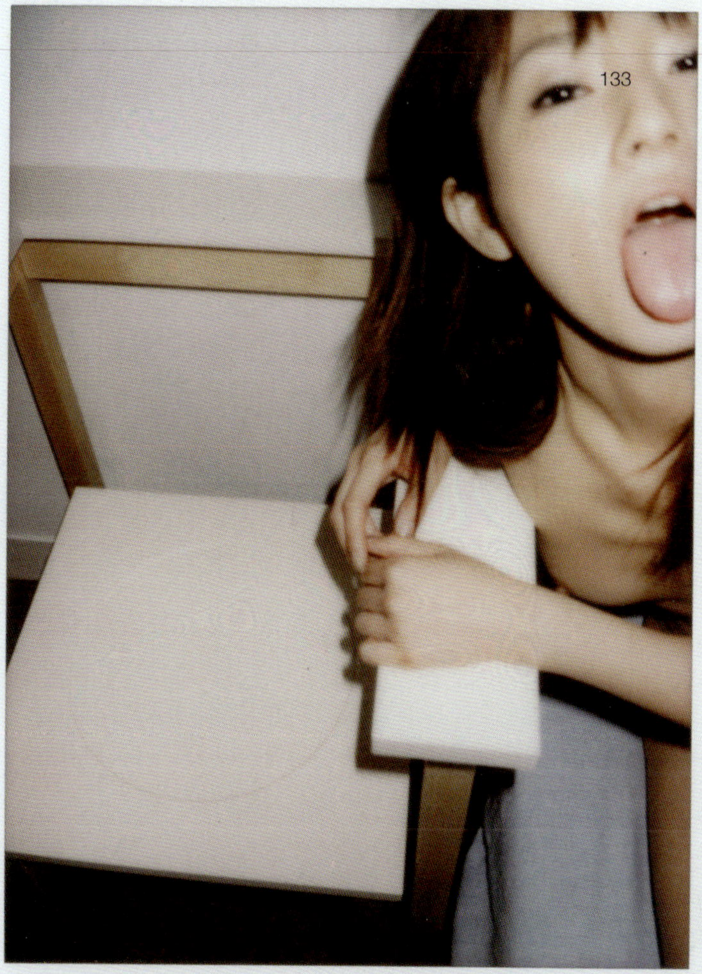

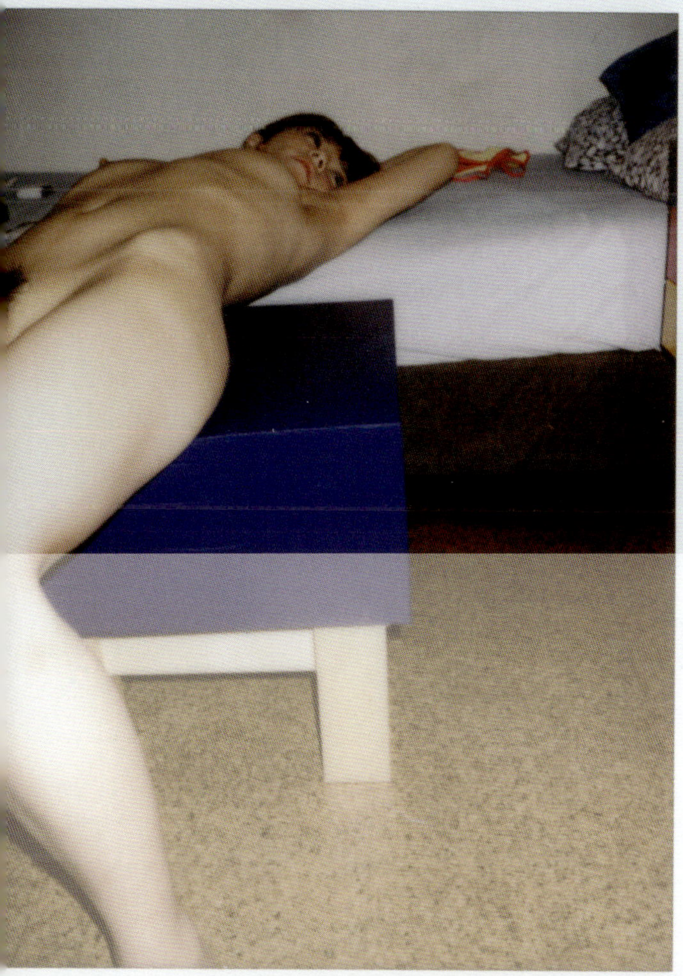

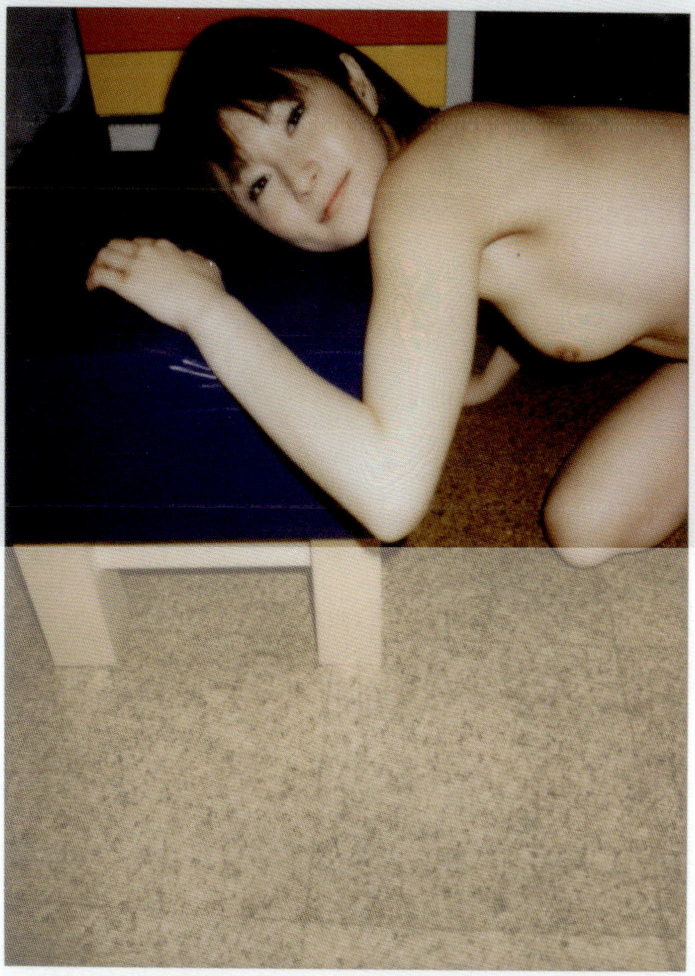

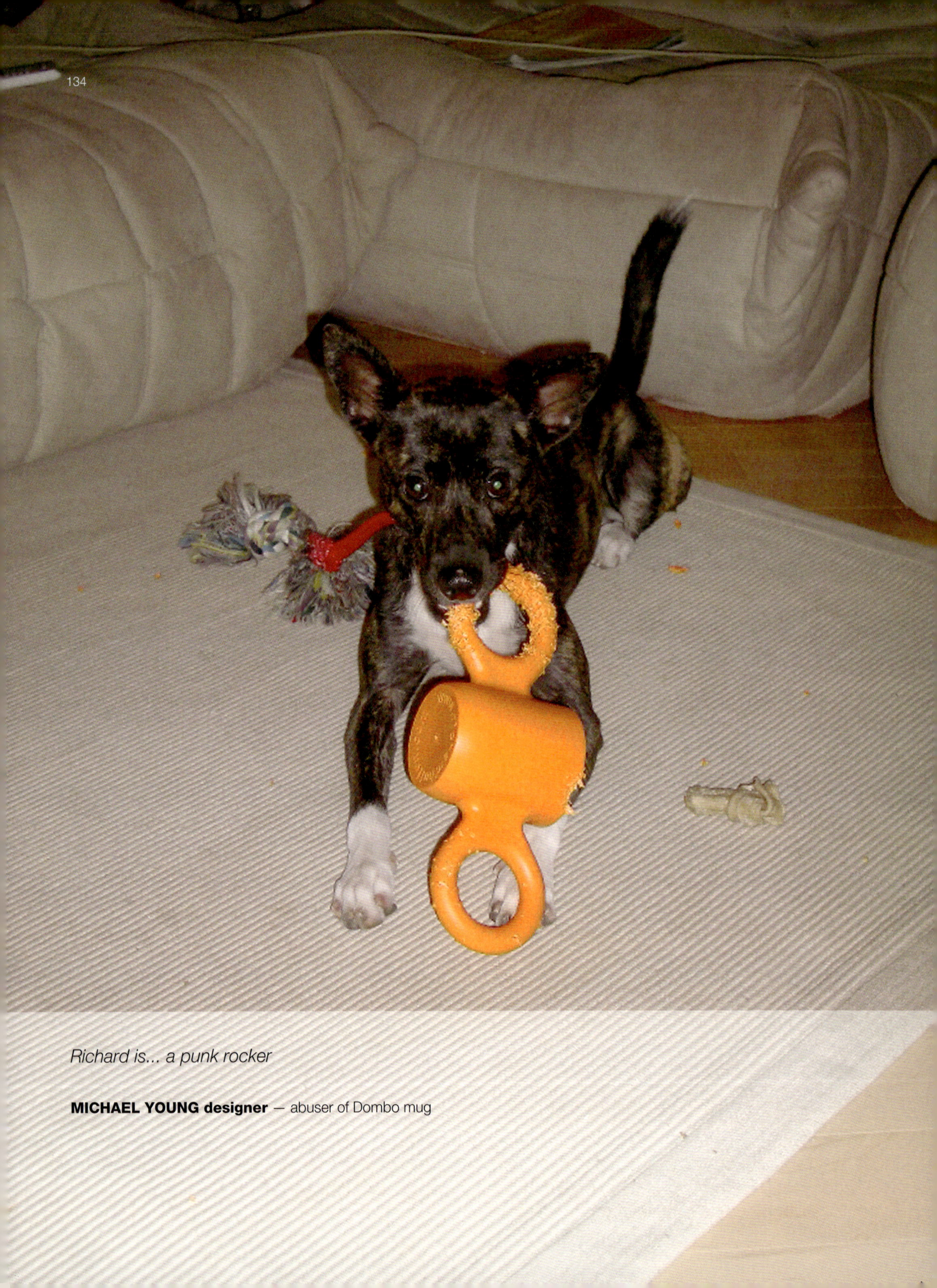

Richard is... a punk rocker

MICHAEL YOUNG designer — abuser of Dombo mug

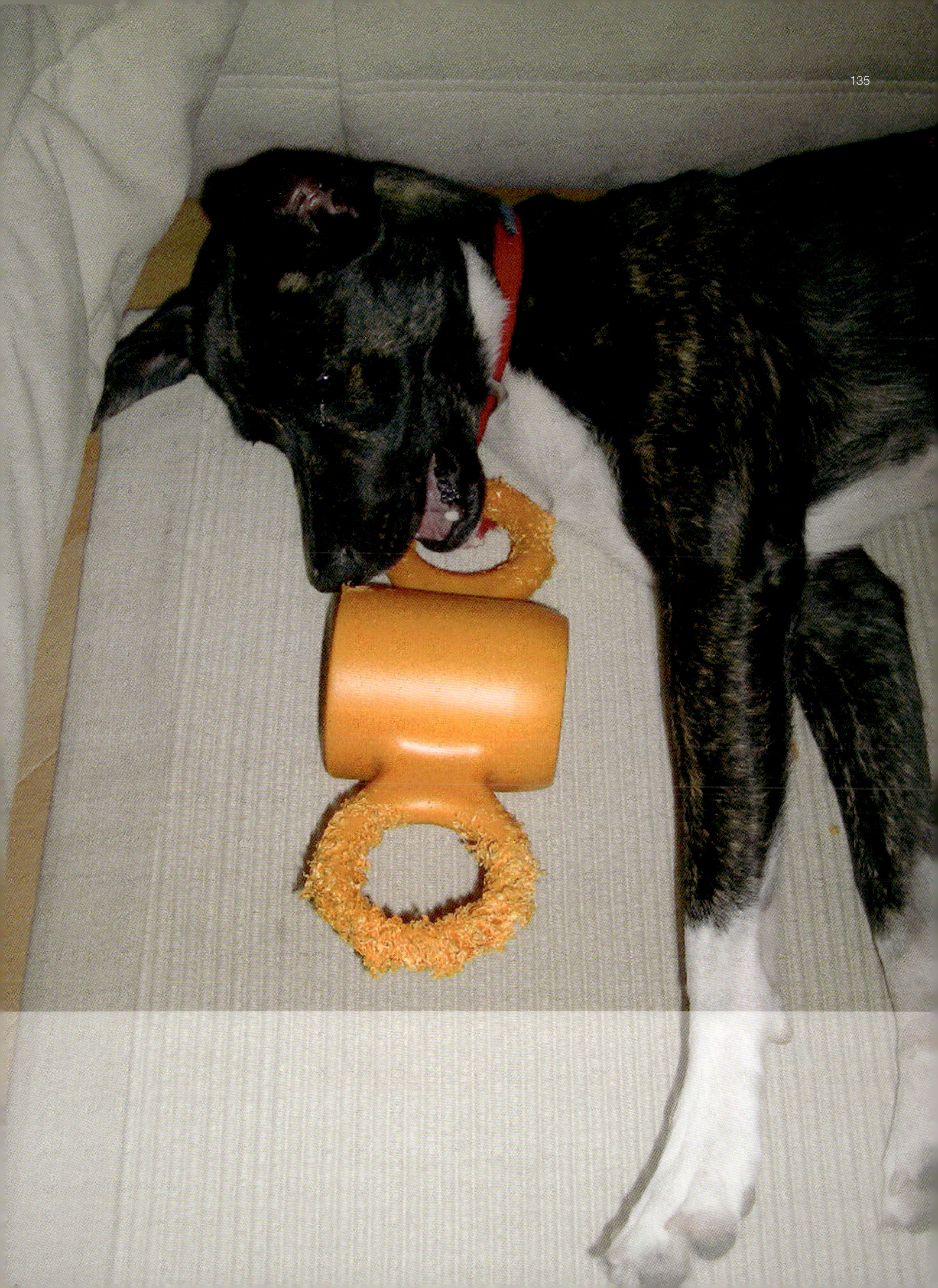

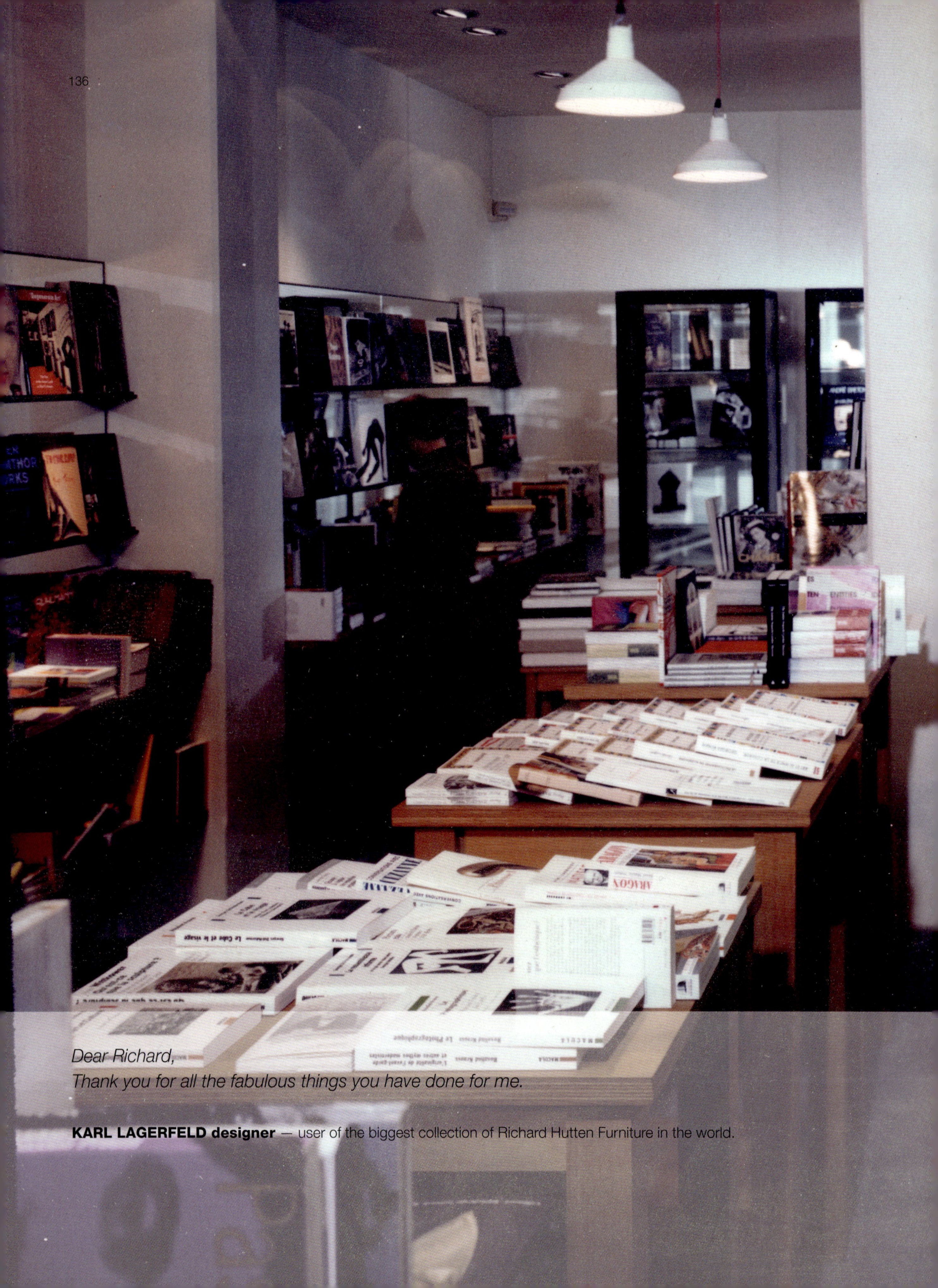

Dear Richard,
Thank you for all the fabulous things you have done for me.

KARL LAGERFELD designer — user of the biggest collection of Richard Hutten Furniture in the world.

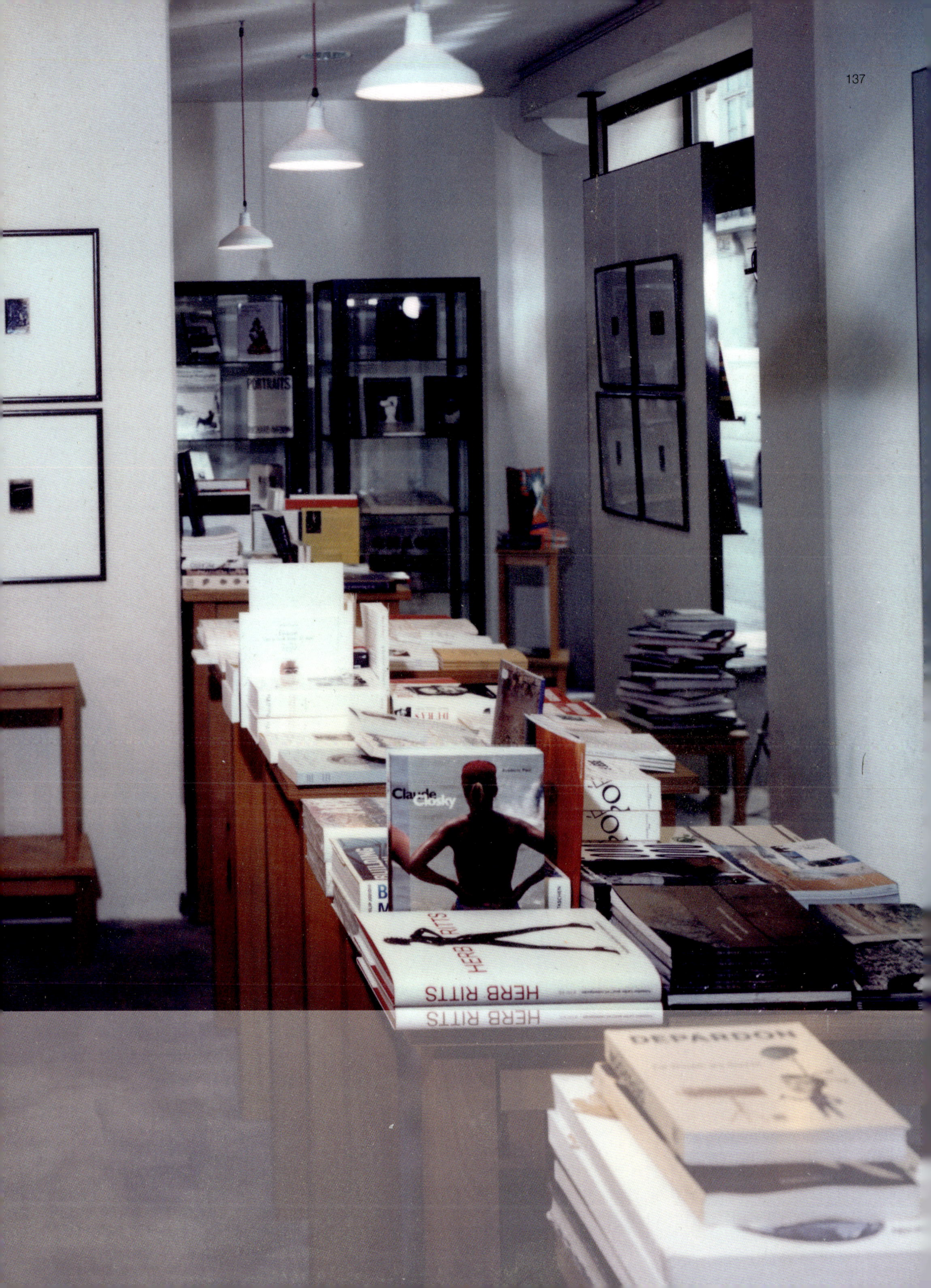
PORTRAITS
Claude Closky
TASCHEN
HERB RITTS
HERB RITTS
DEPARDON

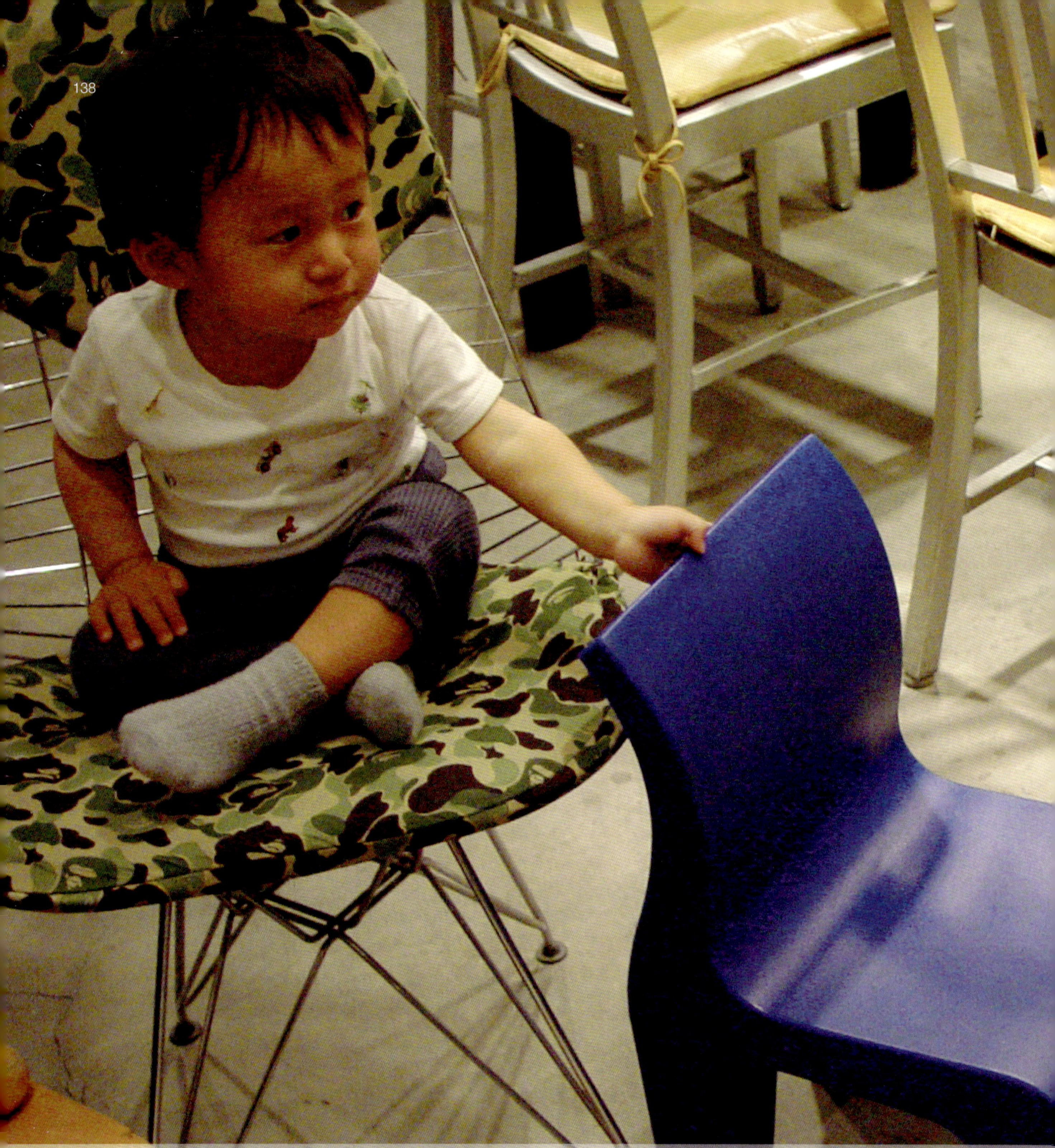

My kids use chairs not only as chairs, but also as friendly toys. They put them on their sides, or jump over it. The chair is no longer a piece of furniture for them. The form is so well thought-out, so even if it is not used as furniture or in a correct position, it will keep the beauty as an object. I recognized this when I saw my kids playing with it.

MASAMICHI KATAYAMA designer Wonderwall — user of the Bronto chair

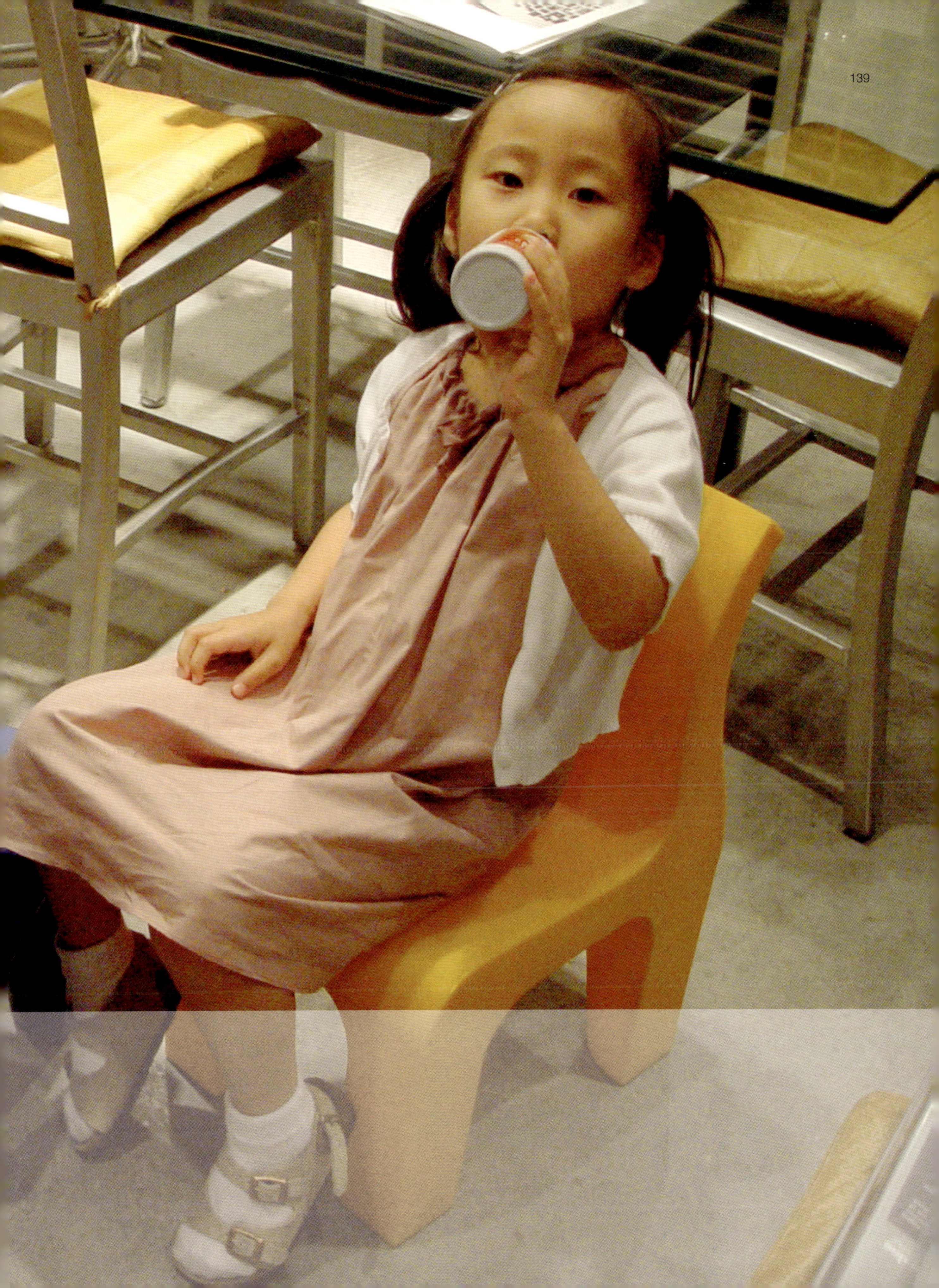

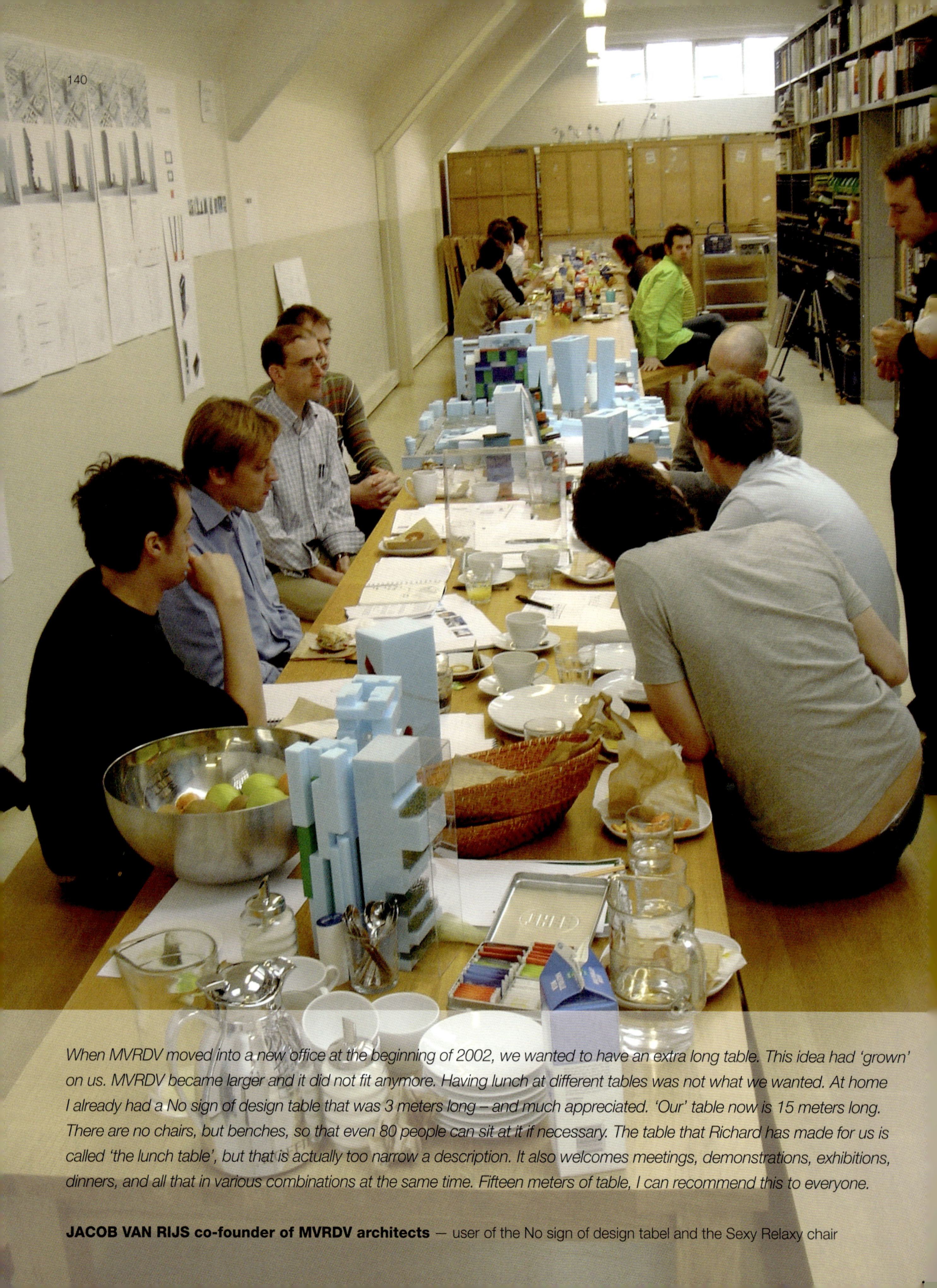

When MVRDV moved into a new office at the beginning of 2002, we wanted to have an extra long table. This idea had 'grown' on us. MVRDV became larger and it did not fit anymore. Having lunch at different tables was not what we wanted. At home I already had a No sign of design table that was 3 meters long – and much appreciated. 'Our' table now is 15 meters long. There are no chairs, but benches, so that even 80 people can sit at it if necessary. The table that Richard has made for us is called 'the lunch table', but that is actually too narrow a description. It also welcomes meetings, demonstrations, exhibitions, dinners, and all that in various combinations at the same time. Fifteen meters of table, I can recommend this to everyone.

JACOB VAN RIJS co-founder of MVRDV architects — user of the No sign of design tabel and the Sexy Relaxy chair

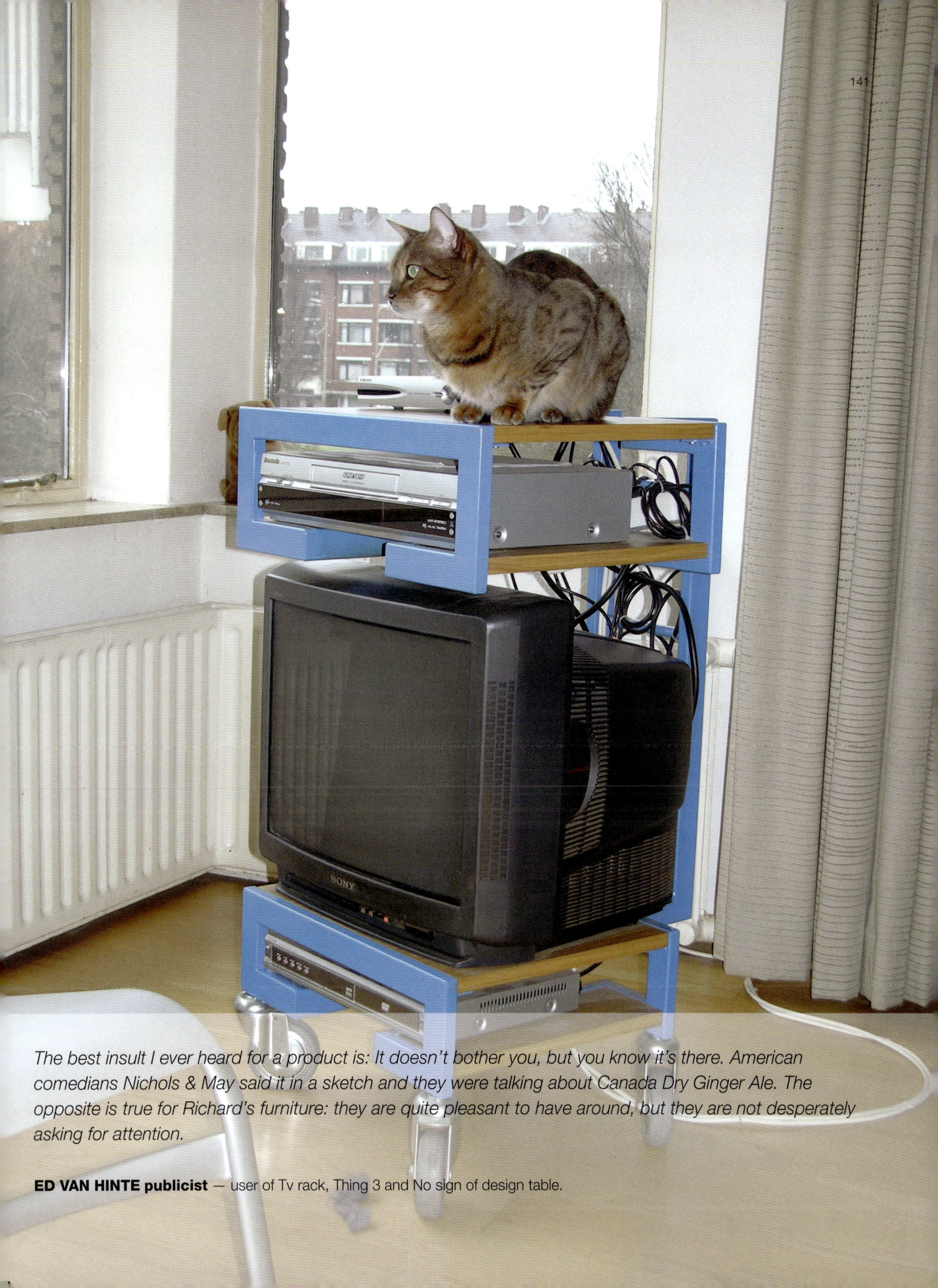

The best insult I ever heard for a product is: It doesn't bother you, but you know it's there. American comedians Nichols & May said it in a sketch and they were talking about Canada Dry Ginger Ale. The opposite is true for Richard's furniture: they are quite pleasant to have around, but they are not desperately asking for attention.

ED VAN HINTE publicist — user of Tv rack, Thing 3 and No sign of design table.

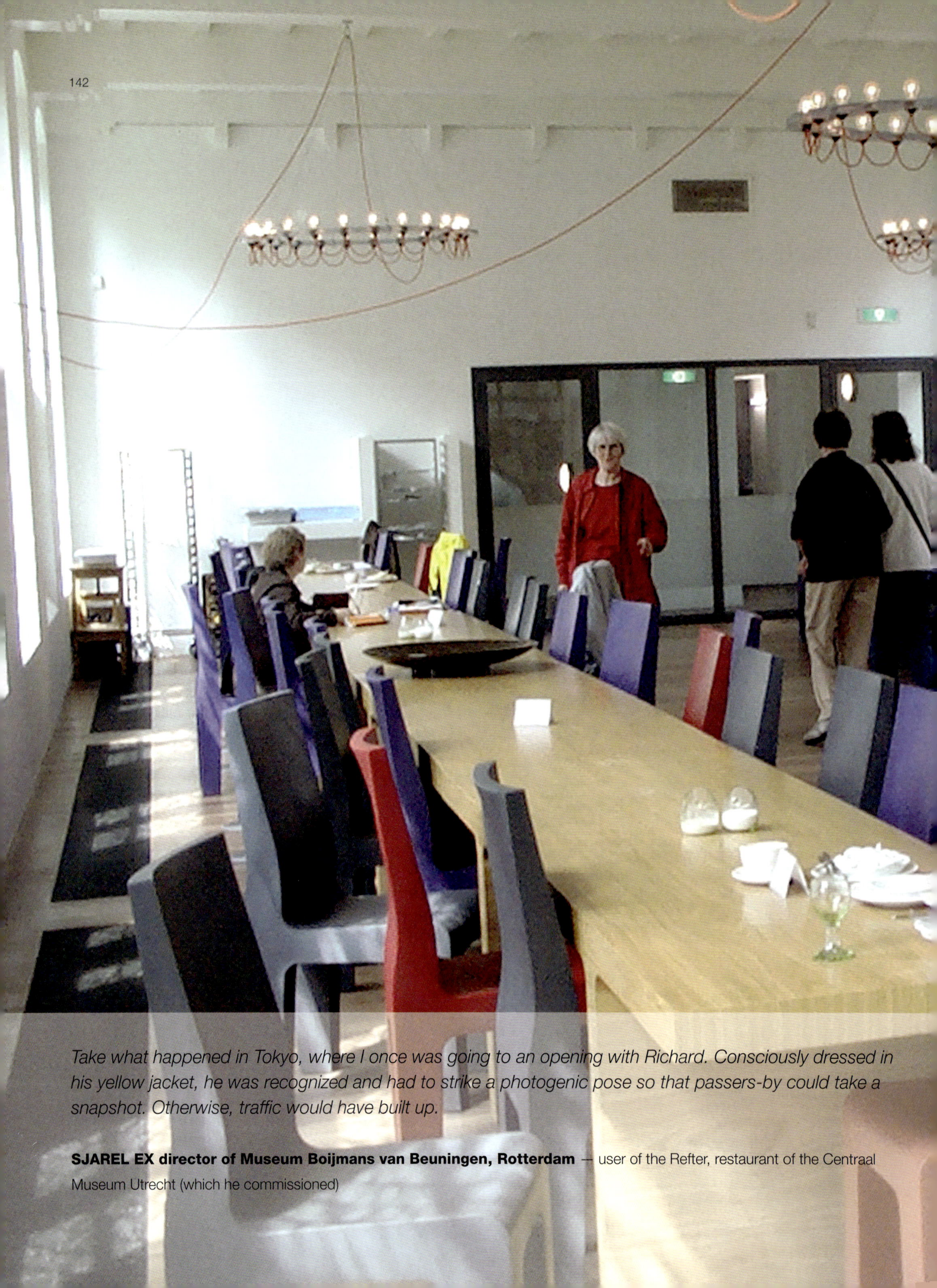

Take what happened in Tokyo, where I once was going to an opening with Richard. Consciously dressed in his yellow jacket, he was recognized and had to strike a photogenic pose so that passers-by could take a snapshot. Otherwise, traffic would have built up.

SJAREL EX director of Museum Boijmans van Beuningen, Rotterdam — user of the Refter, restaurant of the Centraal Museum Utrecht (which he commissioned)

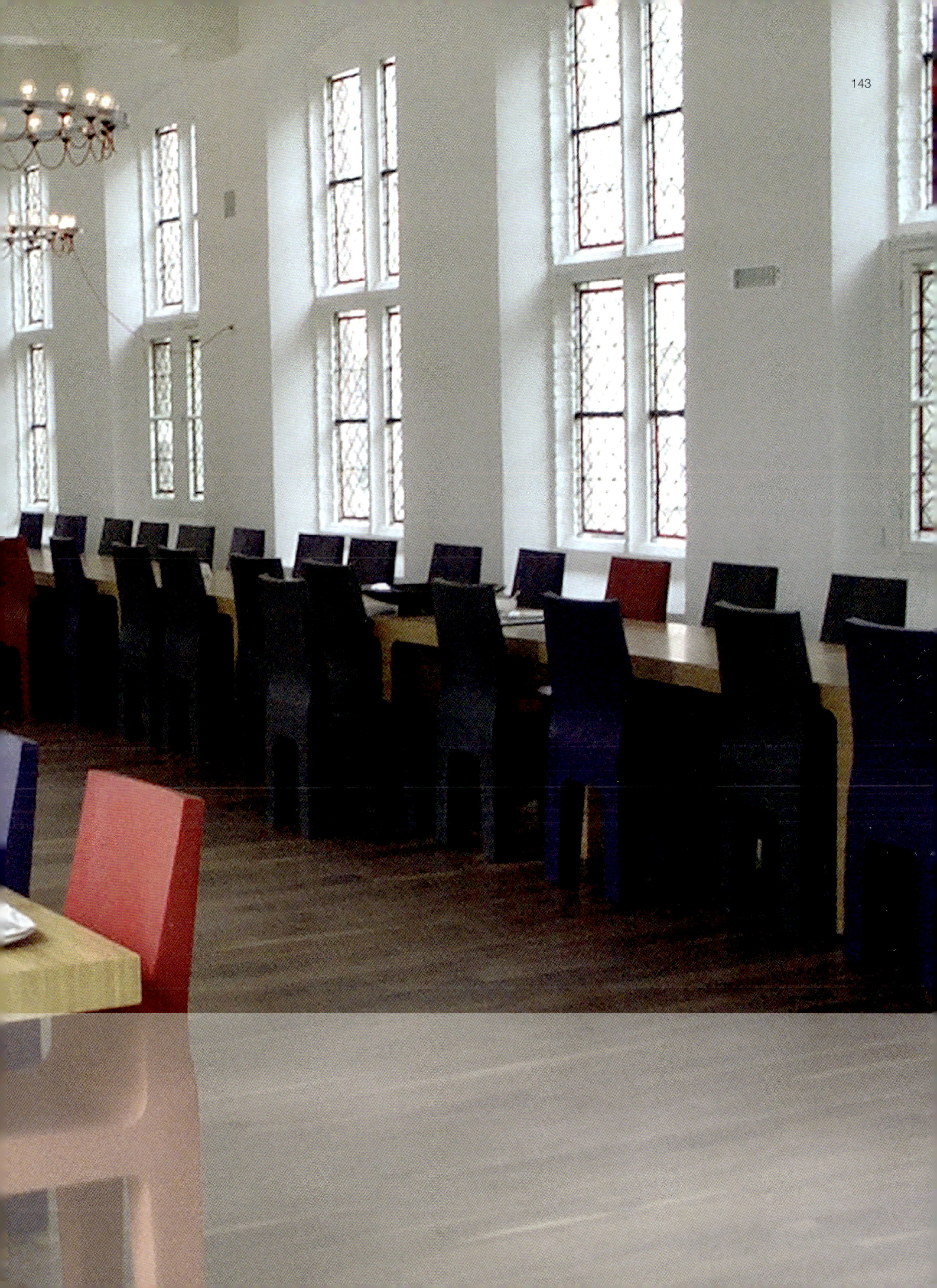

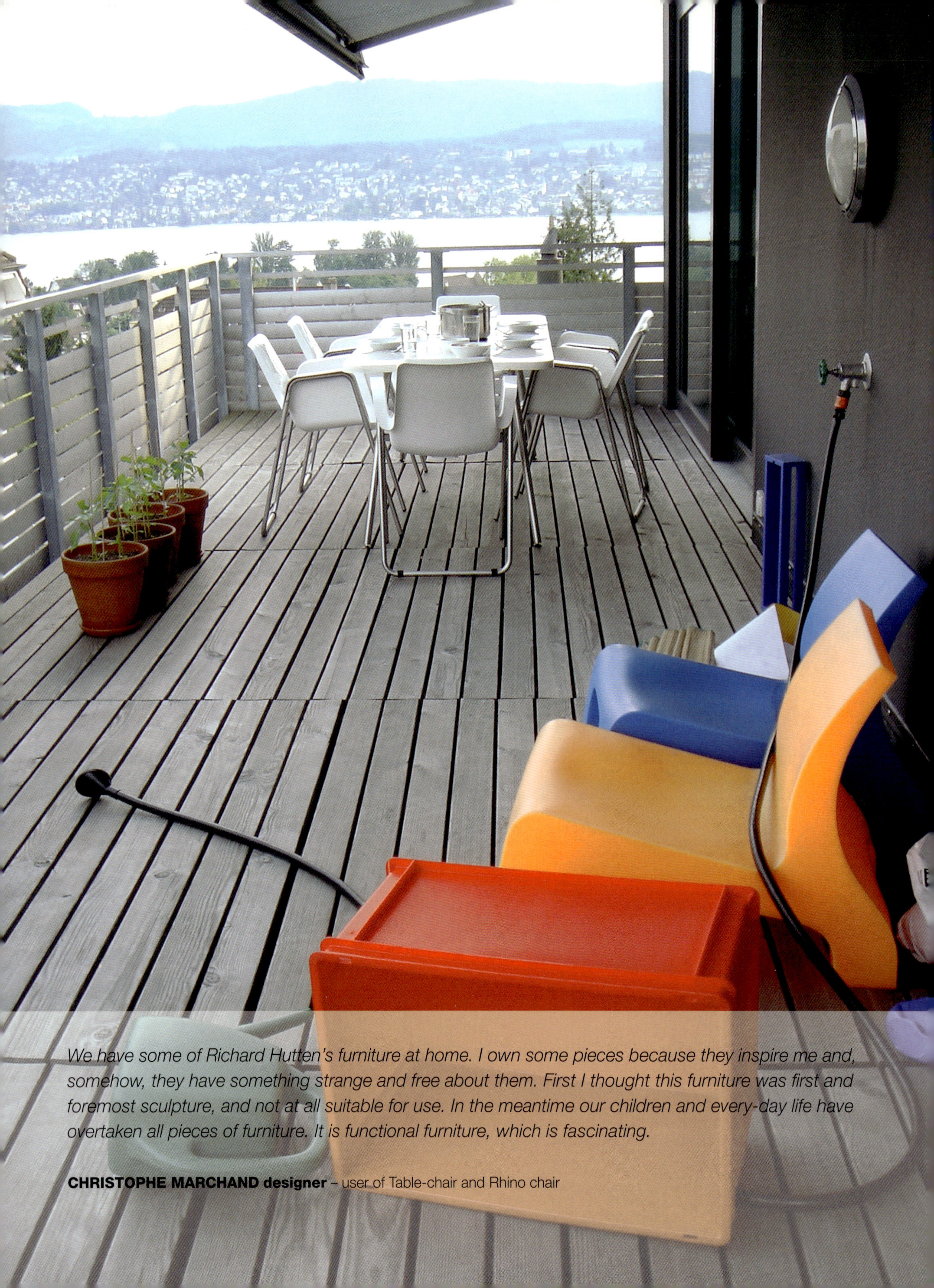

We have some of Richard Hutten's furniture at home. I own some pieces because they inspire me and, somehow, they have something strange and free about them. First I thought this furniture was first and foremost sculpture, and not at all suitable for use. In the meantime our children and every-day life have overtaken all pieces of furniture. It is functional furniture, which is fascinating.

CHRISTOPHE MARCHAND designer – user of Table-chair and Rhino chair

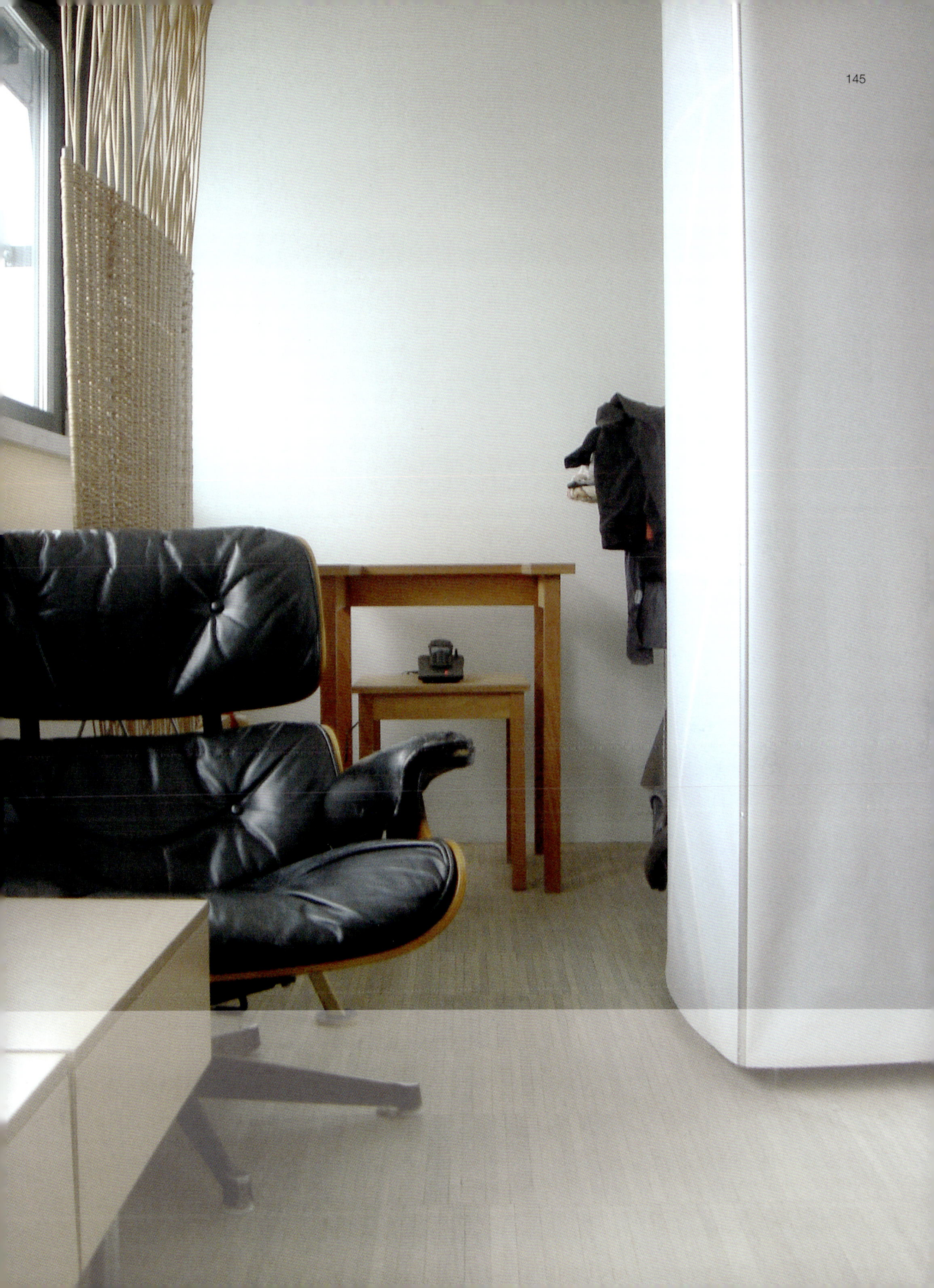

Brigitte Fitoussi

INTERVIEW
with Richard Hutten

Where did this idea of writing a book, the thread of which deals with the circulation of your designs and more generally of their use, come from?

The objects I design are made to be used. The very essence of the designer is to make products that can be a part of everyday life. Since I have been active professionally certain persons have sent me, from time to time, photos in situation of objects that I designed and they bought. It is always very pleasant to see that my designs interfere in their daily lives, that they are even so pleased with them that they take the time to photograph them and spontaneously send me these snapshots. It is a nice way of sharing with me a part of their lives. It is the nicest compliment I could receive. My work does indeed take on all of its meaning through the user. I want to design objects which make it possible to establish emotional ties, which then convey memories…

Most of the photographs that were sent to me (for this book) are of the Dumbo mug. Which seems logical insofar it is the only one among all my products, of which over 100.000 copies have been sold since it was designed. By extrapolating, wouldn't it be amusing if children who are growing up today with this mug should say to themselves in 20 years, 'Did you use a Dumbo mug when you were a child?' For them it would be a means of instantaneously recalling this period.

The meaning of the word 'use' is quite vast. My design projects are not exclusively functional nor are they only designed for everyday use, their use stretches far beyond that. The stakes

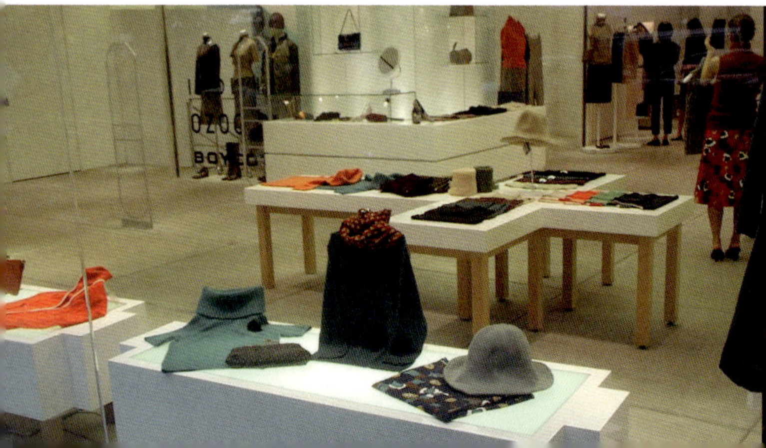

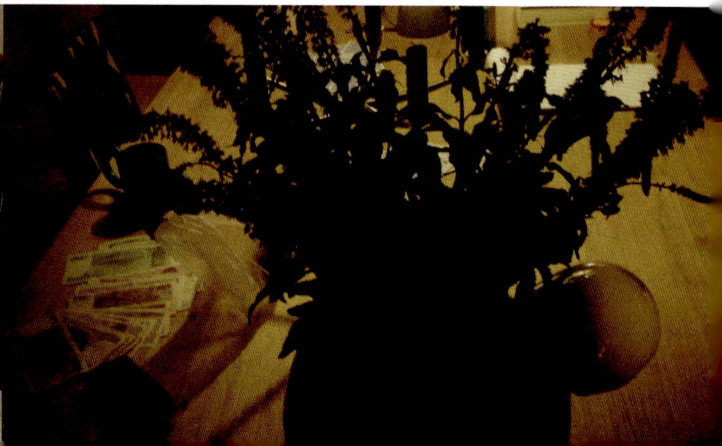

are cultural as well as commercial. The economic aspect of use takes into account the entire designing process: development, production and distribution. My aim is to pay attention to all the individuals involved in this process, to be sure that they are not just content with doing their work but that they should also gain satisfaction from it. The cultural aspect is also very important for me. My work functions, like the objects, in books and exhibitions. It can thus influence people's opinion about design, show them the trends or provide them with the keys. In a nutshell, a design capable of fulfilling these three conditions is a superdesign. Unfortunately this is rare, most of the time only one of these aspects is focused on.

Why release a book so soon after the previous one?

Four years have passed since the last one, which is quite a long time. And since then I have been involved in new projects. My first book, which was biographical, was designed like a catalogue. It came about, in a way, in spite of me. When the publisher suggested it to me, I refused in the beginning. I was 33 years old and I felt that my career was only just beginning. I thought it was premature. Finally, it was published, but I have the feeling that I was only the subject of it. Of course, once it was published I was really happy about it. The book came out really nice. But as a follow up this new book allows me to play a more important role. I purposefully show my work in a real context, with real people. Not just attractive smooth images ready to be published in Wallpaper. I do find it very interesting to show that a product can work in different situations. The context gives it a meaning. Of course, this does not exclude the context from being a magazine article or a museum exhibition. This is not negligible but I prefer to be confronted with real life. Let's take the example of the chair I designed for the Centraal Museum in Utrecht (the Netherlands). It was first created for the museum restaurant. Then, some time later, it was used for the wedding ceremony of our future King and Queen, Prince Willem-Alexander and Princess Maxima. Millions of people in the world therefore saw it on television. It is even currently used in a fast-food restaurant

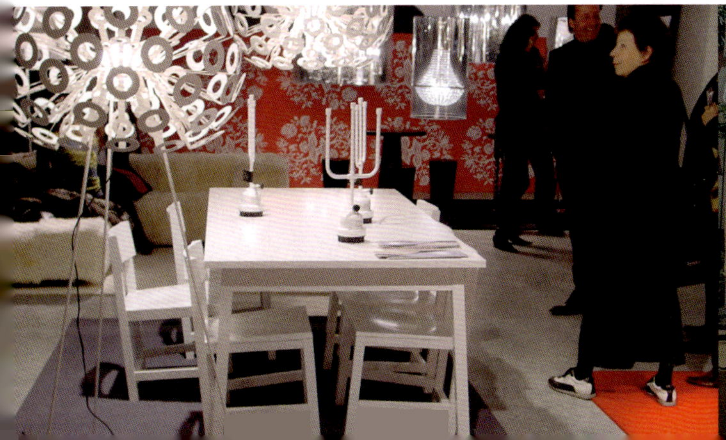

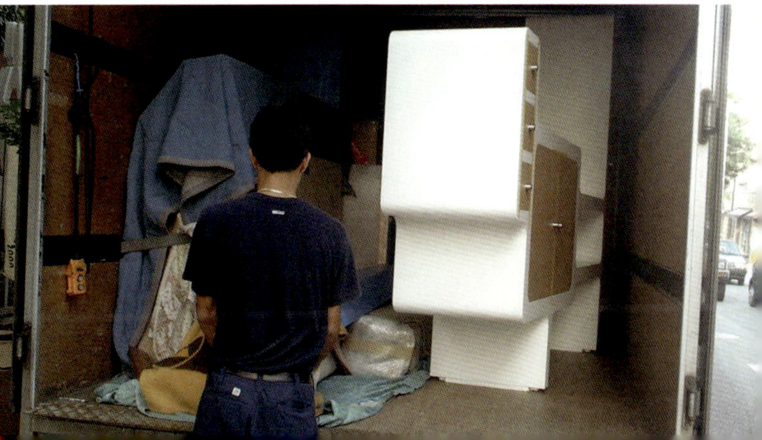

(Kentucky Fried Chicken) near my studio. That's exactly what I like, to see one of my projects as a part of so-called 'high culture' such as in a museum as well as in an ordinary restaurant termed as 'low culture'. In fact, I have no preference for one or other of these places, I appreciate them for different reasons. In the museum because it is a chair which is permanently on display; at the wedding of the prince and princess because it is a unique event, in the fast-food restaurant because it directly intervenes in the everyday lives of people who are not usually in contact with the world of design and who may never have been to a museum.

Beyond the question of use what is the new issue in this book? Is it the sign of a change or a new direction in your work?

It concerns, of course, a change in my work. When I started, the conceptual approach was very strong. Now, this approach remains valid, but functionality has taken precedence. The reason why I became a designer, instead of an artist, is that I like to see my objects being used. Industrial production fascinates me and this has also increased my interest in design. Through my designs, I would like to be of use to as many people as possible, but to accomplish this I need to collaborate with many manufacturing companies. One day I calculated that one person out of a hundred in Holland owned one of my designs. If I could attain this same percentage in the rest of the world it would be fantastic! (There's no harm in dreaming…) The work has therefore evolved from objects that only refer to the field of design to products which come entirely into their own in people's homes. The context has shifted.

Does your catch phrase 'No sign of design' represent a philosophy of design or is it just the name of one of your first collections?

When I was studying industrial design at the Design Academy in Eindhoven, I had chosen the option 'Man and Living'. In 1989, the year I graduated, I decided to do a research project on the

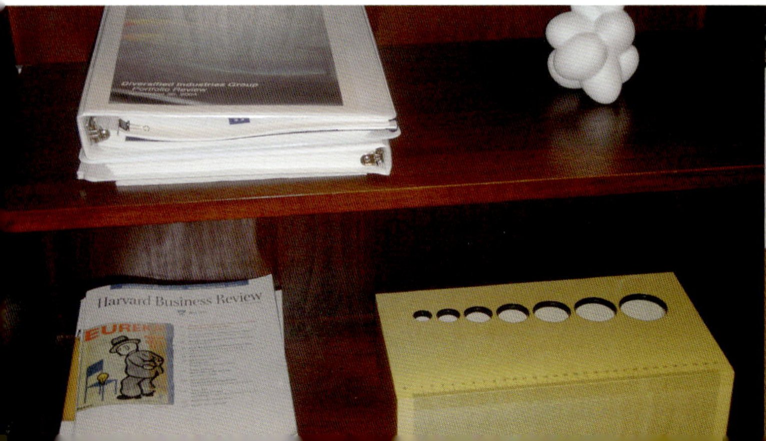

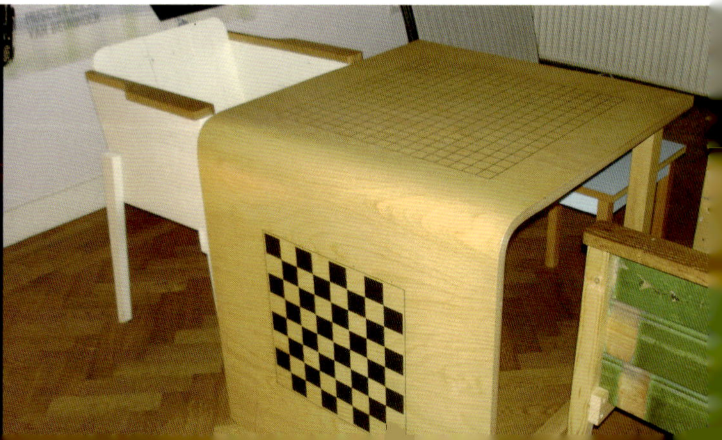

theme 'What does living mean? (Not in the sense of living one's life but as in living in a house.) A very general issue. Through this research, examined from a philosophical standpoint (Heidegger among others) as well as from the standpoint of anthropology, etymology, etc. several questions were raised. I became aware, for example, that being used to these objects was important. First of all, I realised that people lived in several territories: in their bodies, in their clothes, in their houses, in their cities, in their countries, on a planet… I therefore thought that the name of the department at the Design Academy was not appropriate. 'Man and Home' would have been more appropriate given that these courses were focused on the home (living implies a much wider scope including almost everything). However, there are also several territories in the home. Territories which I translated into the archetypal form of the table: a top, four legs and supporting beams between the legs. By adapting the size of the table to the human dimension, I rendered it significant in a different manner: the table became the floor, the bedroom, the bed, the wardrobe, the dining room table or the stool/chair. In reality, this design showed the link between architecture and furniture, the relation also between different types of furniture, for example between a sofa and a bed. In the end this demonstration no longer represented design for me but a form of science. A science which provided me with the tools to create objects. Taking existing forms and not inventing new ones, has enabled me to find new solutions. The advantage of using these forms is that people are already accustomed to them. As I said in the introduction, 'being used to' is another important aspect of life. The Dutch word for 'design' is 'form-giver'. But in this case I have not given a form, I have taken forms. Thus 'No sign of design' was born. Like a new way of designing products. No one had used this expression before me in the world of design, and this was several years before Droog Design was born. I didn't invent new forms but I sampled existing forms. Whereas the world of design, as we knew it at the time, was not very receptive to this kind of work. It is partly because of this that I called it 'No sign of design'. It was essentially a frame of mind, a conceptual approach, which provided me with the keys to a different approach to design, beyond form.

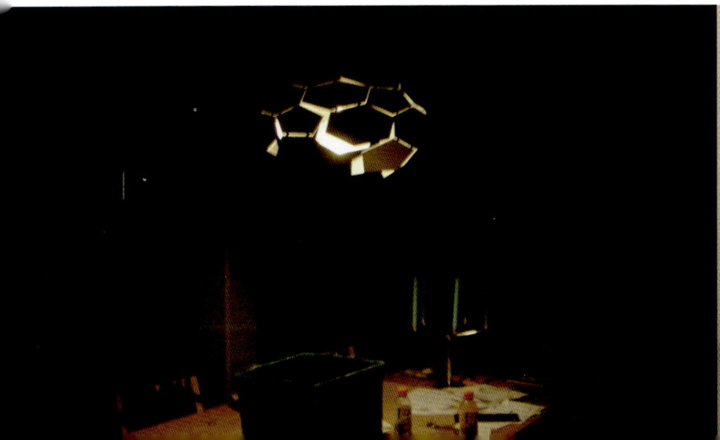

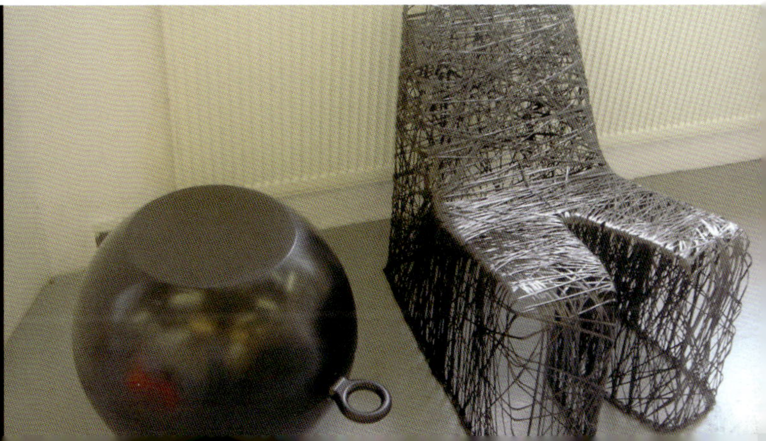

What are your main sources of inspiration?

I don't like the word 'inspiration' because it contains a certain element of chance or randomness. Something like 'I saw a pretty flower so I decided to design something in the shape of that flower'. This type of inspiration doesn't interest me. My work is based on knowledge and a feeling in relation to the period in which I live. But if I had to rely on inspiration I would say that the people for whom I design are the greatest source of inspiration: their needs, joys, constraints, curiosity, sexuality, habits, humour, limits, beauty and the whole spectrum of human emotions are a constant source of inspiration. The context in which I have to create also contributes to this inspiration. It might be a business or a place, a situation, a problem... Basically, anything that may have a connection with the project.

Does your workshop enable you to design special orders or is it a response to a lack of manufacturers in Holland?

The lack of manufacturers was the main reason for starting it up. The reason why I kept it going was its function as a laboratory. I can make prototypes in it, special design installations or objects that have no mass production potential but which, I feel, should be developed. Experimentation is indispensable for any designer worth his salt. To be able to make mock-ups or life-sized prototypes, in spite of the wonderful designs that can be done in 3D, remains essential. To see, feel and touch; a context is essential to be able to assess a new design.

Do you always approach design in the same way?

Fundamentally I always adopt the same approach. For me design is a thinking process. Being a designer implies several crafts or functions. It means being a technician, a philosopher, a scientist, an inventor, a politician, a sociologist, a diplomat, a romantic, a rebel, a mediator, an engineer, a craftsman, a revolutionary and finally (to quote Philippe Starck) a 'creator of forms'. For me,

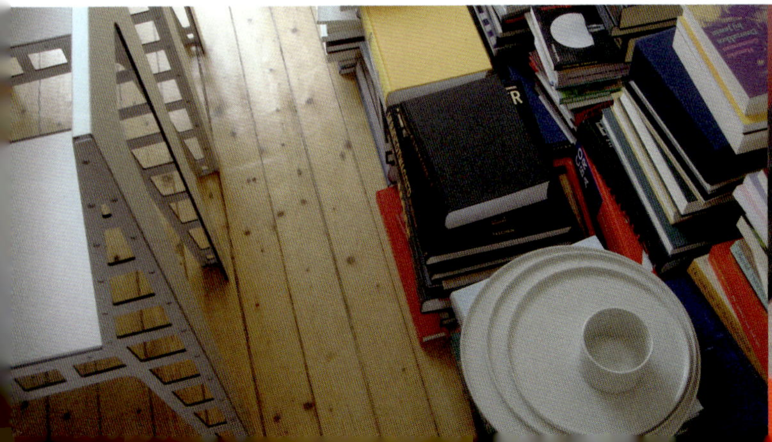

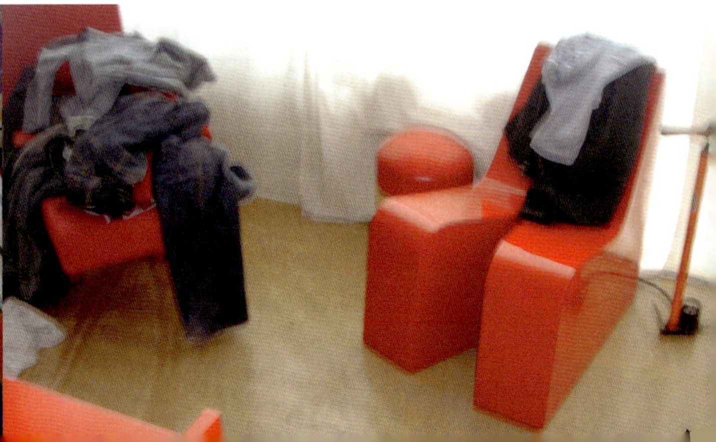

form is almost accidental, it is an outcome. Things must have a final form, it is inevitable, but it is not an aim in itself. The client has a problem, and according to the context, I translate this into a solution.

In terms of this approach, how do you begin a project? Is it based on materials, technical processes, the function, the programme...?

Design is a dialogue. The client plays an important role. What does he want, what does he need? Is it really what he needs or is it something else. Answering this question is already half the job. The project starts therefore with curiosity and involvement. I try to explore the limits of the task, its restrictions but above all its possibilities. The thinking process can then begin. It is a solitary process. I cannot think in my studio where I am distracted by a thousand things. I can think more easily while riding my bicycle, or when driving (without music, just to the sound of the car engine), in a plane (it's perfect, no one can call me on the phone!) or even in the evening at my kitchen table. During this process I may talk with my assistants, but it's rarely the case. I like this whole creative process, but the beginning, the conception of an idea, always remains very special for me. When, from this mental process a clear concept results, when I have given a definition to the constraints with which I would like to work, I explain to one of my assistants what I want. This is often the moment when I do my first drawings. Then, with my team the visualisation process begins. Most of the time this is done on the computer but we often start in the workshop by producing mock-ups or small 3D models; this process requires constant discussion. I am fortunate in having a good team who understands my way of thinking and often thinks in the same way as I do. (When I am on the other side of the world I explain my ideas simply over the phone or in an e-mail and they are often able to make the initial drawings!) The process, as a whole, is almost rational. Each aspect of the design has its own reason, can be explained (even in the case when we take a decision based on emotional reasons). It is the type of rationalism

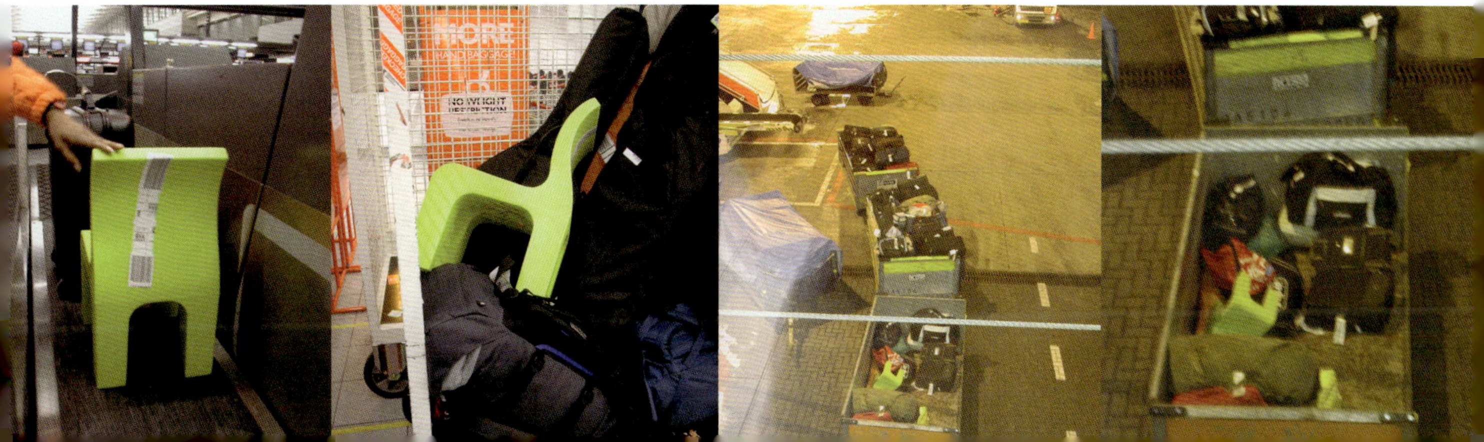

that can be seen in nature. Each aspect of a butterfly, for example, has a function, a reason. The result is always extraordinarily beautiful. And there is a huge variety of butterflies! It is the rational approach which leads to different types of butterflies! Could one possibly imagine needing only one type of butterfly? What is wonderful in nature – and this is where the analogy with design comes in – is that all of these variations function very well side by side. Obviously, in nature there is this need for diversity that is also found in Man.

Let us return to the dialogue with the client...

After having finished the first phase of the design, we present it to the client. Once again, this is a very important moment. The dialogue with the client continues. It calls on my skill as a designer but I also need the client's expertise and know-how. We therefore develop the projects as a design team, from the sketch to the finished product. (This dialogue, as I see things, is what opposes art to design. The artist creates a work in his studio and once it is finished he shows it to the world. Art is like a monologue. The designer has a much wider palette of investigation to conduct, which involves him to a greater extent in contemporary issues such as mass communications, mass production and globalisation. Contrary to the artist, he does not only react according to this reality of daily life, he also has the capacity of transforming it, thus conferring on design a cultural impact that is much stronger than art.)

Style is never an issue. Besides, I have no sense of taste. Taste does not go with creating something new or something that has never been seen before, on the contrary it blocks all possibility because it is based on established parameters. One of my objectives is to introduce an element of surprise not to be complacent about a certain taste or style which could prevent me from doing so. Generally, design is a matter of seduction, but I don't seek to seduce people through my designs, I prefer to amuse them. 'The sexual aspect' of design, seduction, is not my principal objective. I could compare that to the attraction between a man and a woman. For

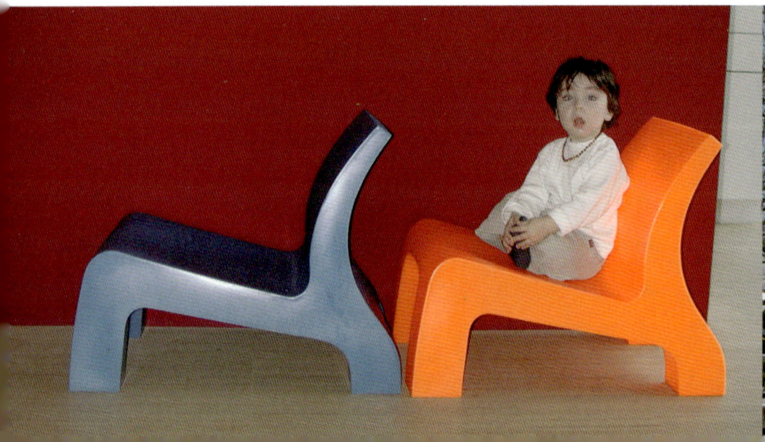

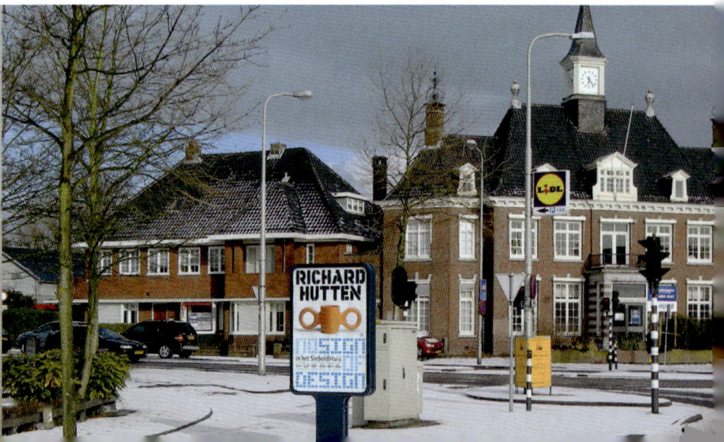

me this type of love – because design is about love – has an impact on everyone's background. It is, I believe, closer to the type of love a parent feels for a child. It has more to do with joy, surprise, composition, dreams... For many designers, beauty, in the aesthetical sense, is the main objective. This is not the type of beauty I am looking for. I try, in my projects to move and to affect people in several ways, for example with humour and by surprising them. Naturally this approach is not unique, if one refers, for example, to the work of Gaetano Pesce, Ingo Maurer or Achille Castiglioni, but it is the way of working that suits me best. It resembles me.

Let us come back to the starting point, which was (is?) the part played by Droog Design in your career?

Very early on Droog understood and appreciated my design approach, my spirit too, providing me with an international platform. It was practically at the beginning of my career. At the time I gave general answers to general questions (usually my own questions). How people react, what is the importance of objects in their life...? From these answers original projects were born. Today, I provide specific answers to specific questions which are almost always posed by someone else. I seek much more to collaborate with companies which manufacture industrially or who possess real expertise. An important part of my job consists of always observing. I always feel this need to connect with what exists, with the same attitude as when I started with Droog but with a different approach. Droog continues to focus on the cultural aspect of design. I feel, on the contrary, the need to expand more widely the scope of design, namely to the notion of use which has become more important for me.

What does modernity mean for you?

For me modernity makes me think of something from the past. Many (too many) architects from this period have been incorrectly interpreted. The magnificent buildings designed by Le Corbusier,

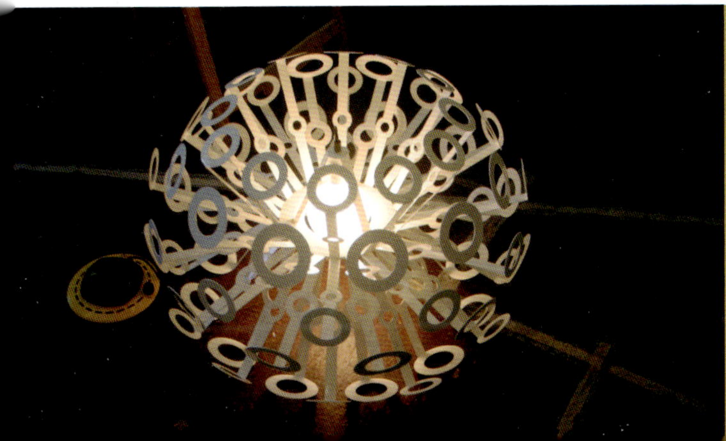

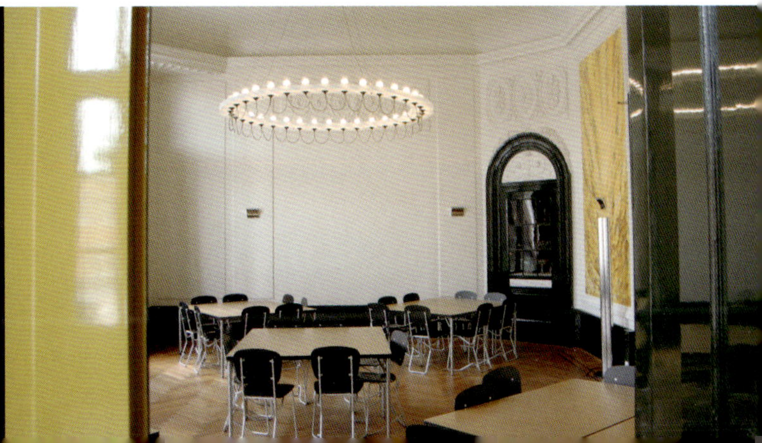

Rietveld or Mies van der Rohe were used as excuses by many architects to build boring and insipid buildings, soulless places. If a project has no soul, whatever the discipline involved, architecture or a product, it has no reason for being. Modernity has been interpreted as a style instead of a philosophy. It is very difficult to design a simple project that has a soul…

At first sight your designs appear to be very simple or based on the archetype, but it soon becomes apparent that they always have something more (irony, humour, a story). Is this difference one of the keys to your creativity?

If I use existing forms and/or archetypes there are several reasons for this. Above all, they make it possible for the user to recognise them immediately; to appropriate them more easily. Using existing forms, from history, from nature or from existing crafts always seems to me to be a good starting point. I also like to create a surprise effect, hence this quirkiness often found in my designs. To make people smile by surprising or even shocking them. Humour is a good way of trying to make people happy. For me, happiness is not a trivial thing but one of the most important aims in life. 'Stepping out of context', or using quirkiness, is also a way of encouraging reflection. The work is often extreme, radical in its point of departure and in the limits I set myself, but the result always leaves a door open to interpretation. Traditionally design is about solving a problem. I don't solve problems, I create possibilities. Like a story the end of which would be free to interpretation or that might be invented by the reader…

Can you give three words of advice for being a good designer?

Listen, see, hear or… think, try, make.

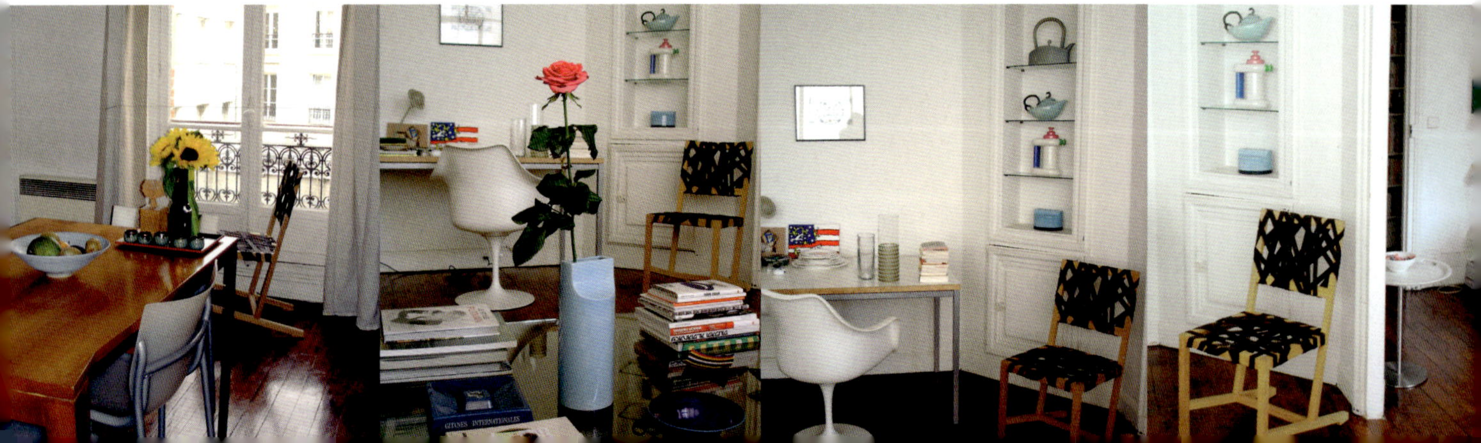

Your plans for the near future?

Some of them will be in this book with drawings for Christofle, Muji, the interior of a building in Amsterdam signed by the architects MVRDV, the offices inside the Boijmans Museum in Rotterdam, a house for Ytong, a lino floor for Forbo, and a dozen other designs…

What motivates you most in this profession?

The most important thing for me is to try to please the greatest number of people possible. The more I please them, the happier I am. I would like to make a small contribution to their daily lives. This book is therefore in homage to the people who use my designs. I would like to celebrate every day of my life, that's why I make tools to celebrate it with. But beyond that, this profession also allows me to travel very often, to meet many wonderful people and to have a lot of fun…

An interview conducted by Brigitte Fitoussi

January - February 2006

Brigitte Fitoussi is the art director of Christofle since June 2005, in charge of the corporate image of the great Parisian silversmith. Trained as an architect, she formerly worked as a journalist, a curator and consultant. Editor of the design column for the French review NUMERO since 1999 (for which she continues to write), she has regularly contributed to French and foreign publications. She is the author of ten books on design and architecture.

User of Dombo mug, Berlage chair and Zzzidt object.

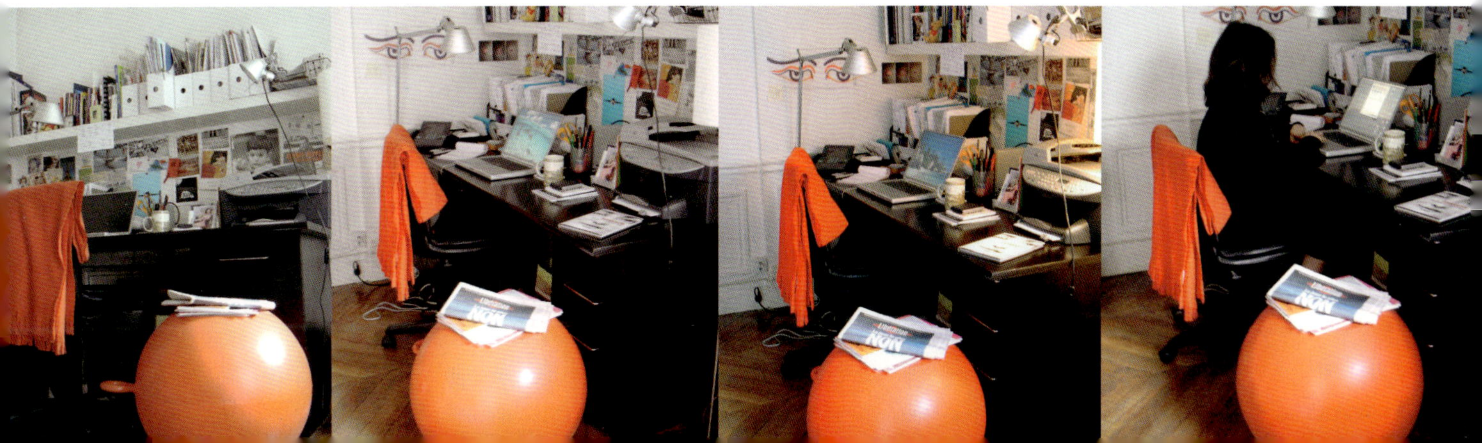

Traditionally design is about solving a problem. I don't solve problems; I create possibilities.

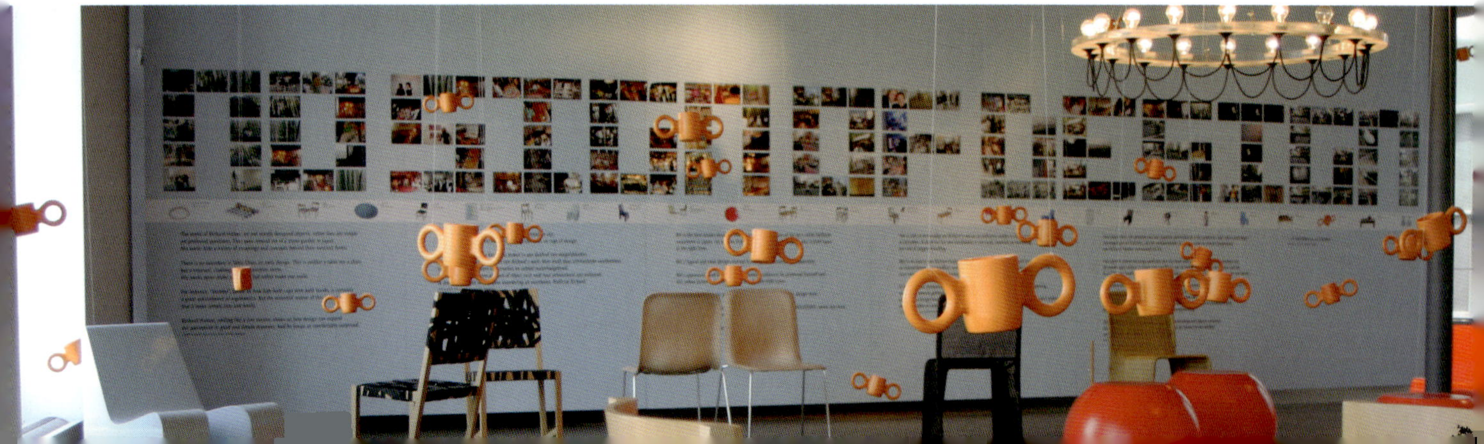

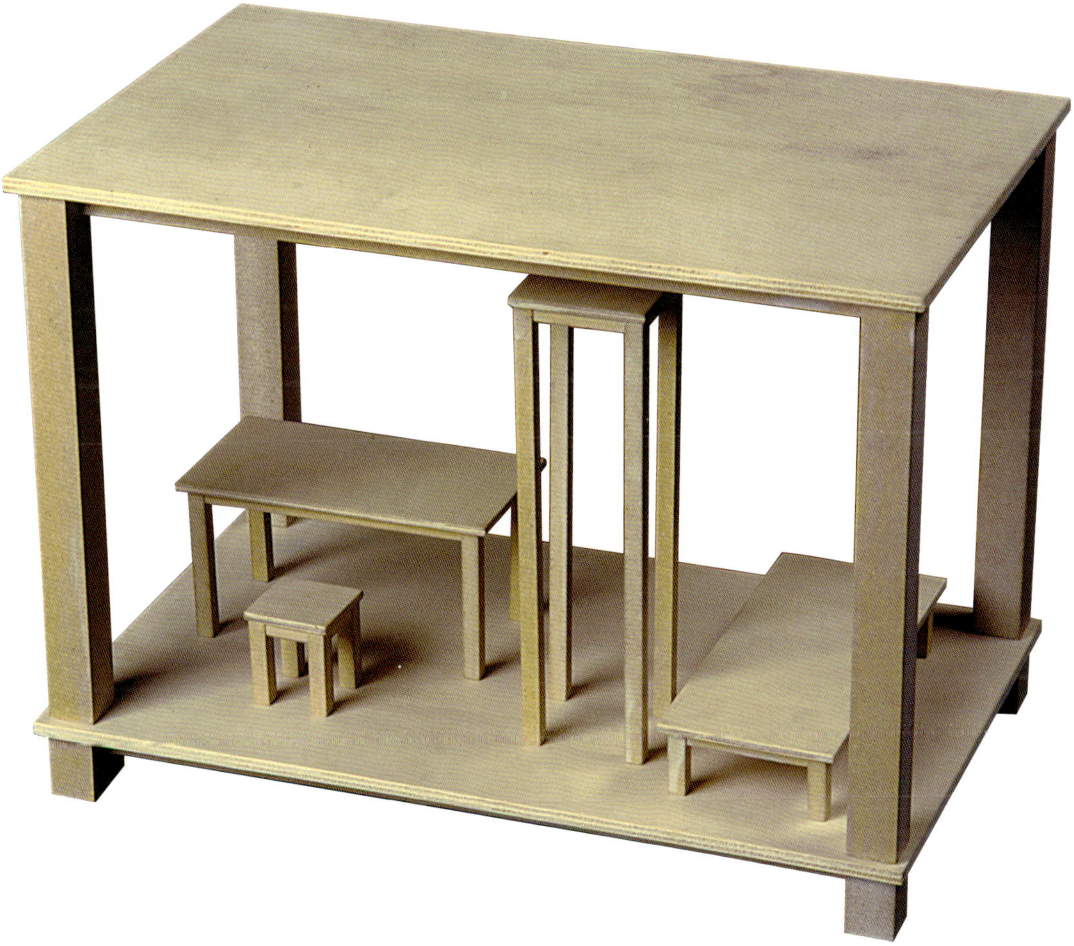

Table-upon-table concept — 1990

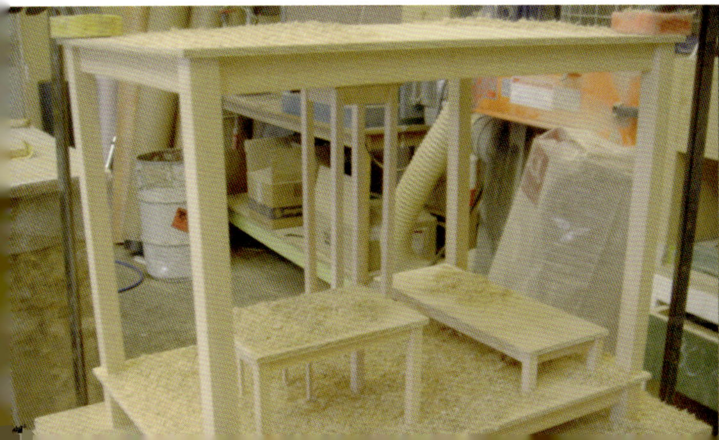

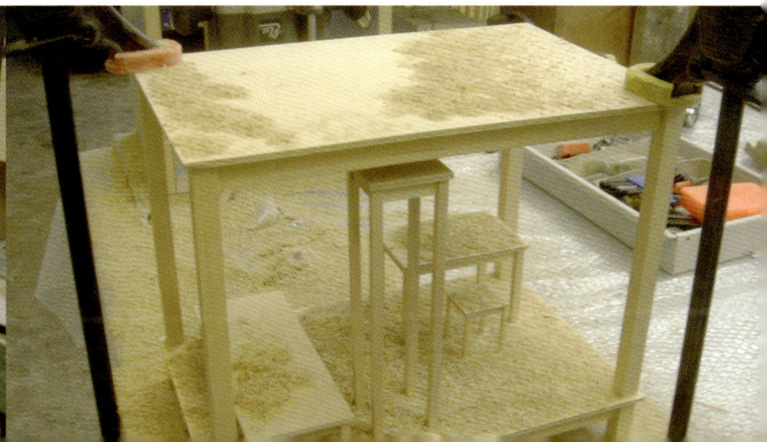

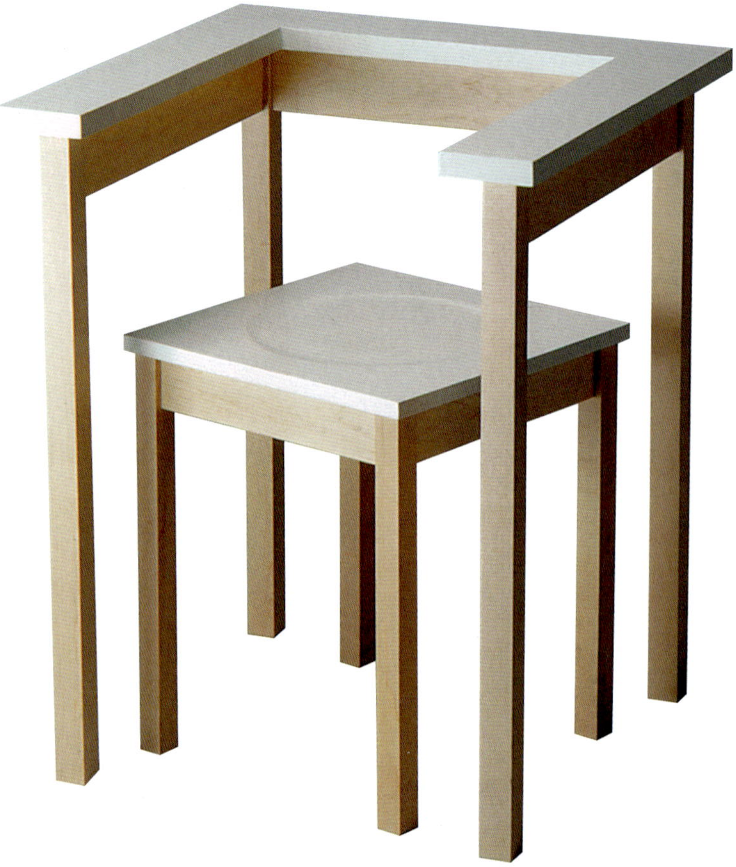

Table-chair — 1990

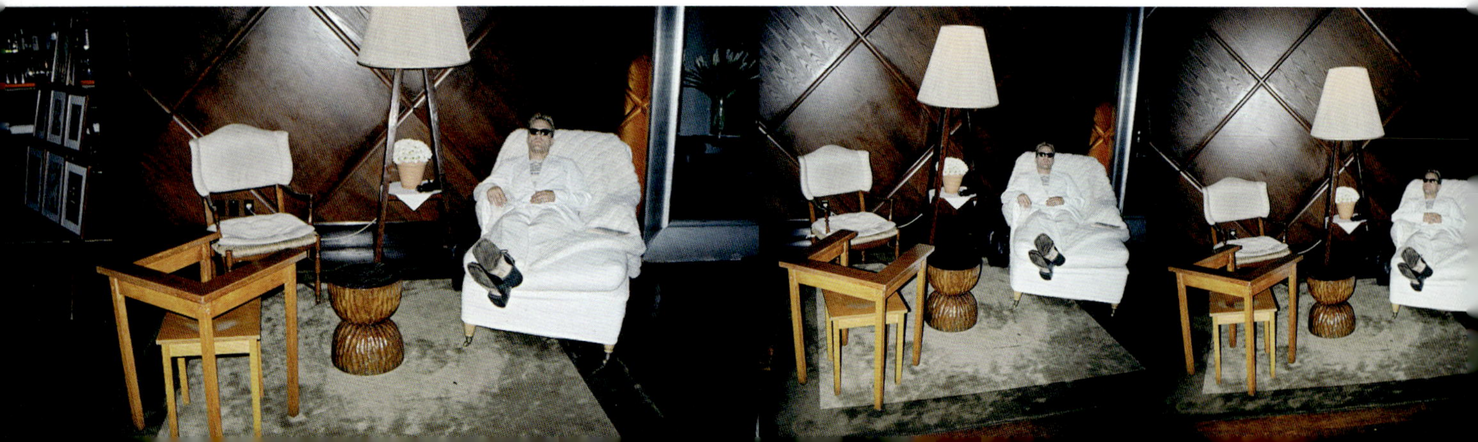

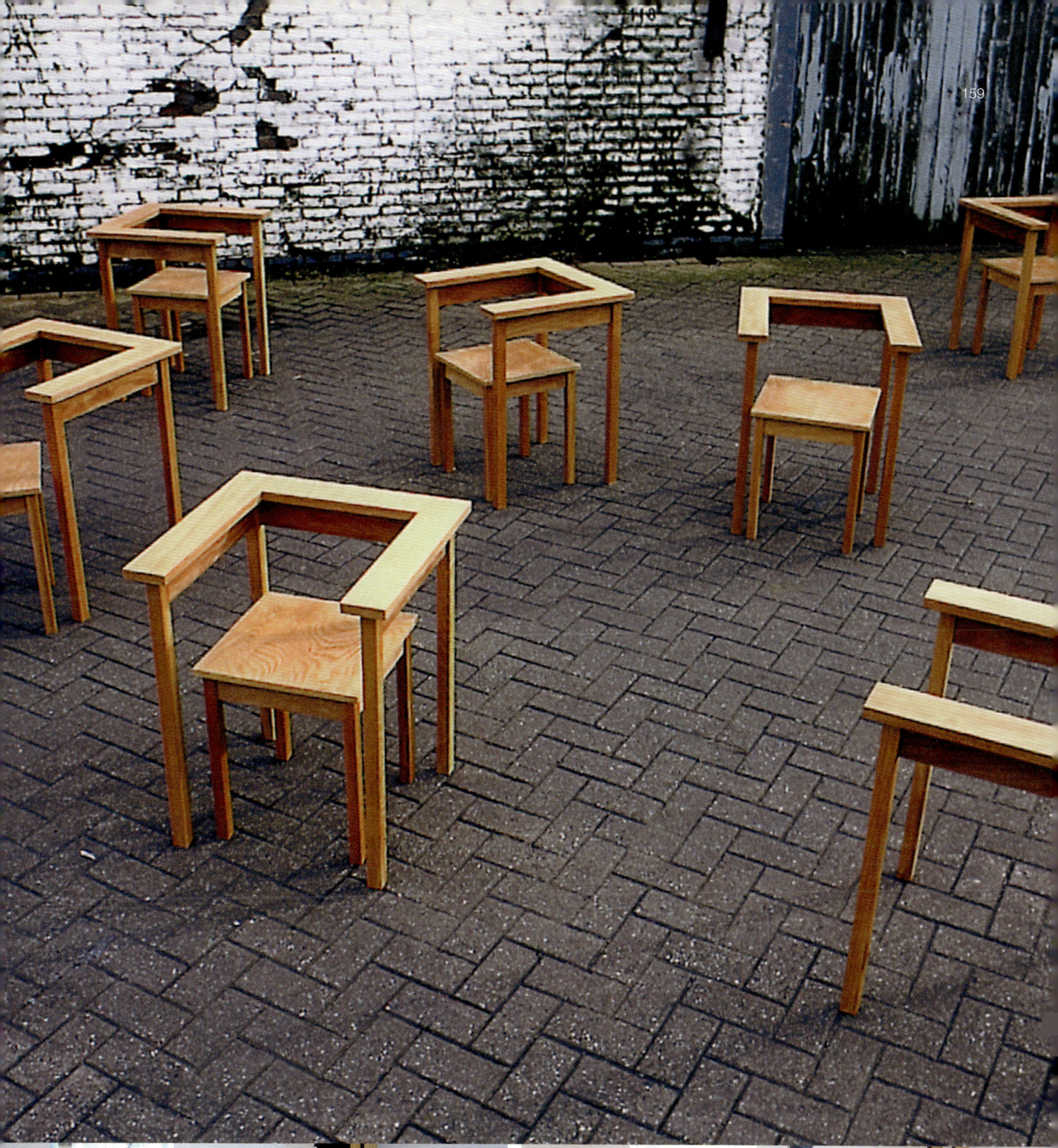

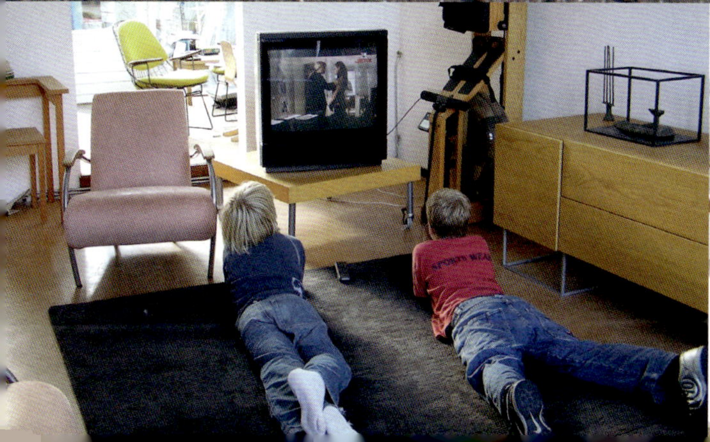

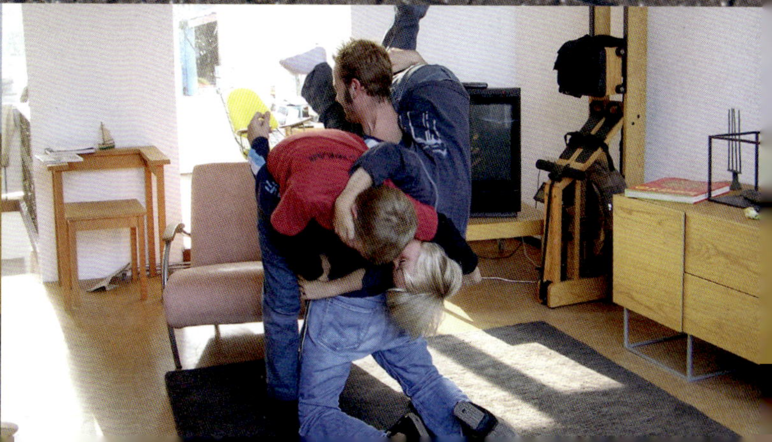

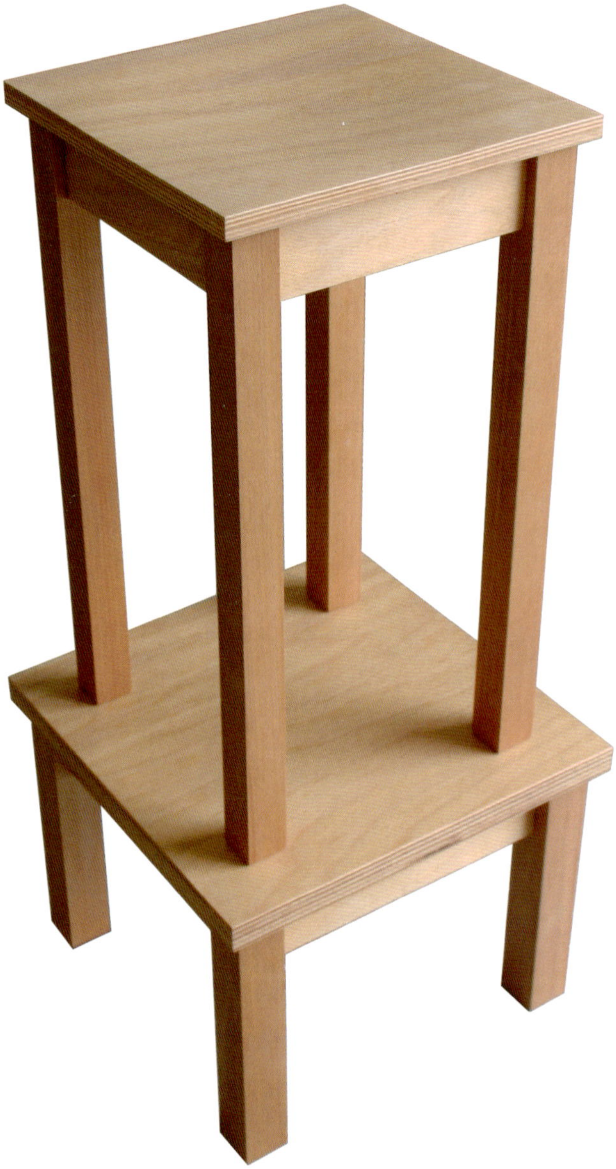

Table-upon-table barstool — 1991

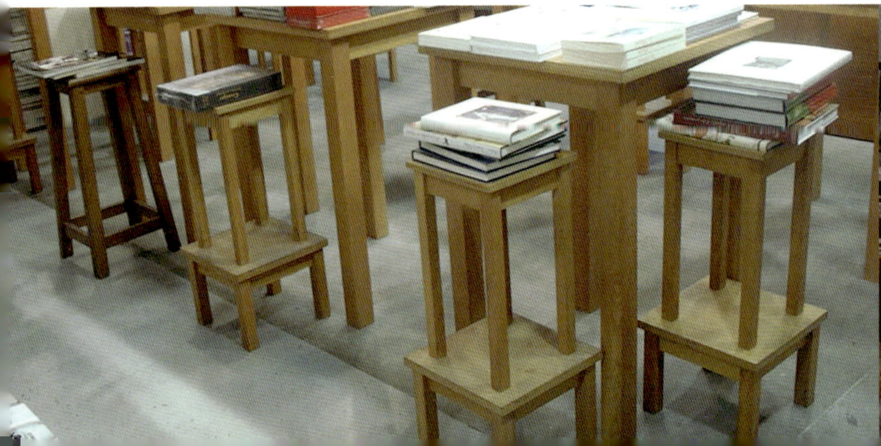

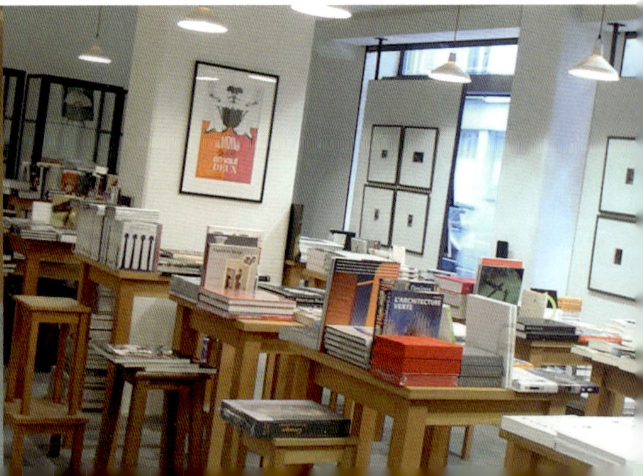

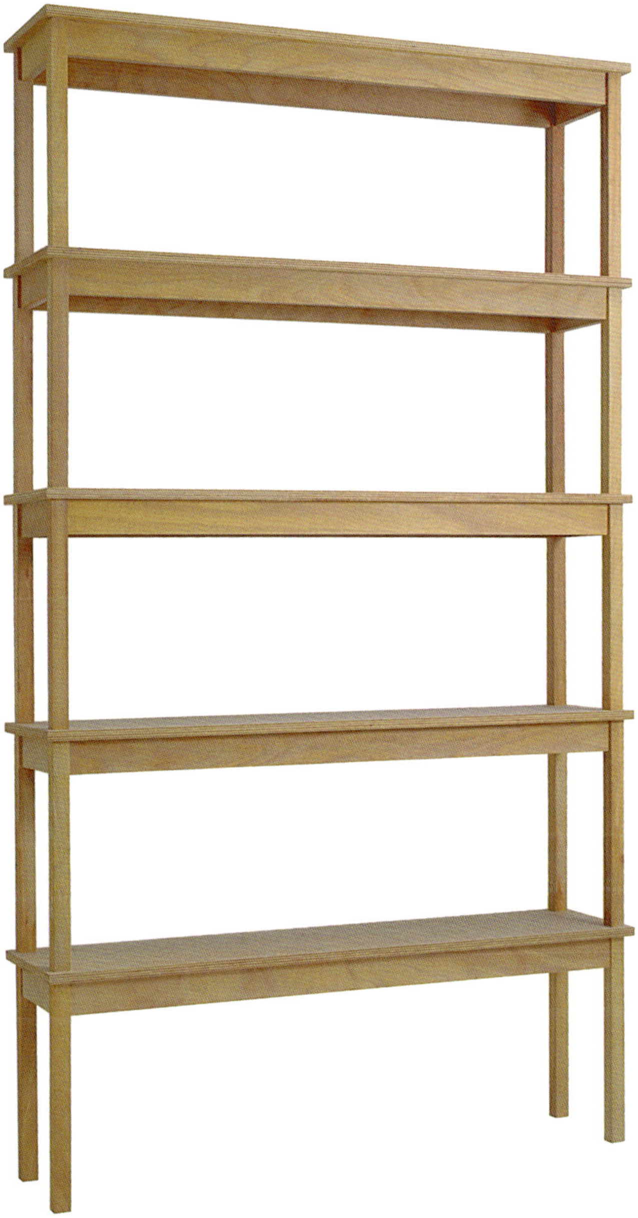

Table-upon-table bookshelf — 1992

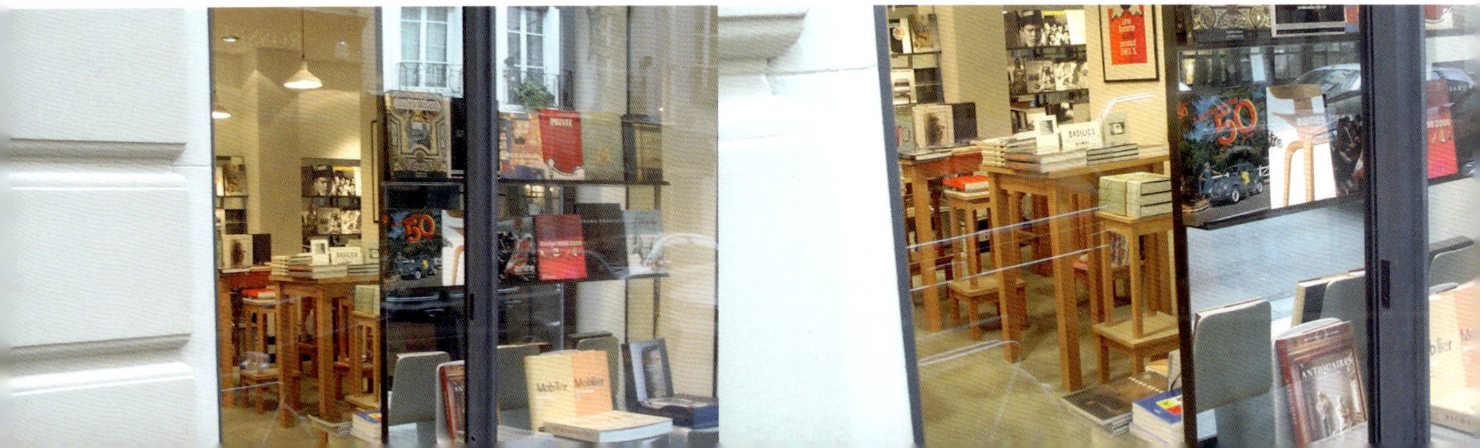

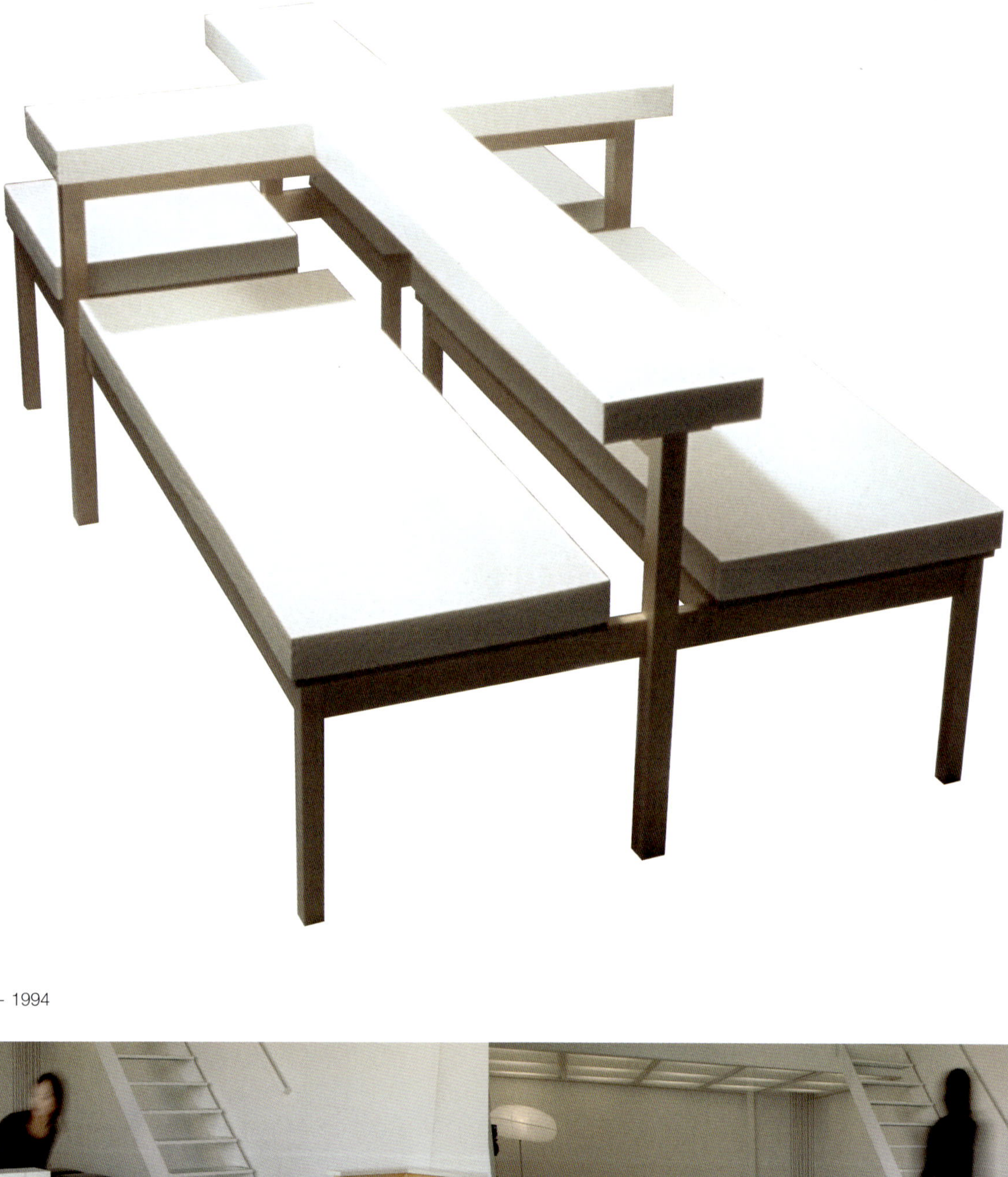

The Cross — 1994

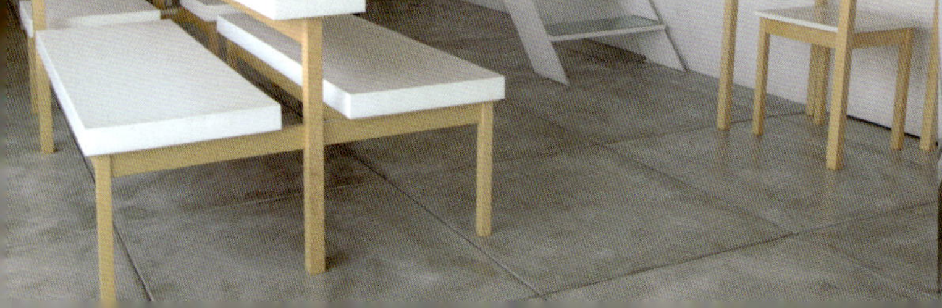

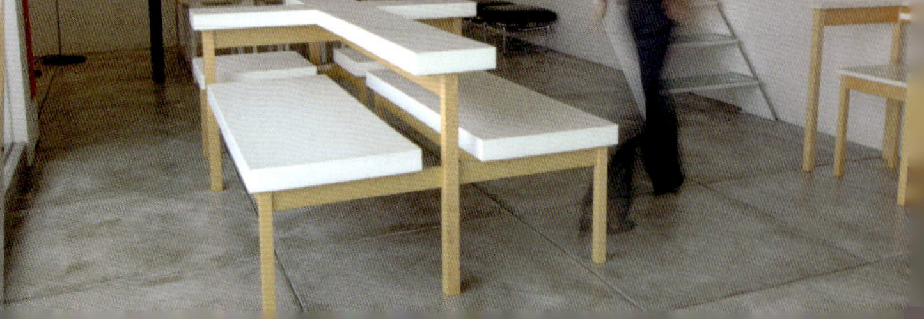

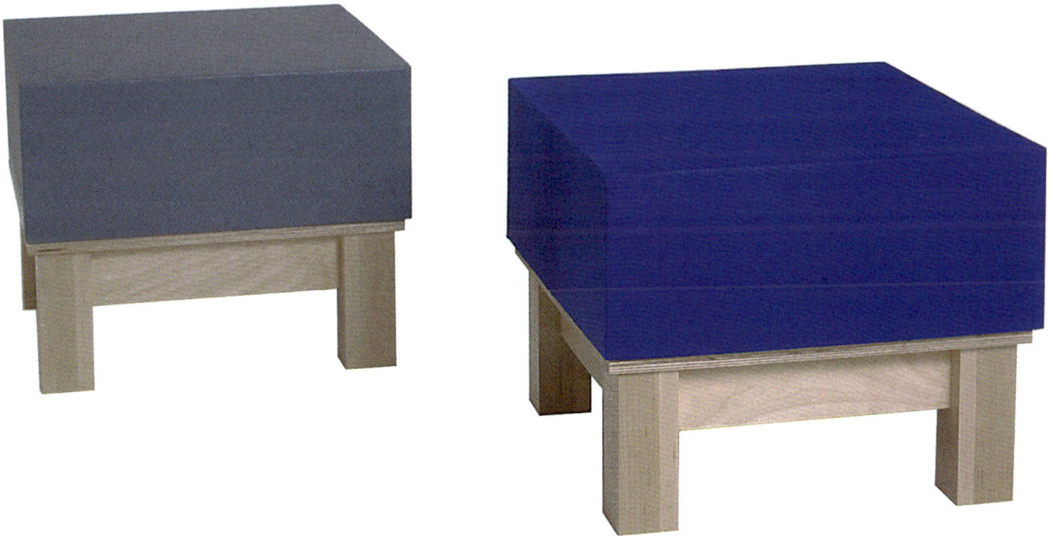

Poef-pouff — 1994

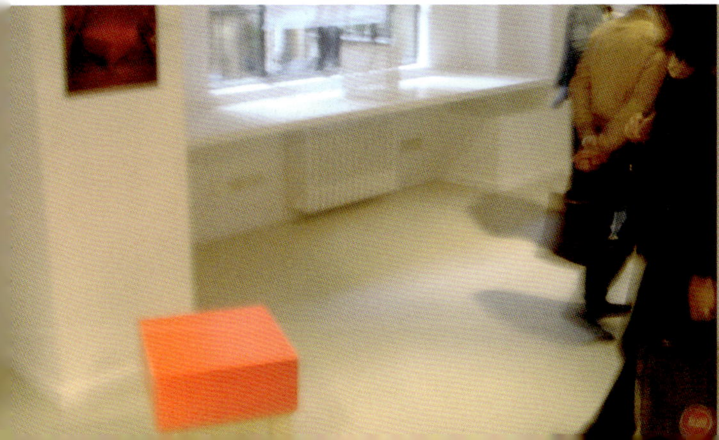

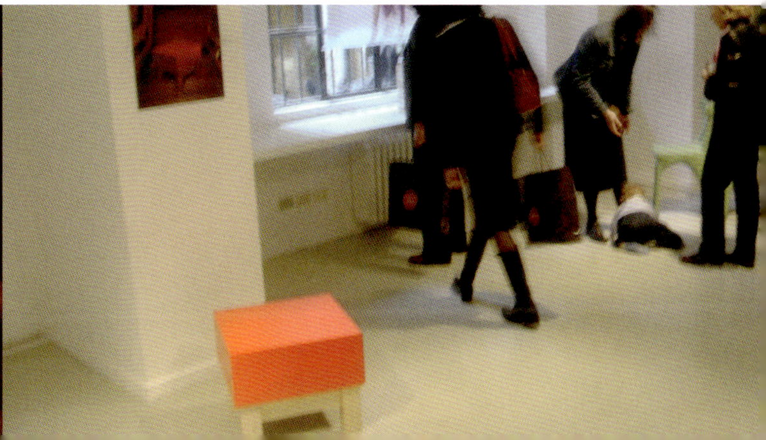

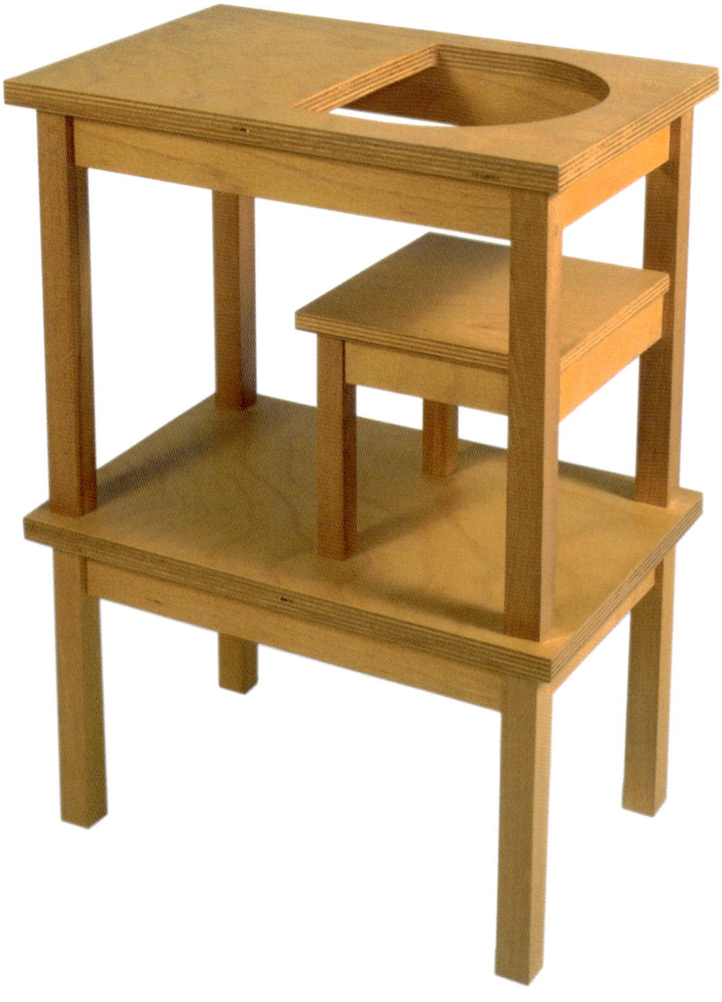

Table-upon-table children chair — 1998

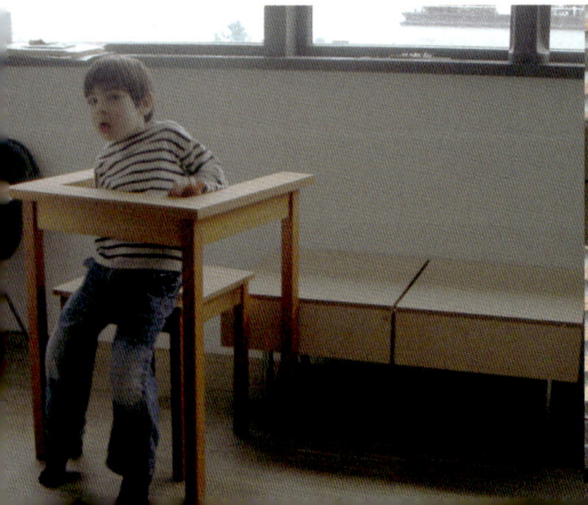
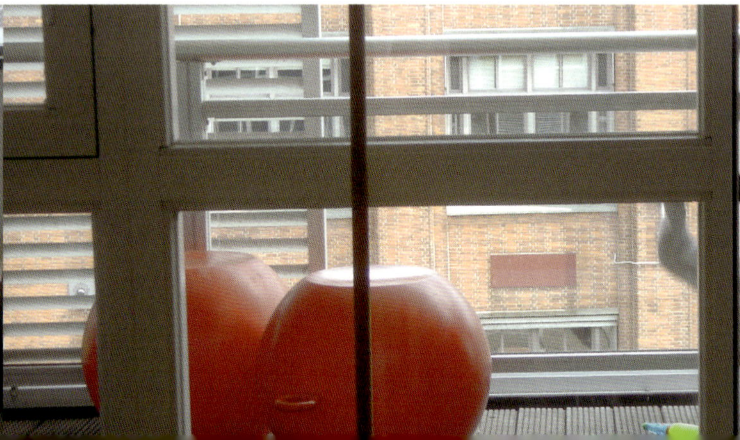
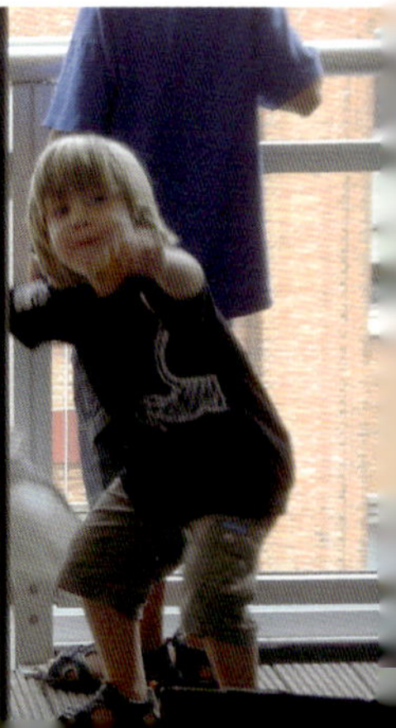

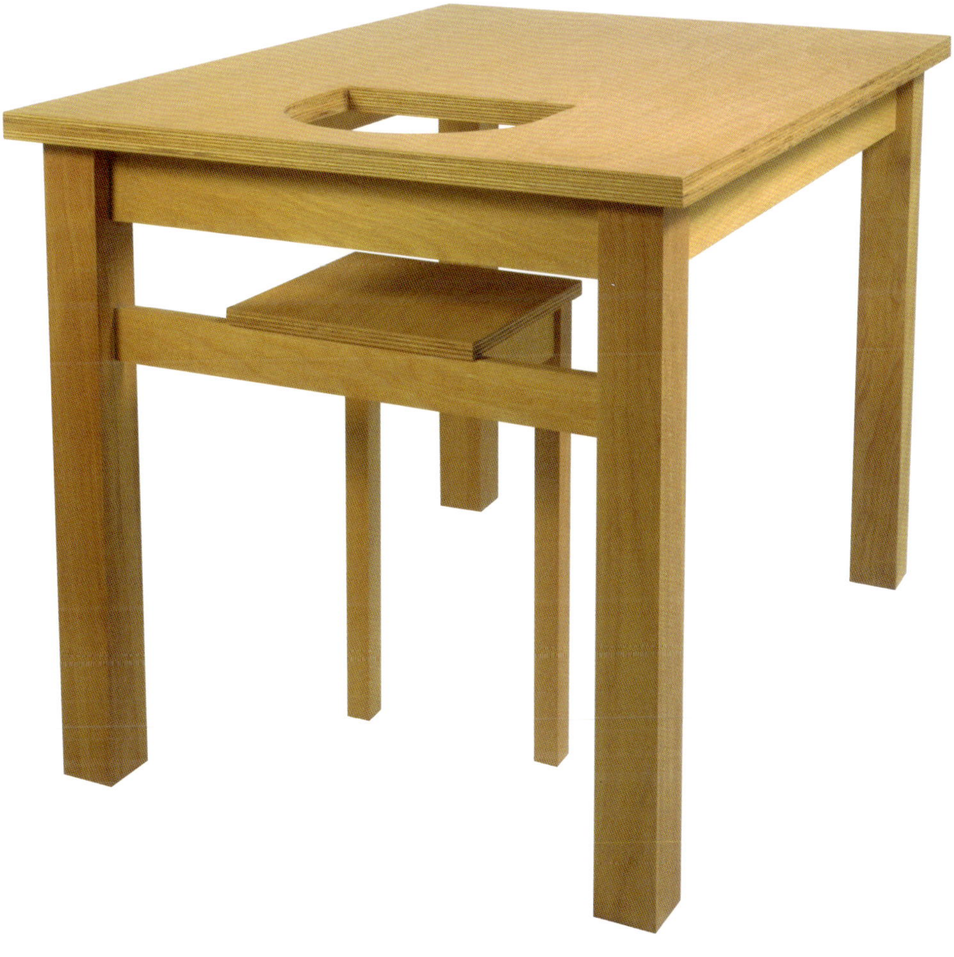

Who's the boss — 1998

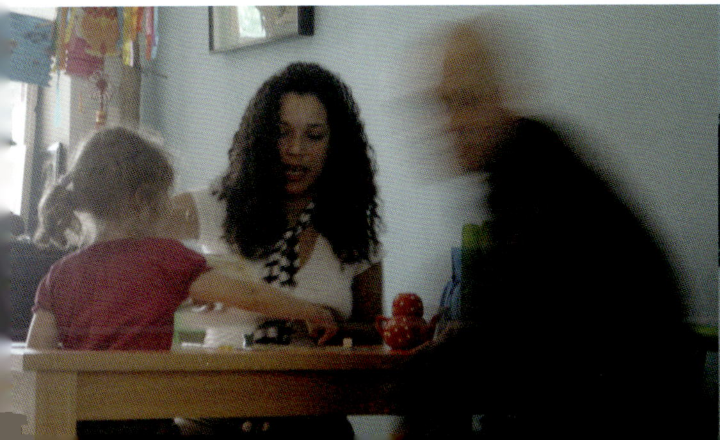

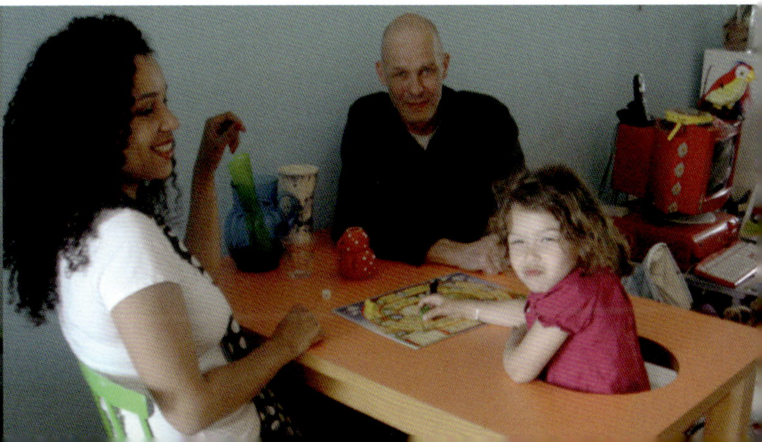

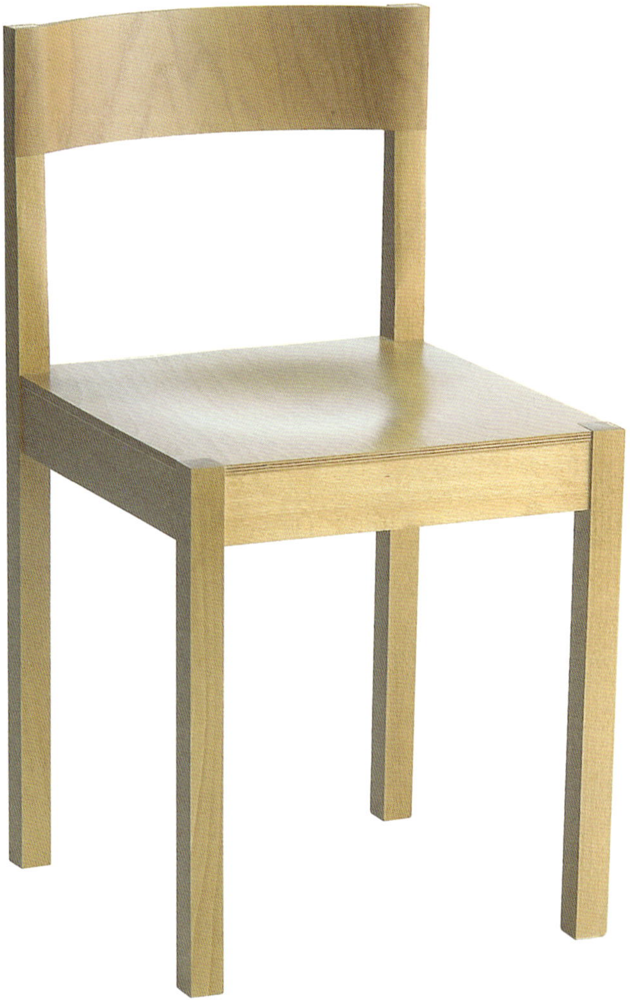

No sign of design chair — 1992

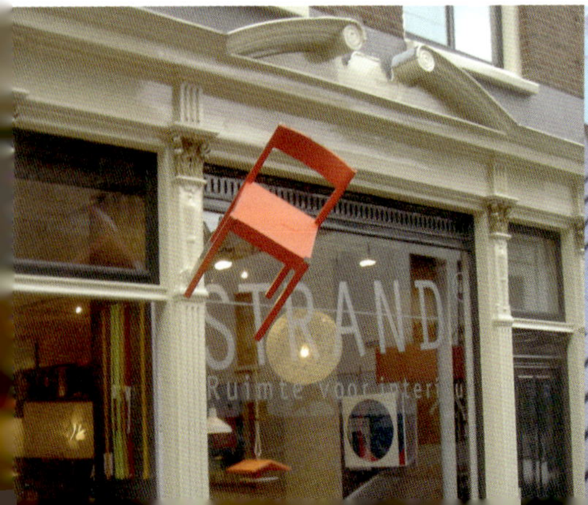

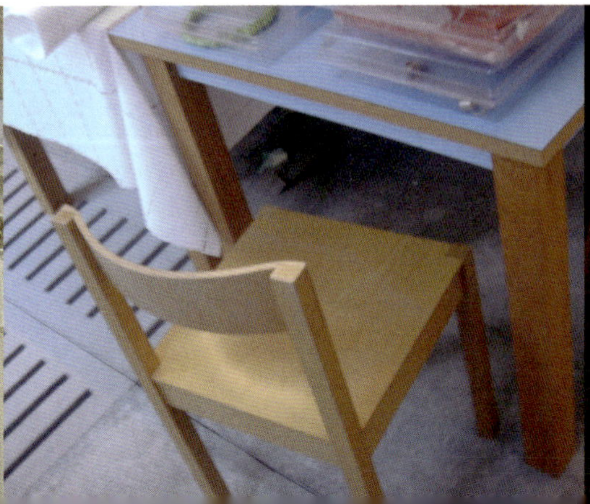

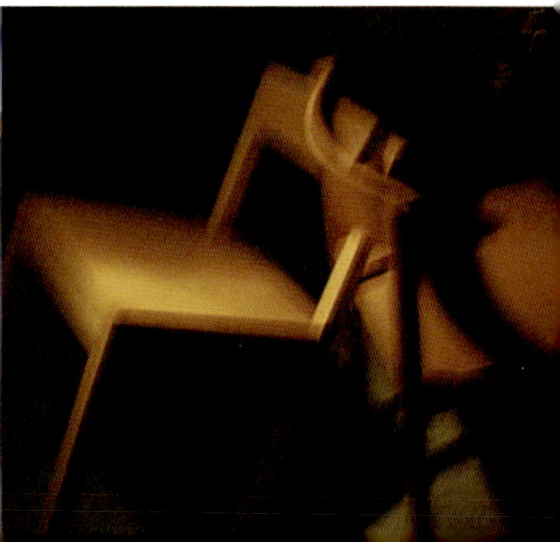

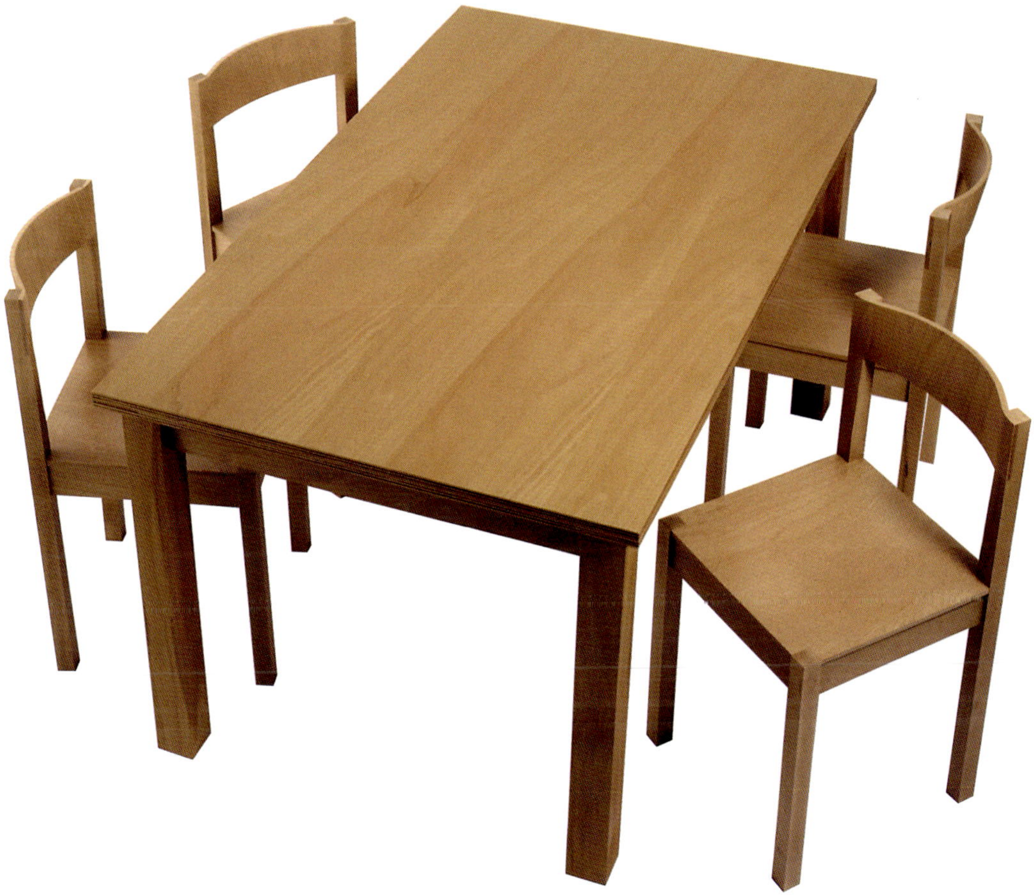

No sign of design table — 1990

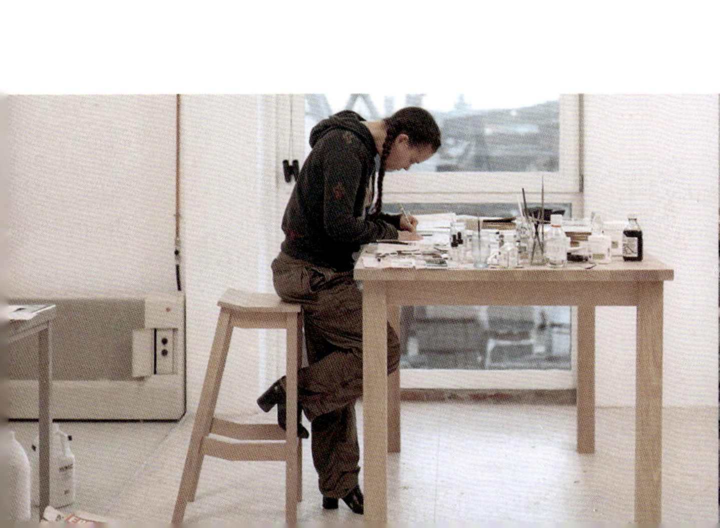

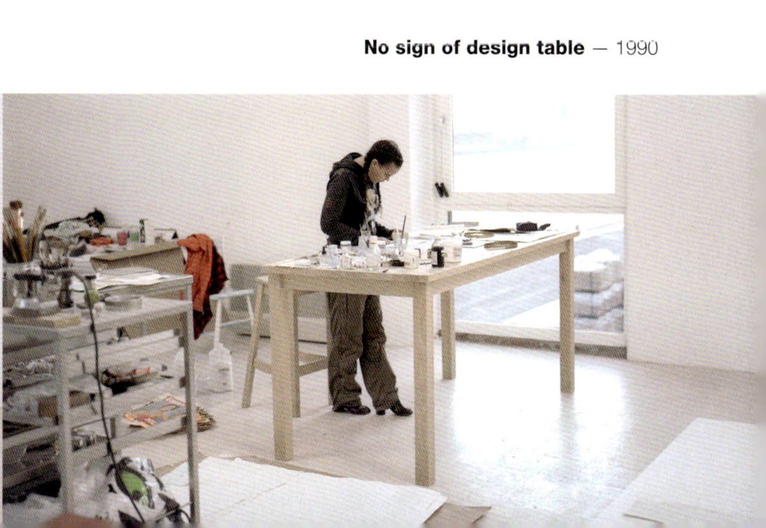

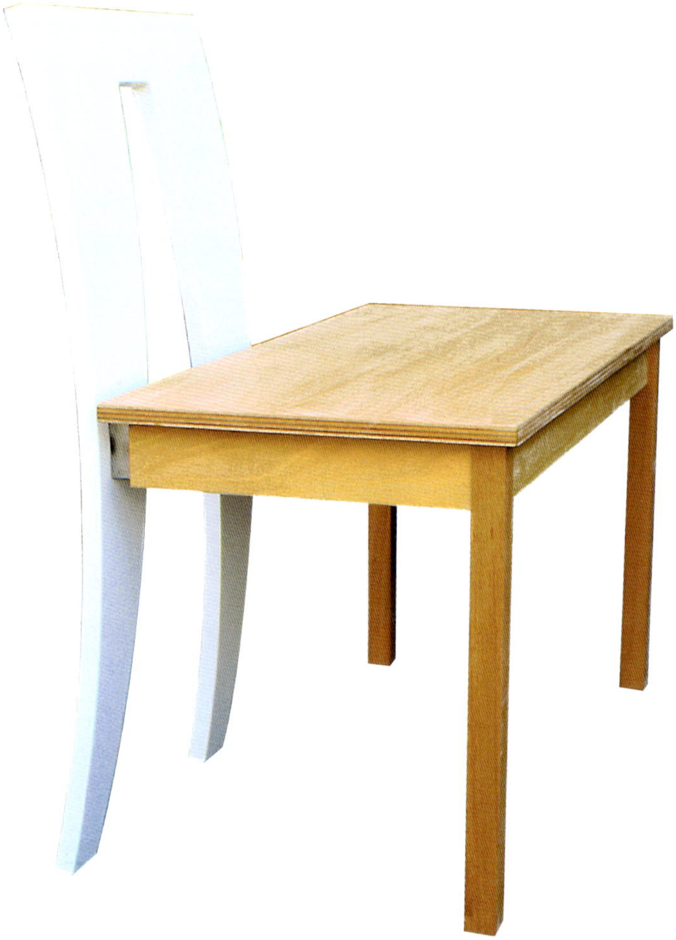

Thing 1 — 1995

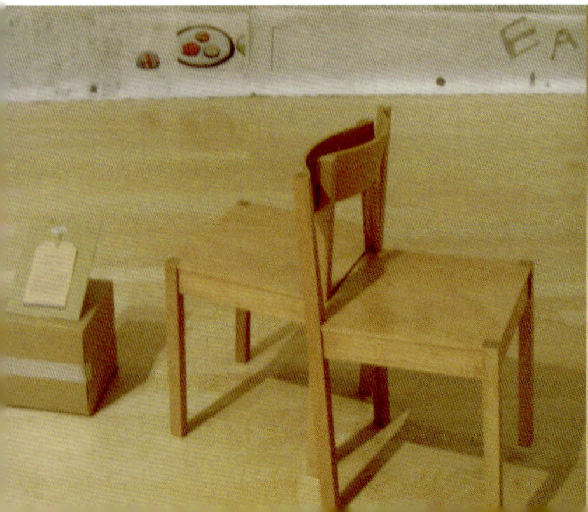

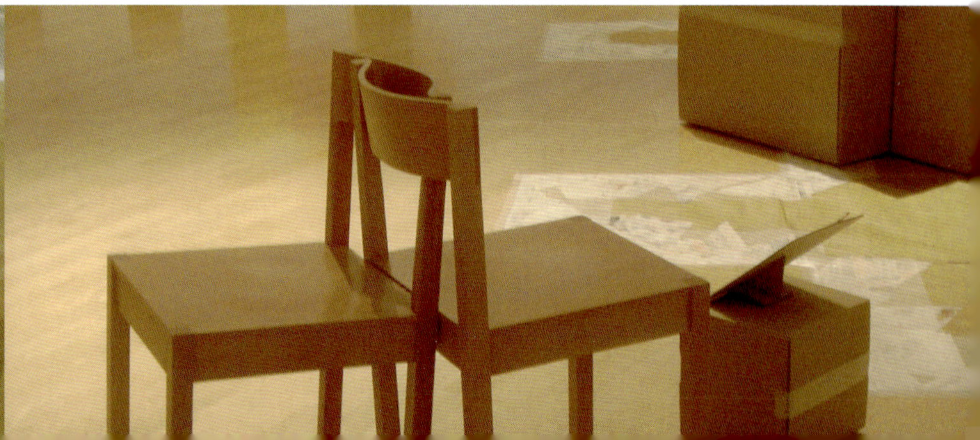

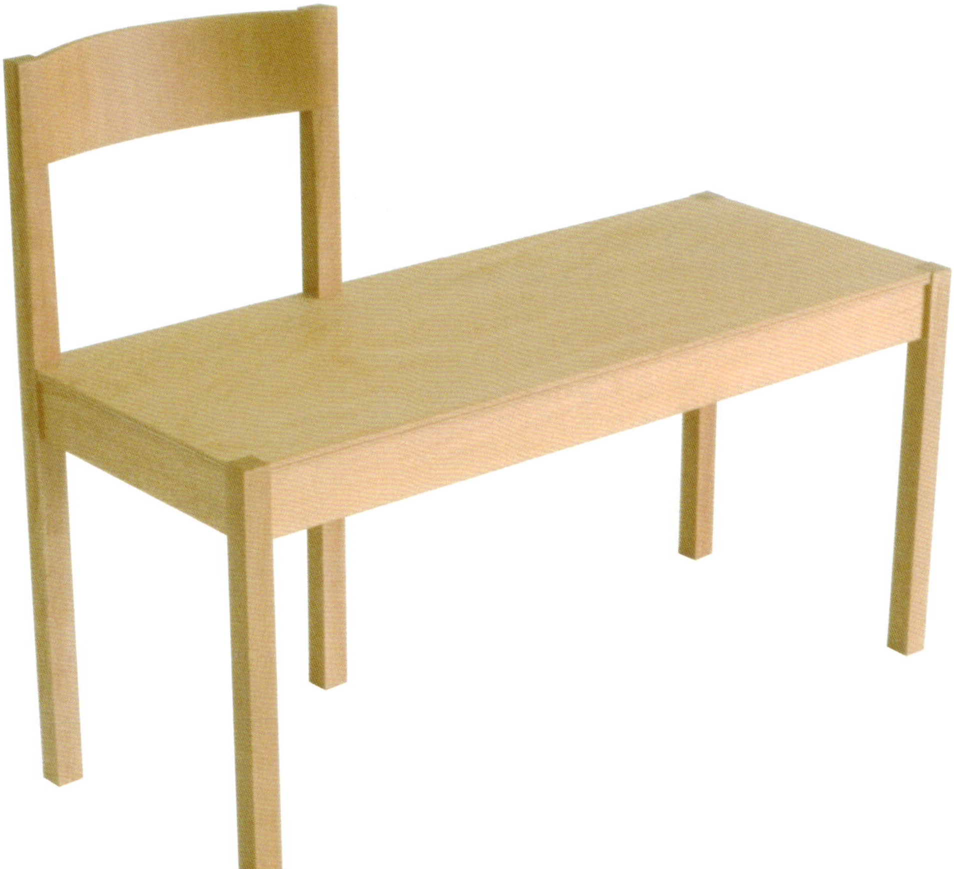

Thing 2 — 1999

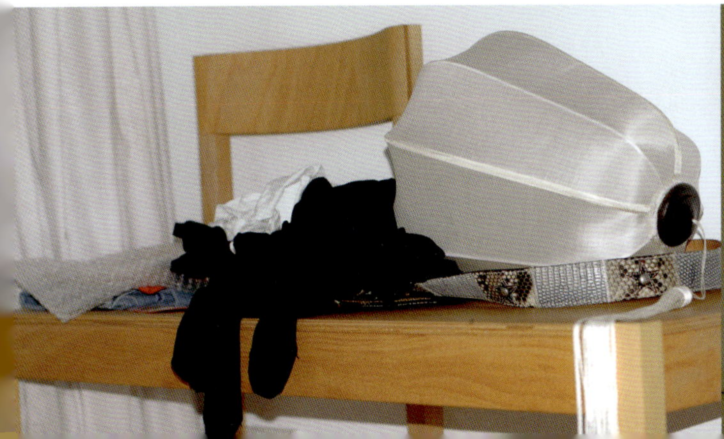

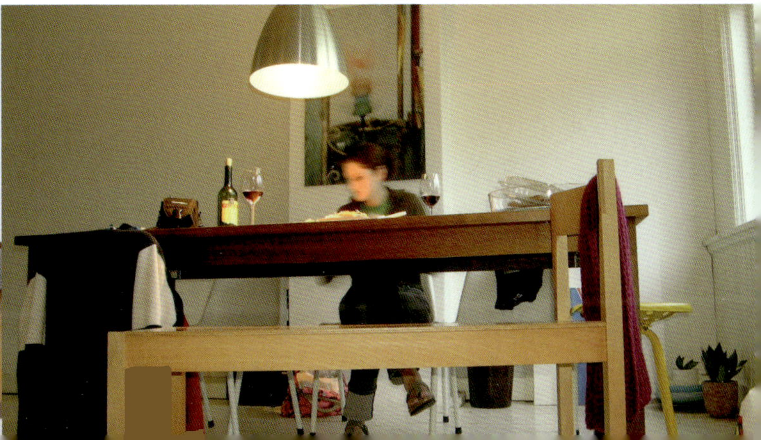

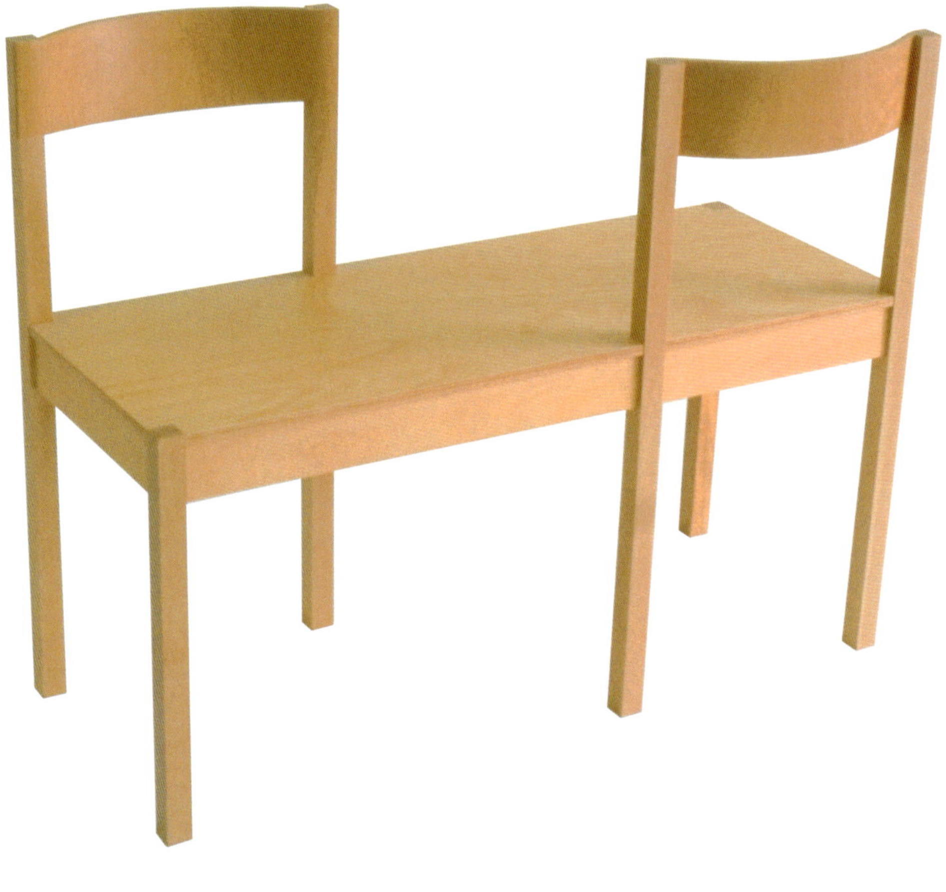

Thing 3 — 2000

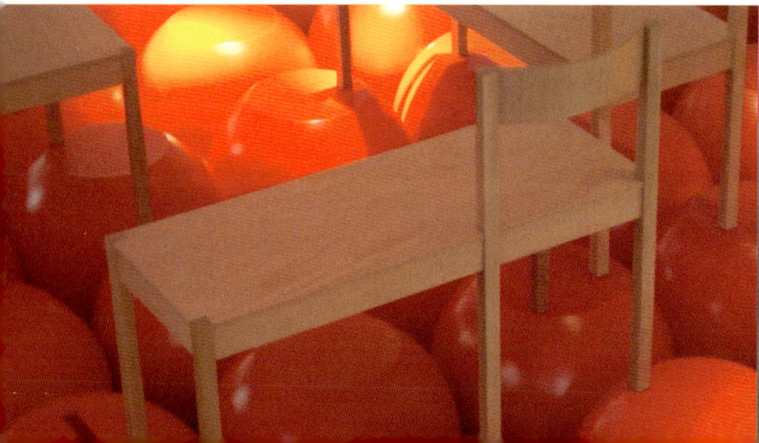

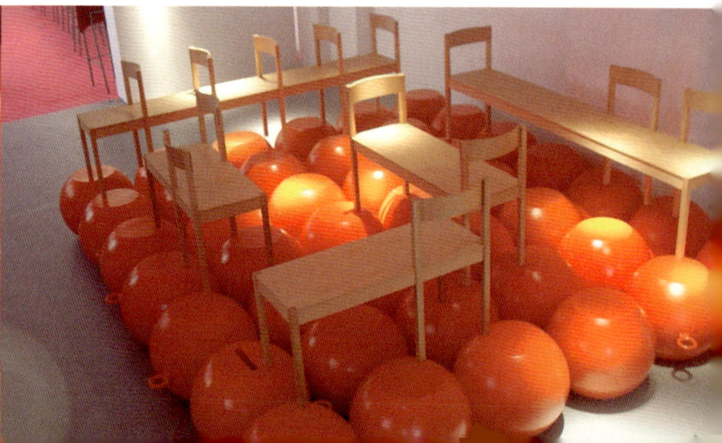

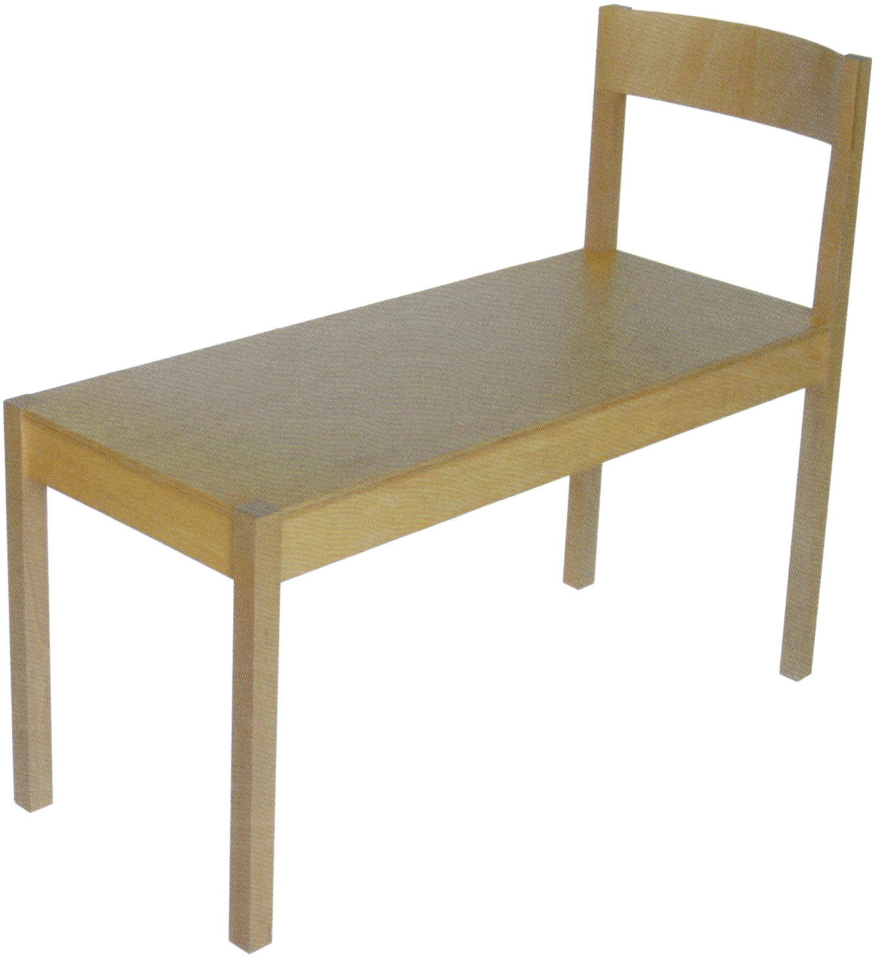

Thing 5 — 2000

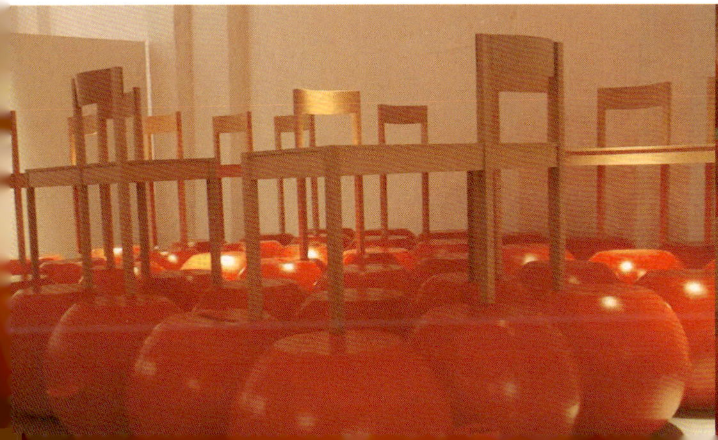

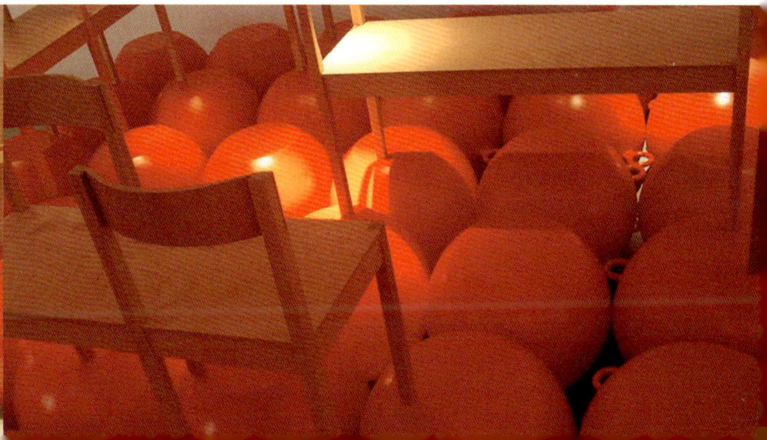

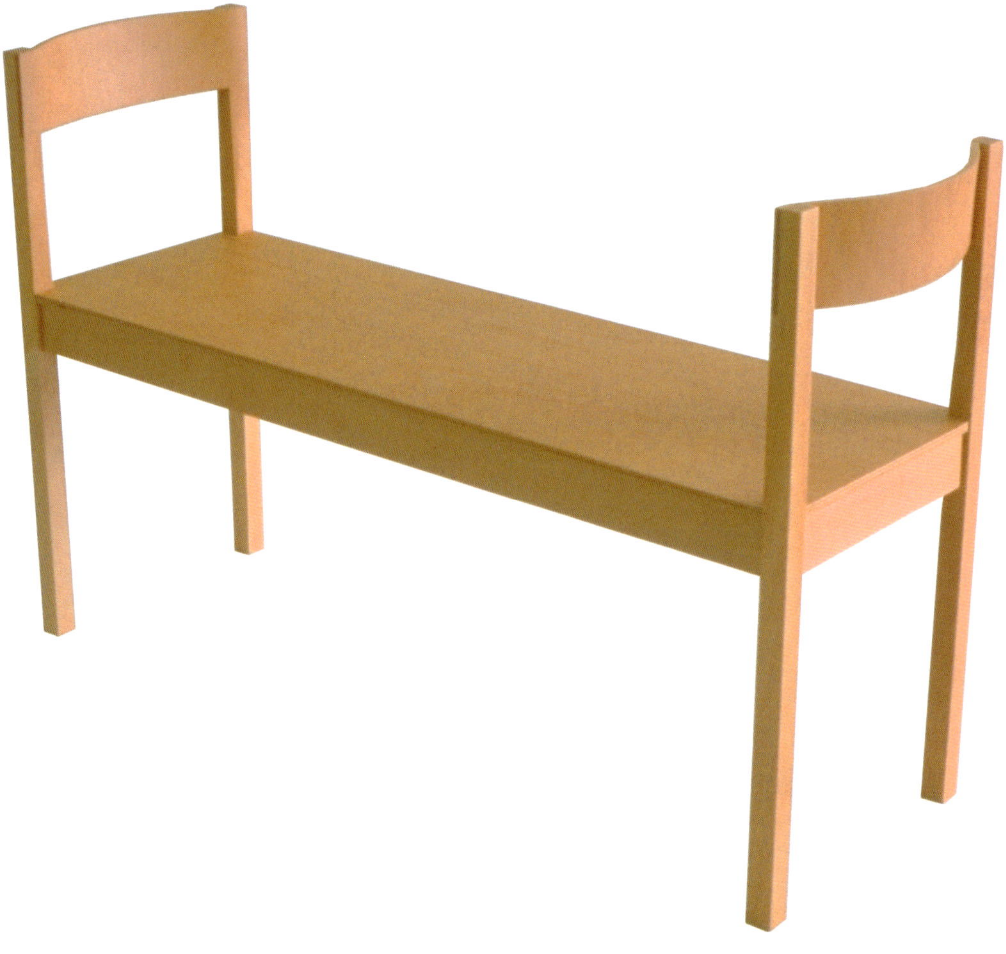

Thing 6 — 2000

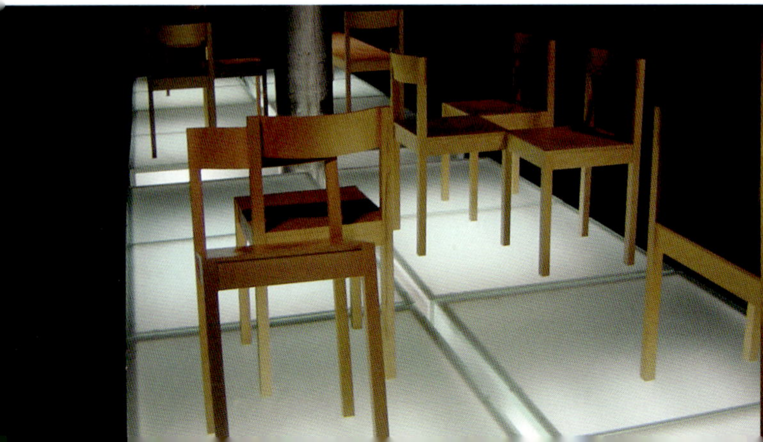

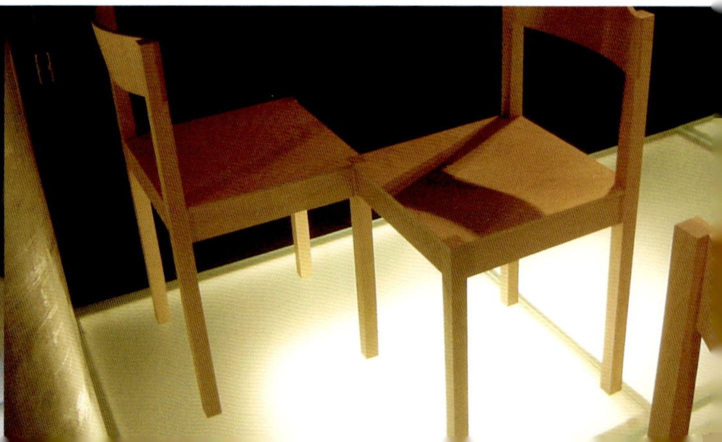

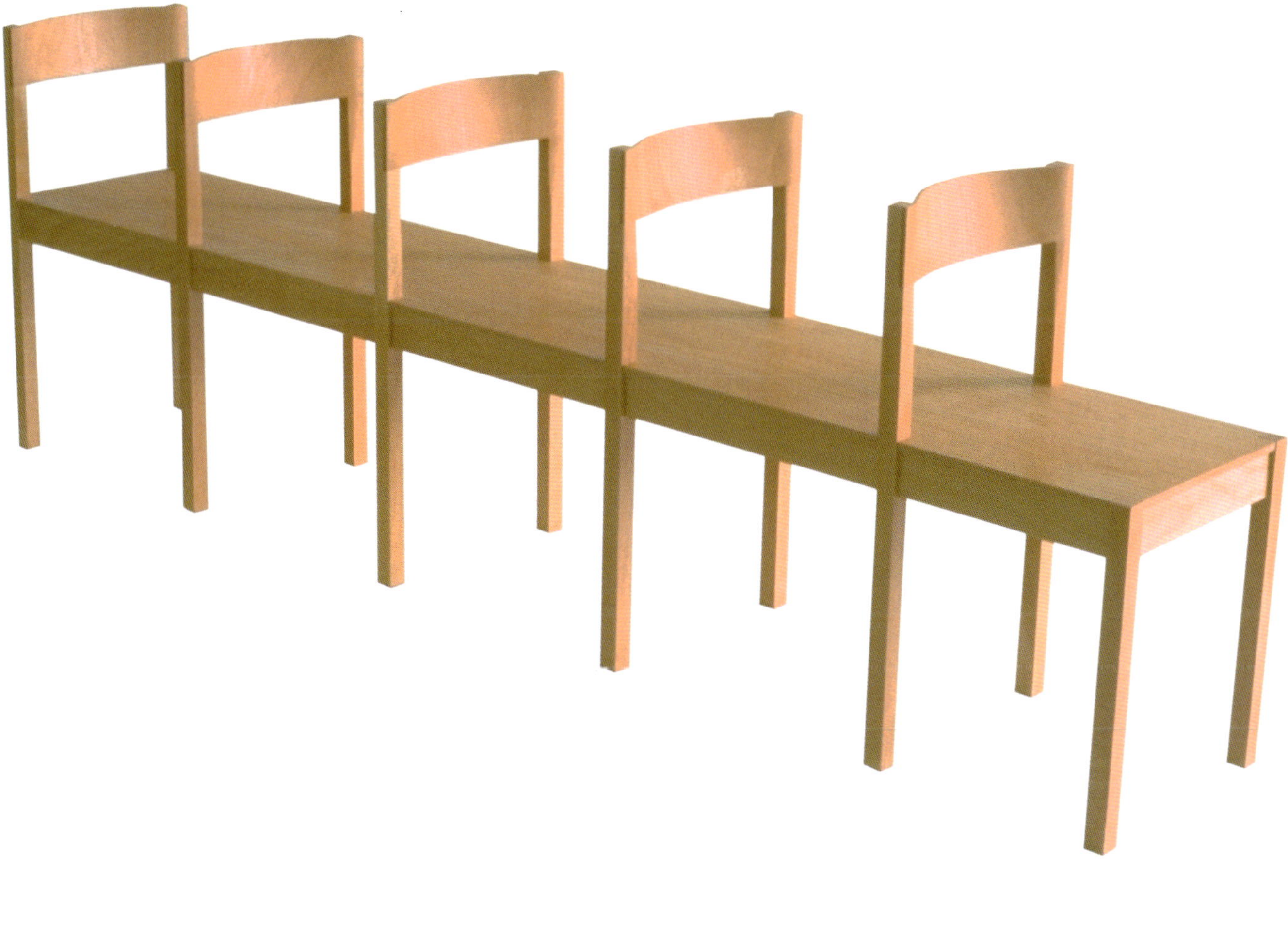

Thing 10 — 2000

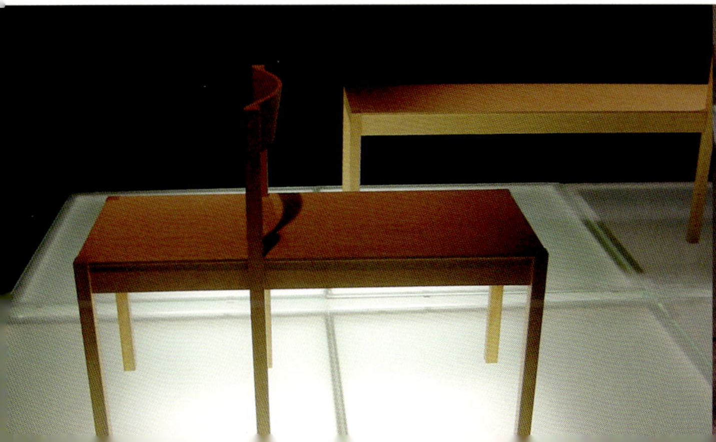

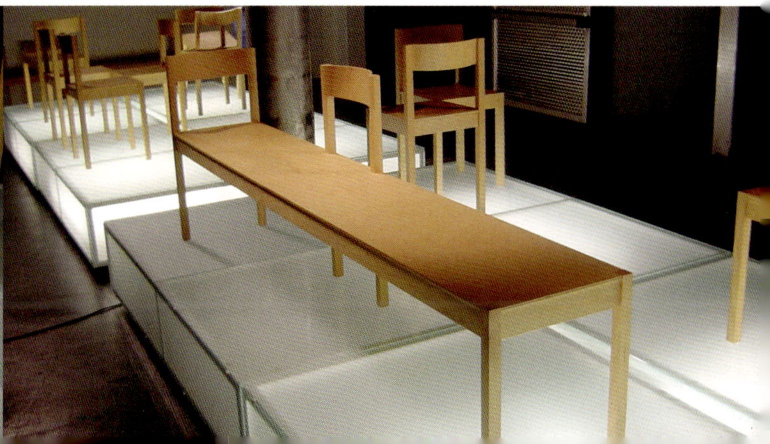

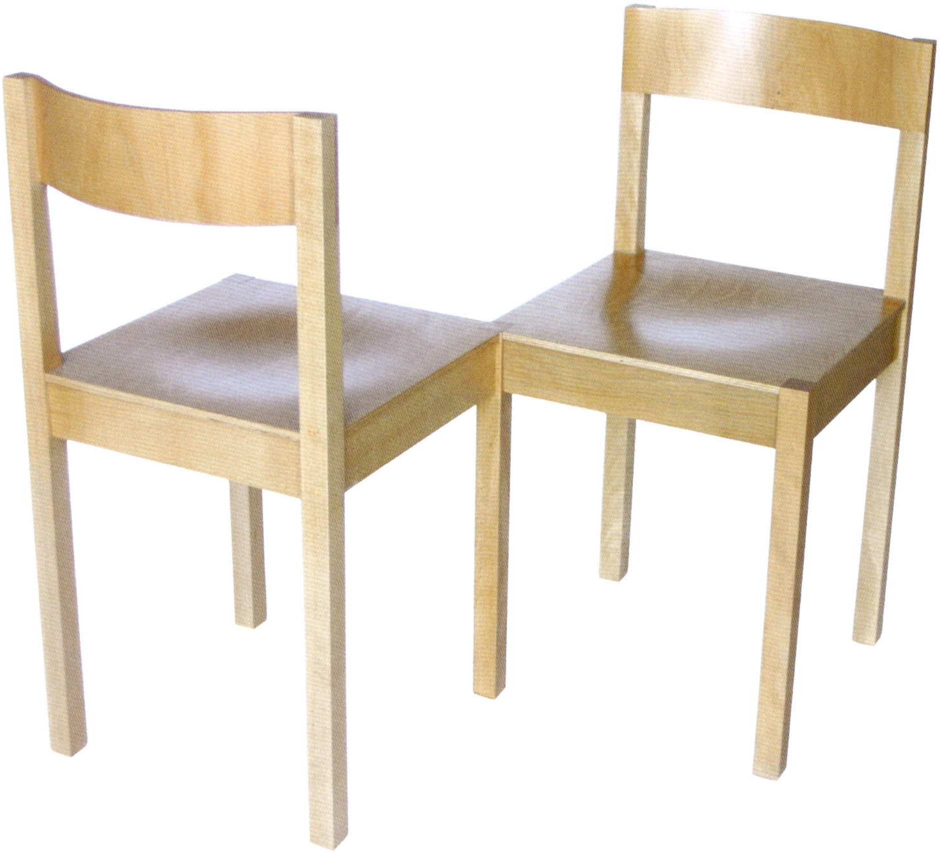

Thing 11 — 2002

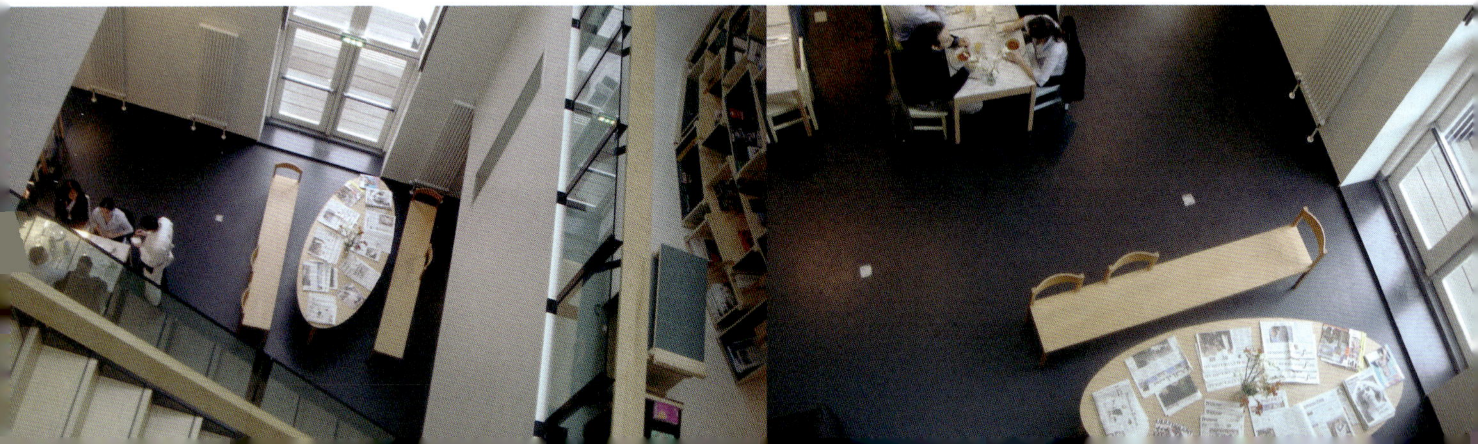

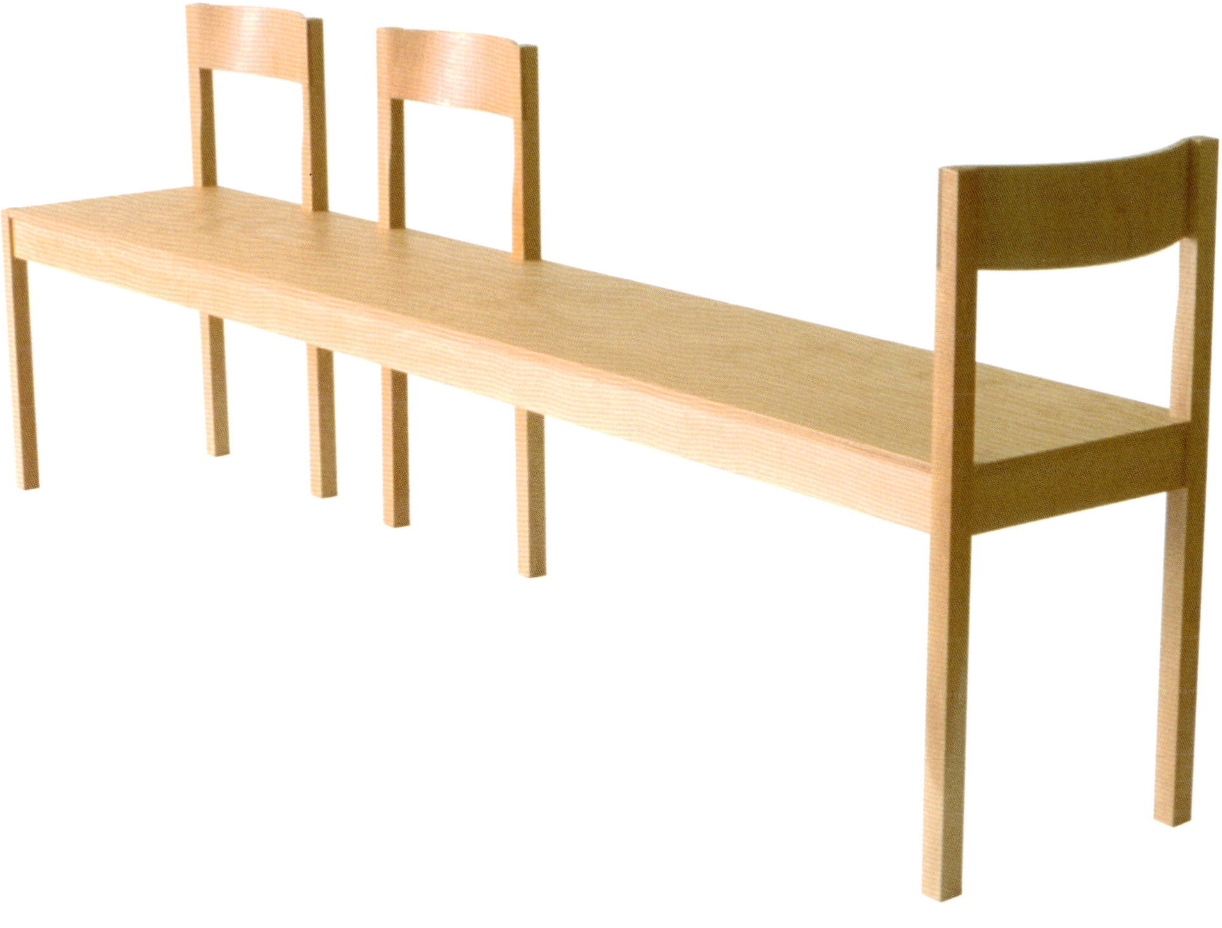

Thing 14 — 2003

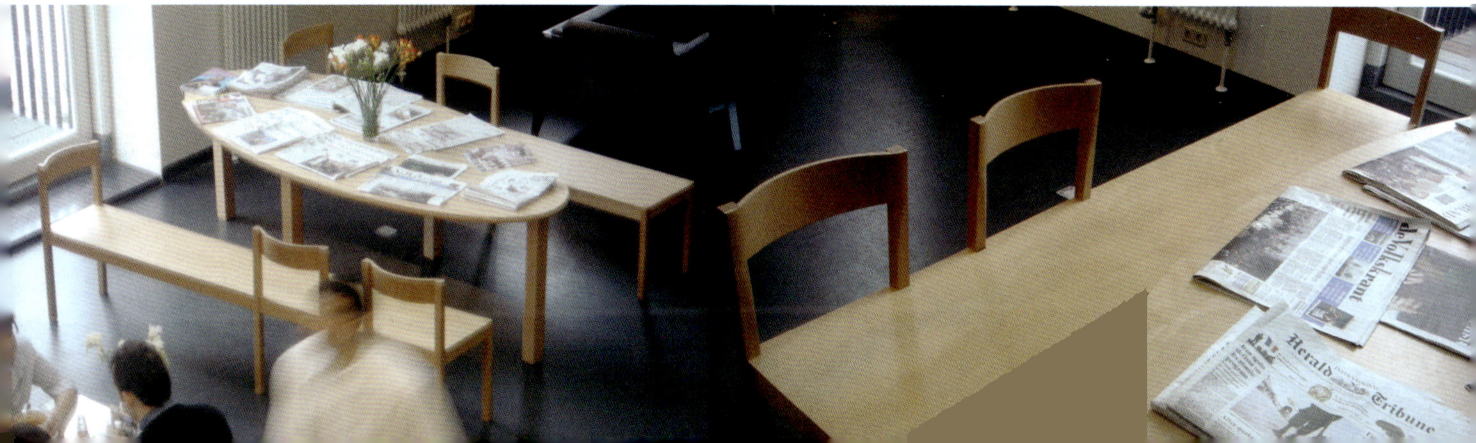

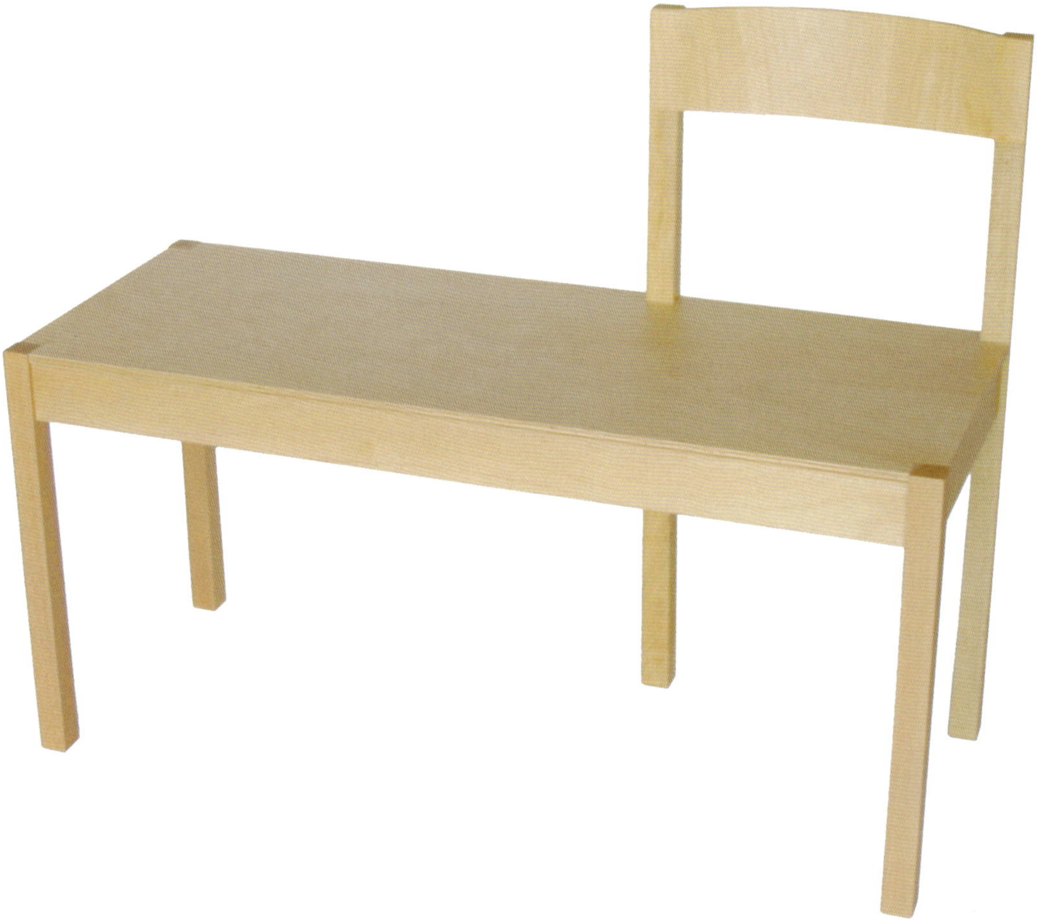

Thing 22 — 2004

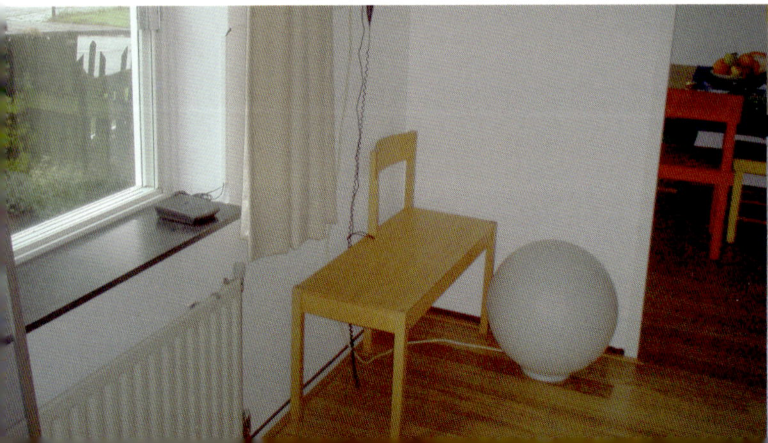

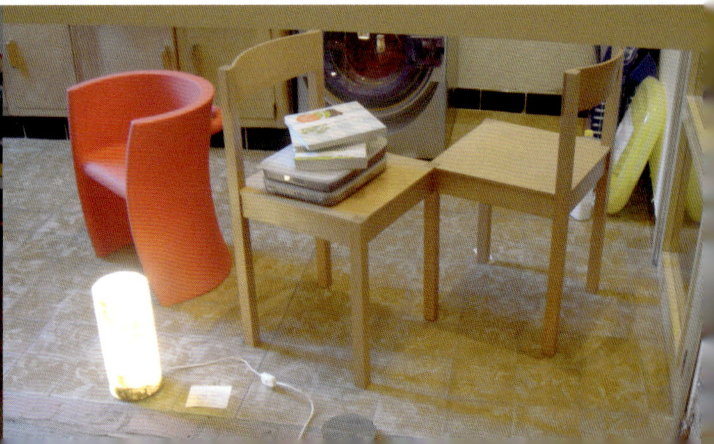

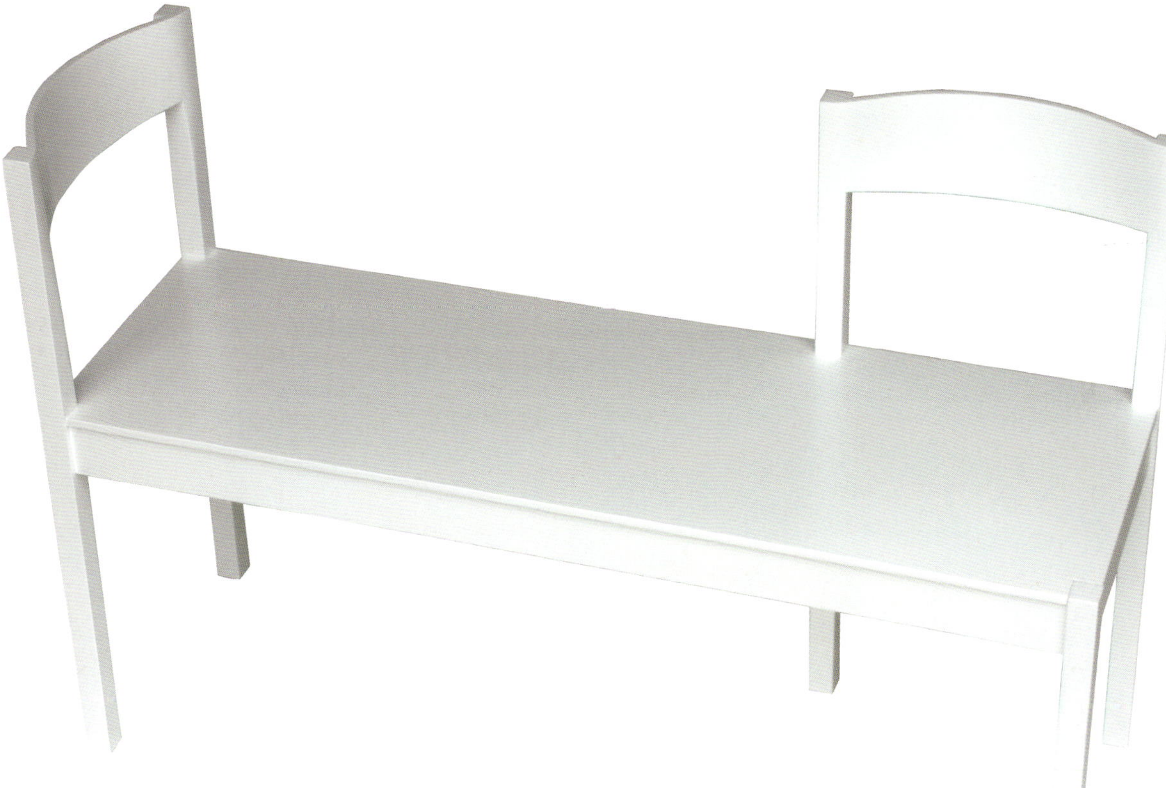

Thing 24 — 2001

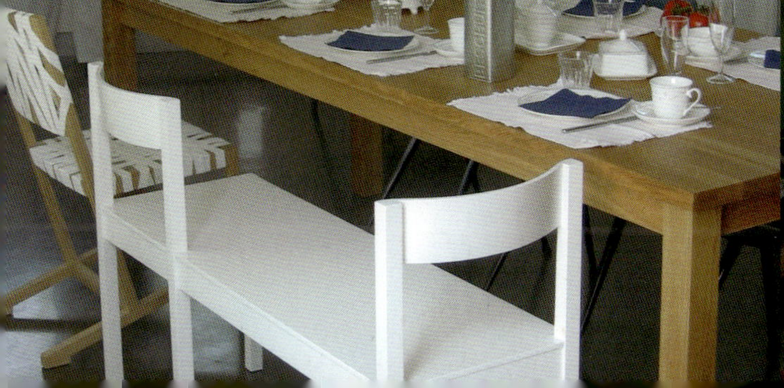

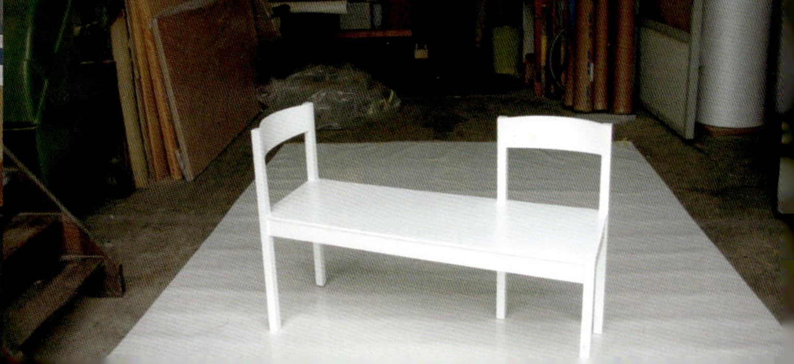

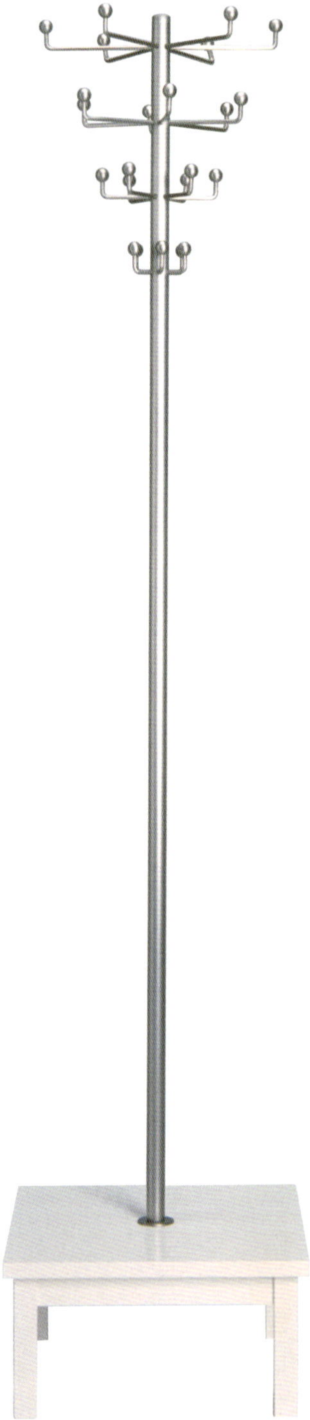

Table-rack — 2000

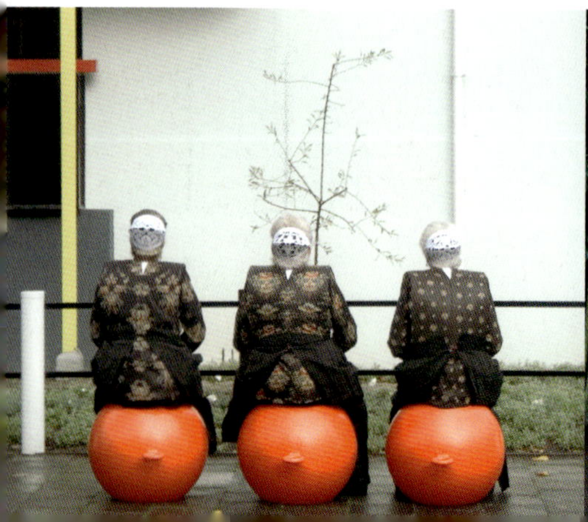

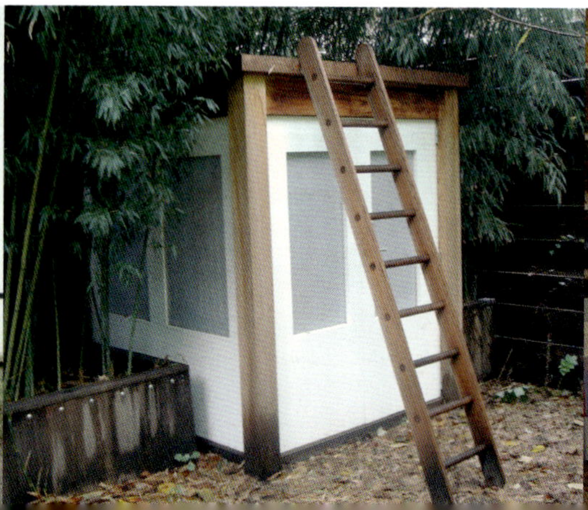

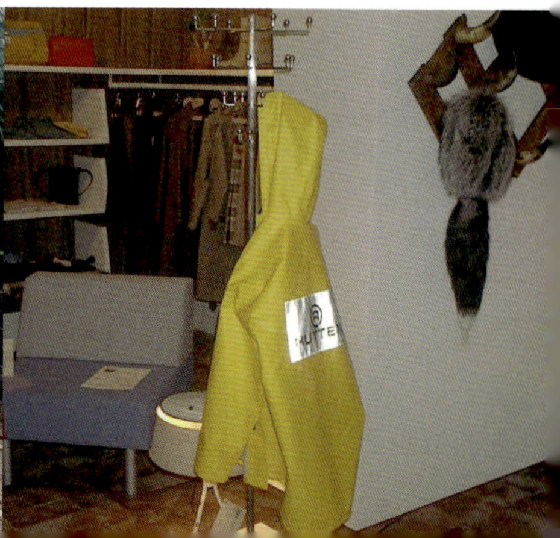

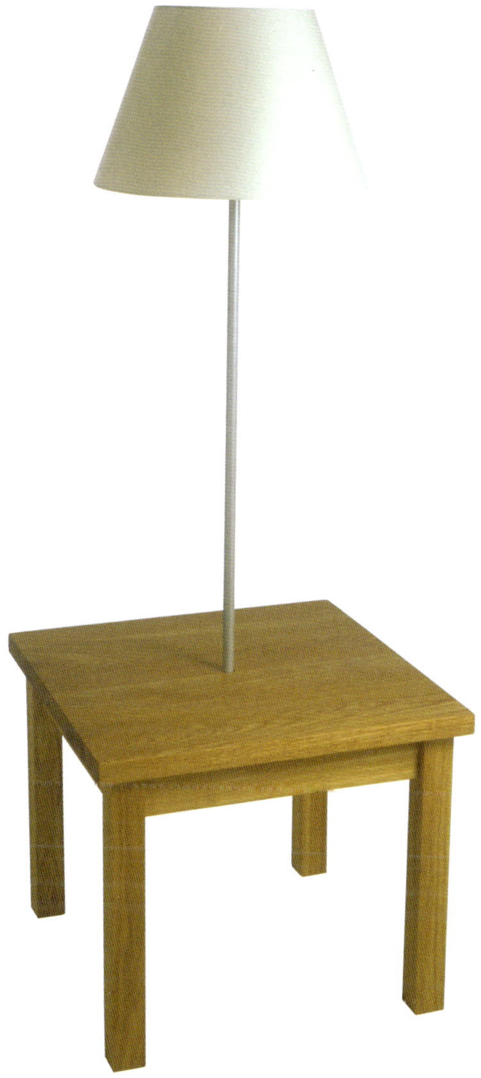

Table-lamp — 1998

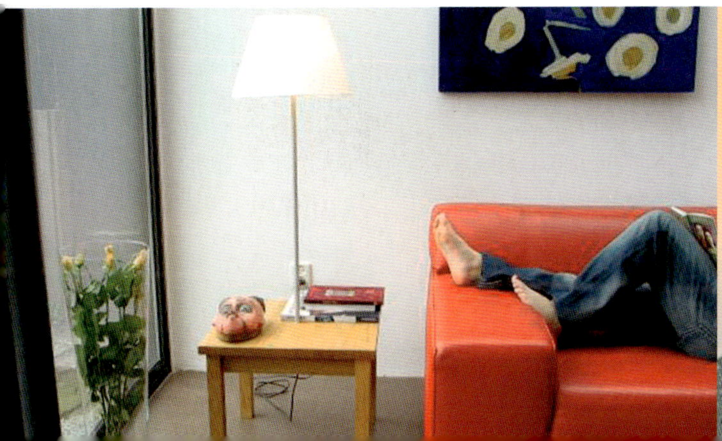

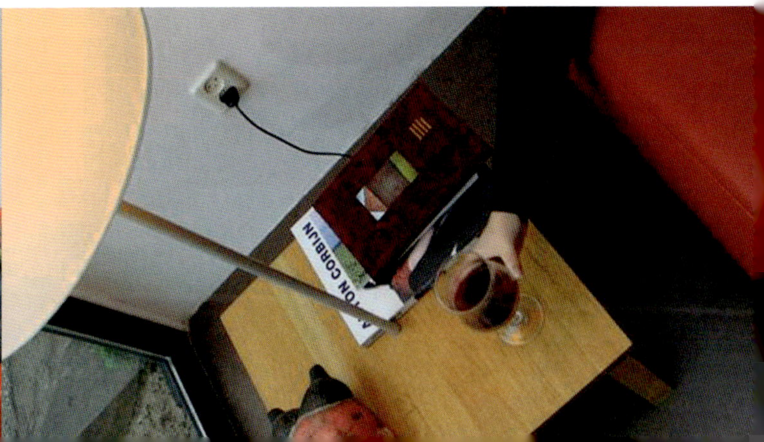

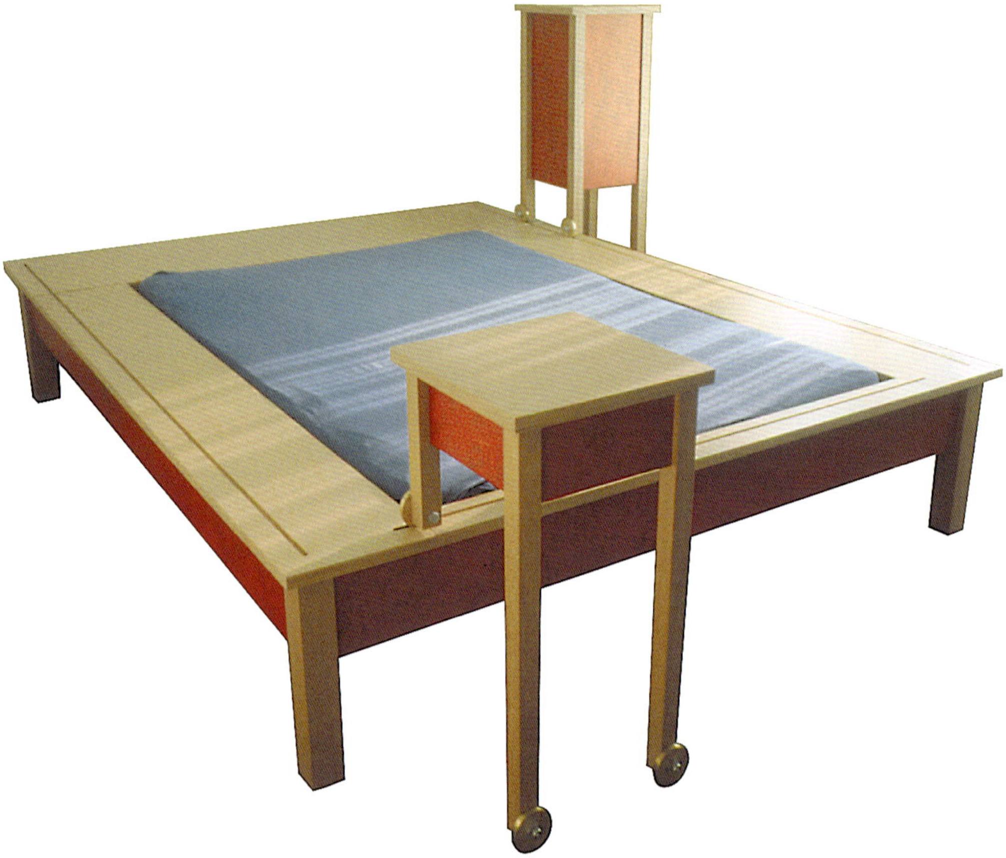

Bed=room — 1993

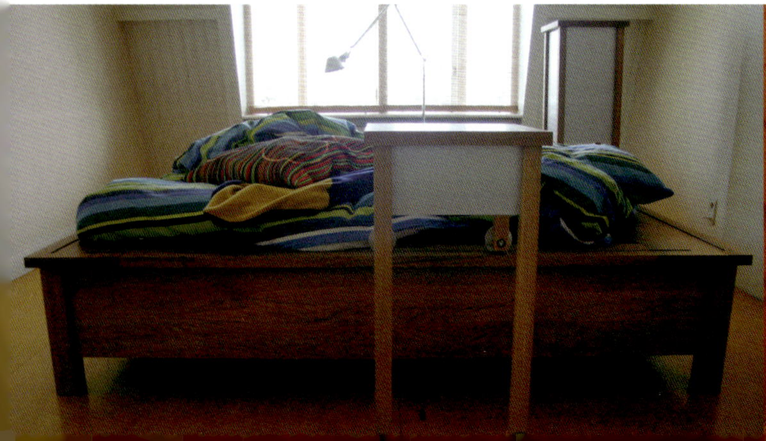

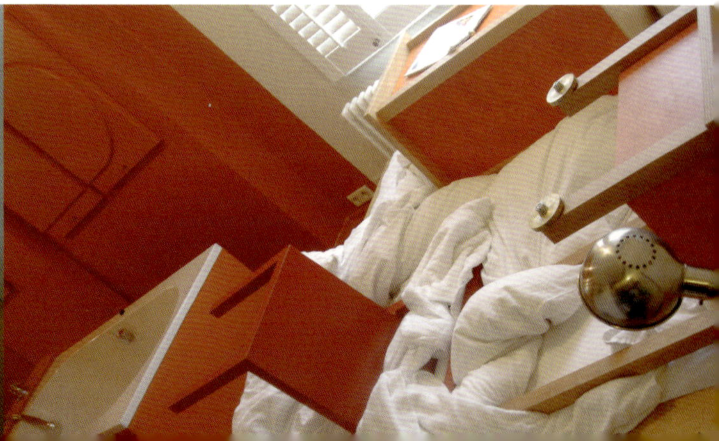

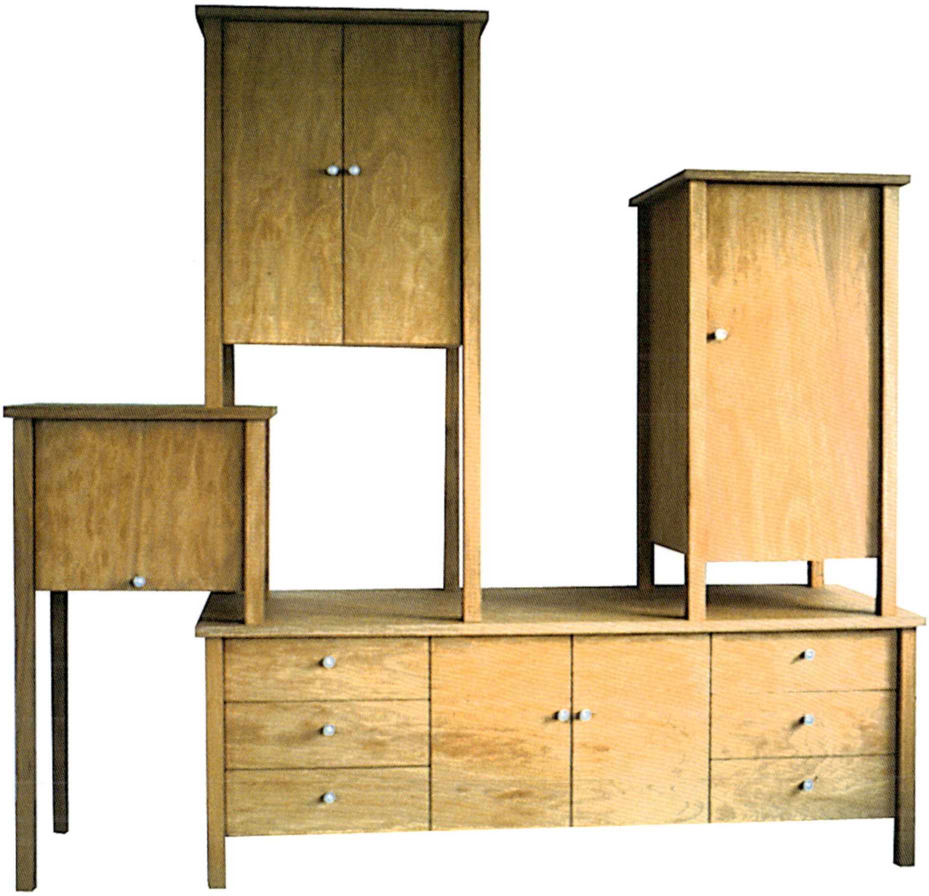

Skyline Rotterdam — 1994

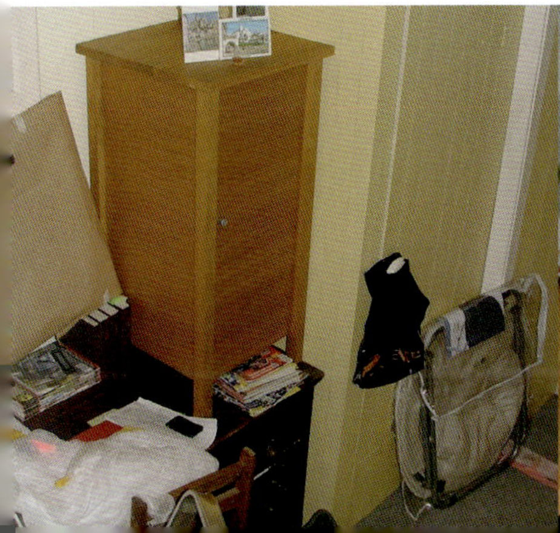

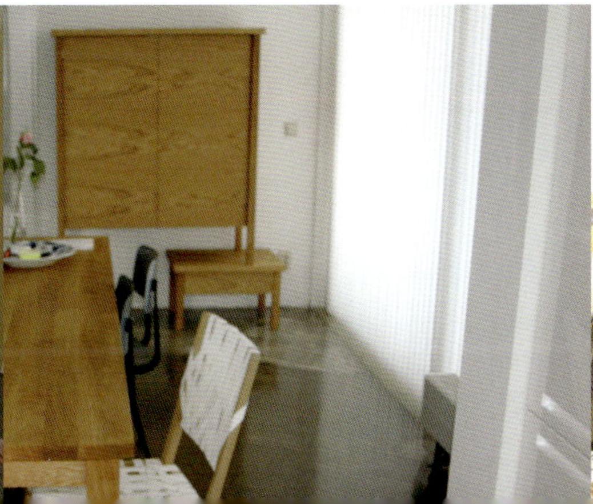

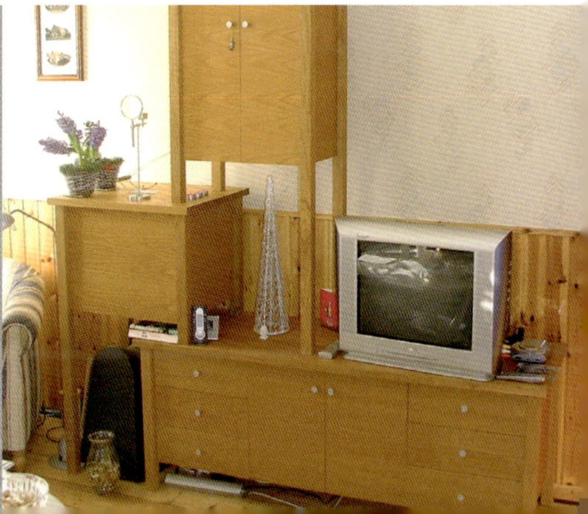

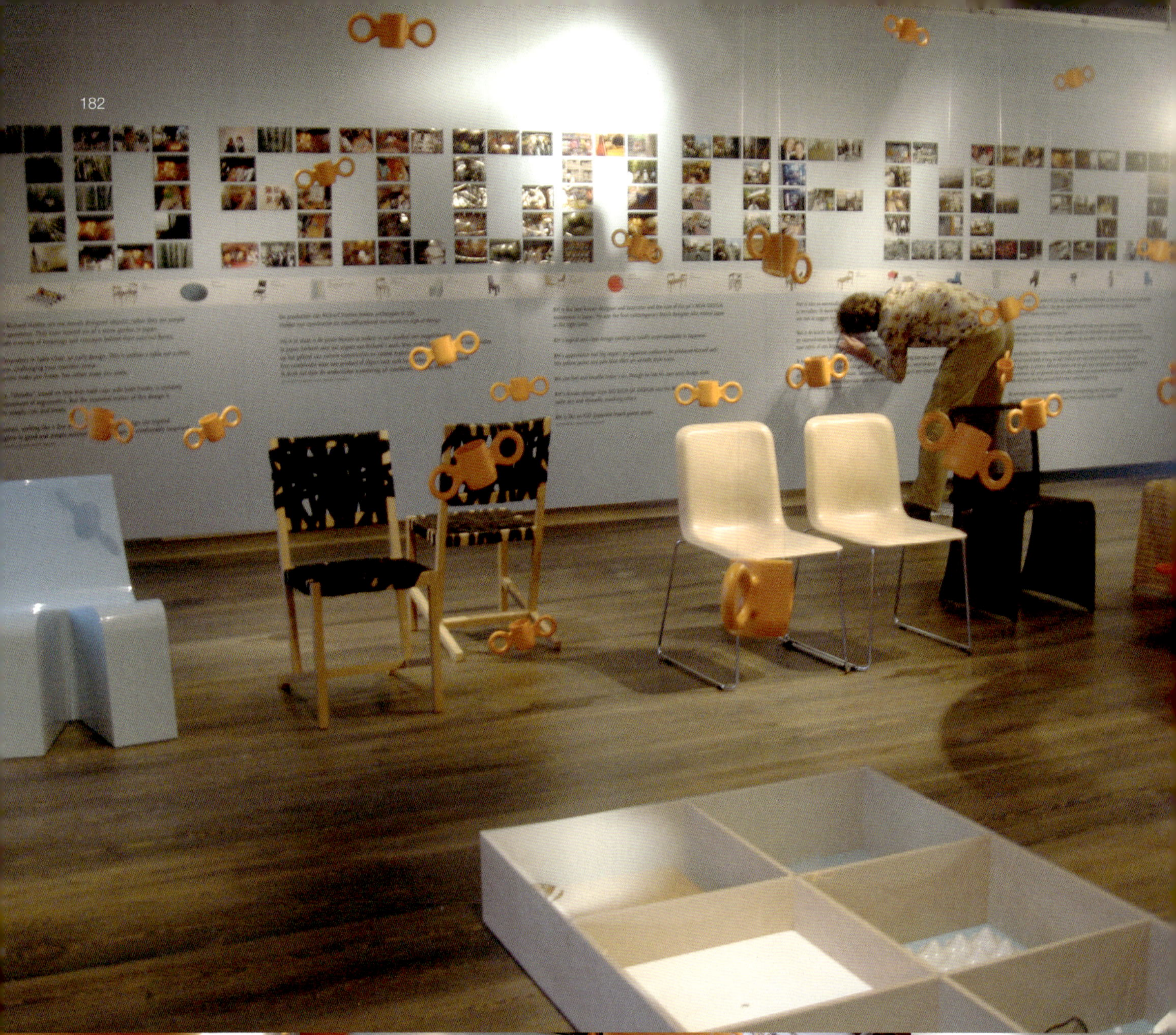

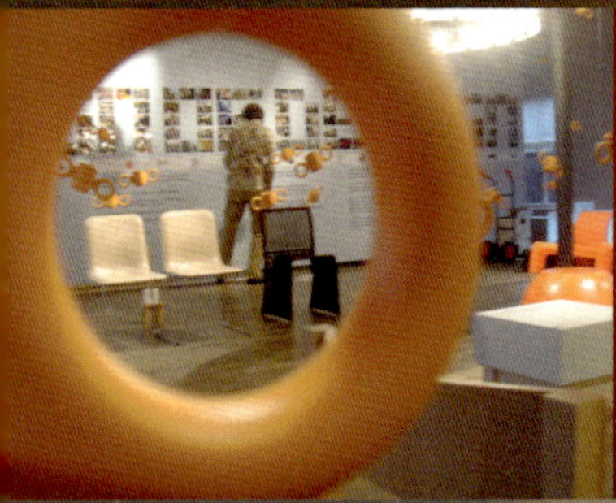

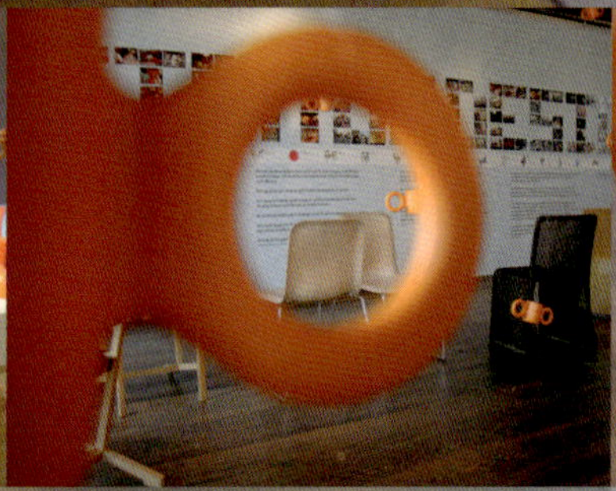

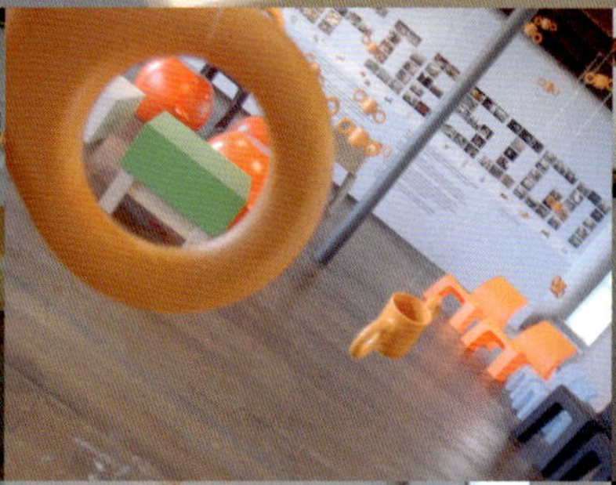

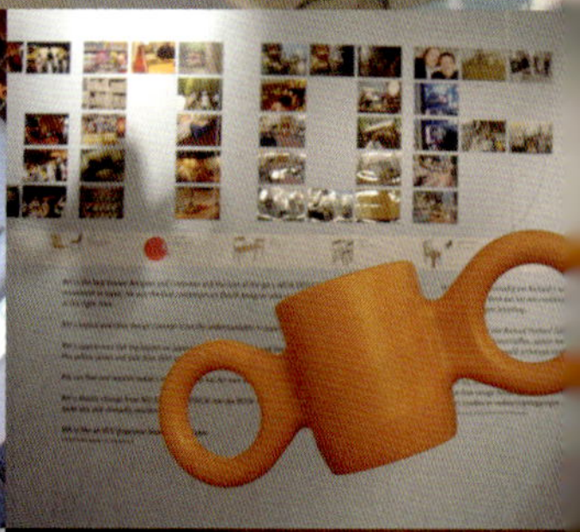

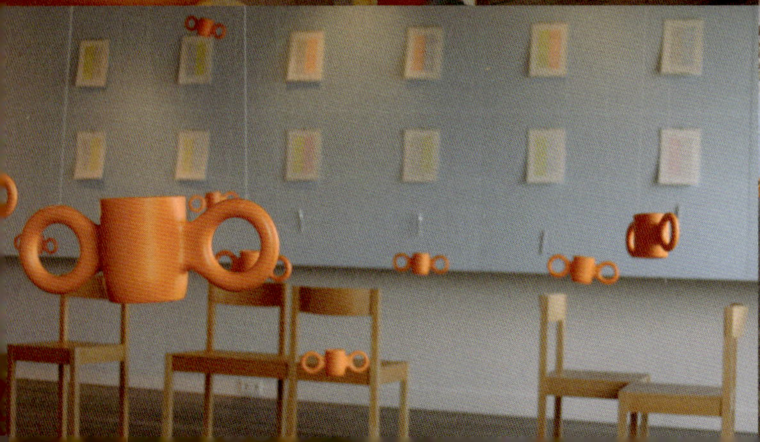

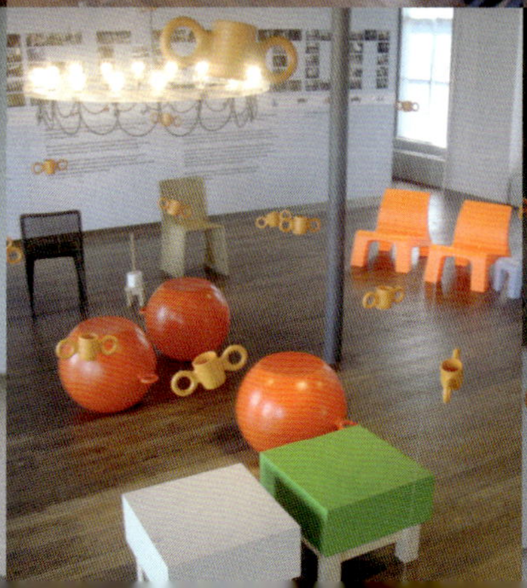

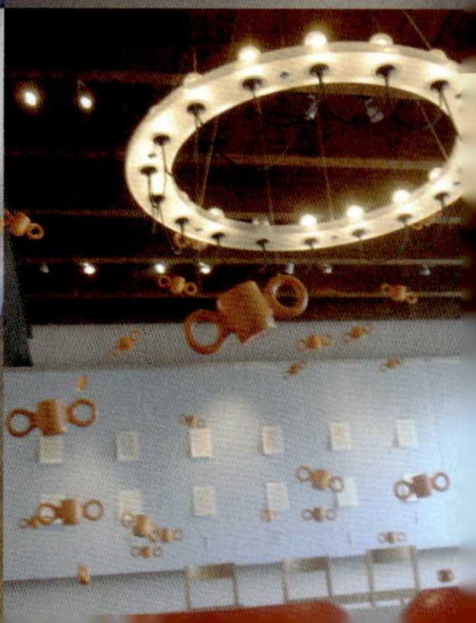

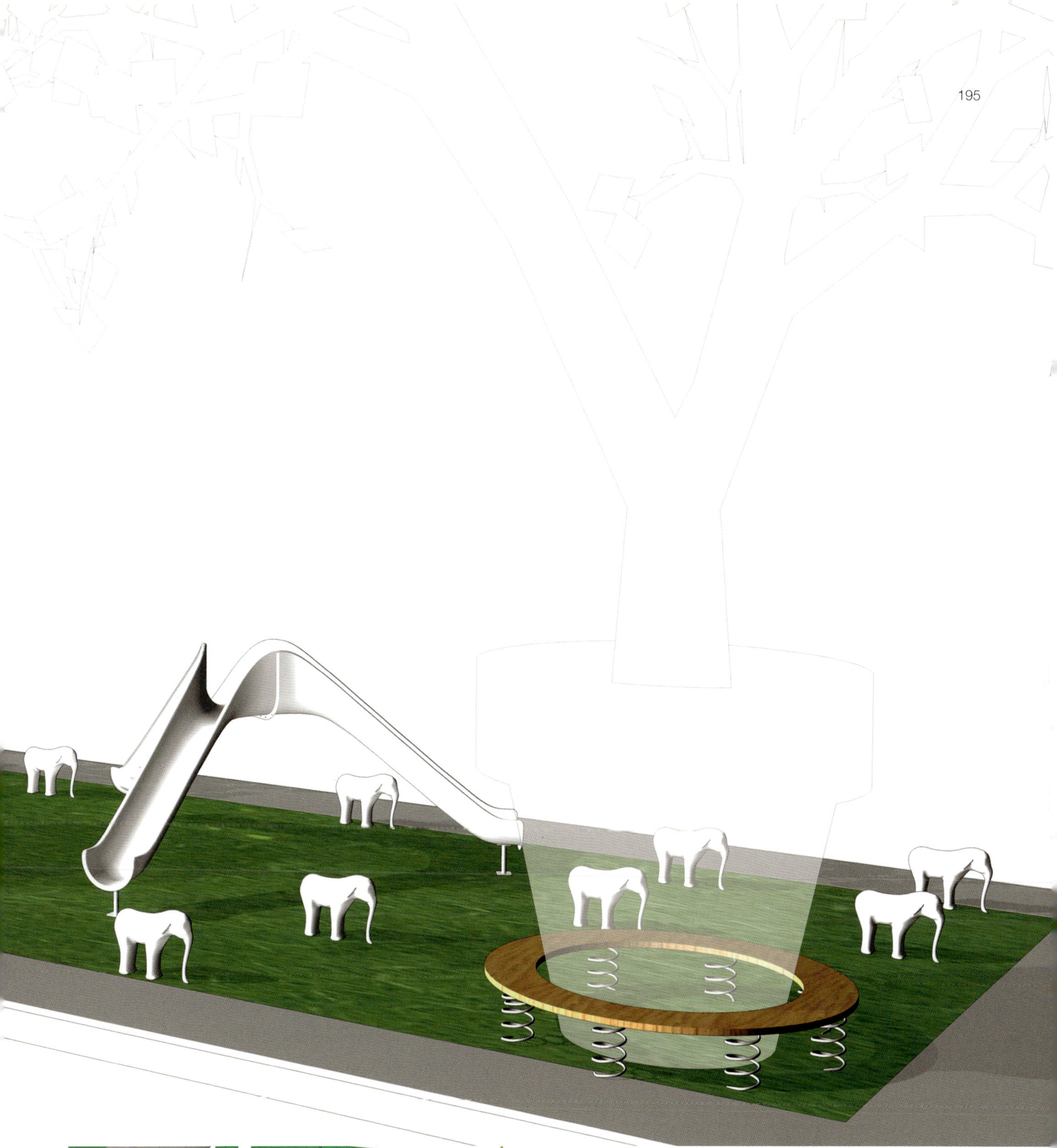

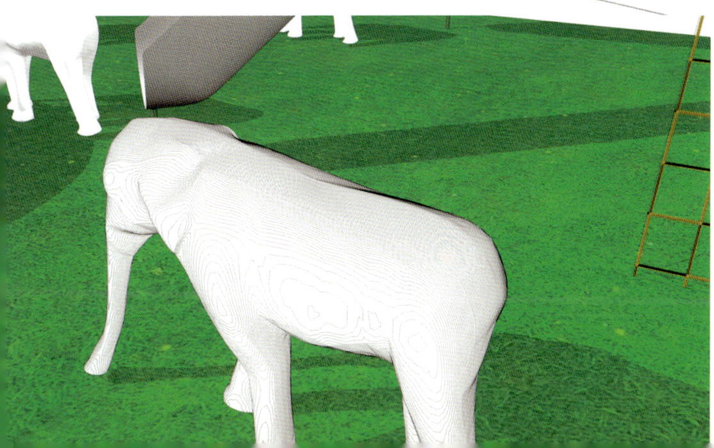

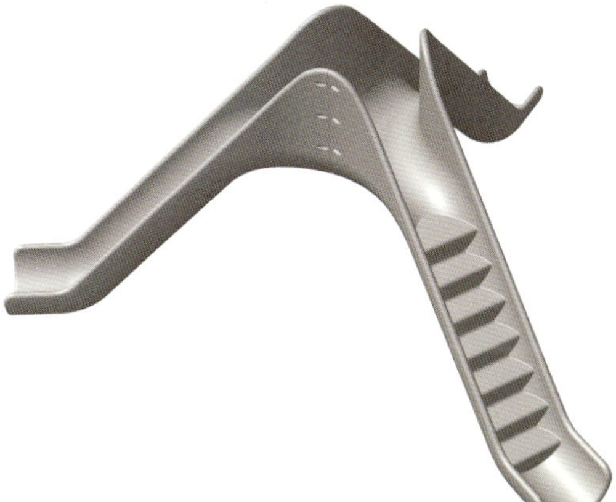

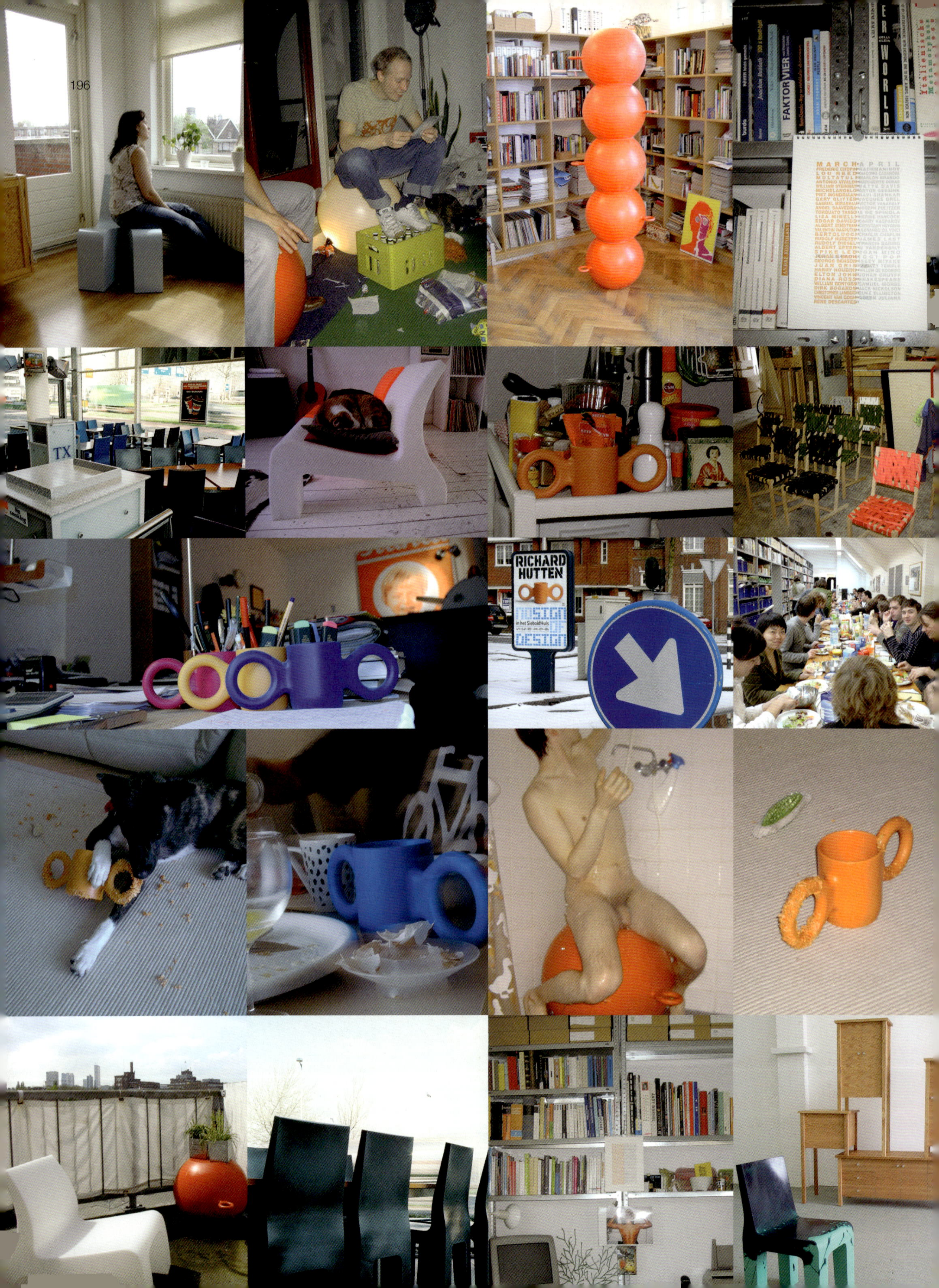
TX
RICHARD
HUTTEN
in het SieboldHuis
OF
DESIGN

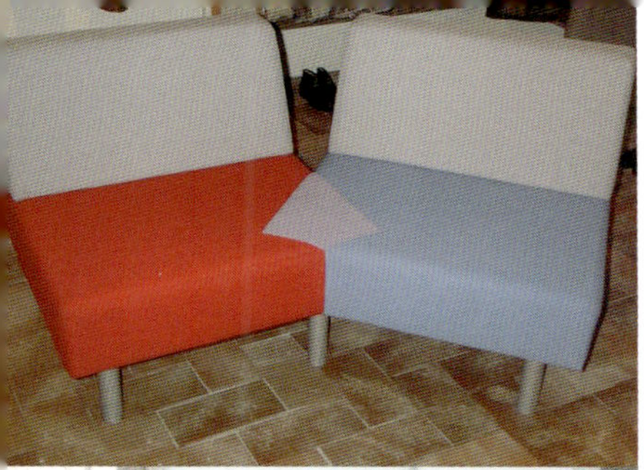

Entity & identity — 2000

E&Y, Japan.
Materials: steel, upholstery.

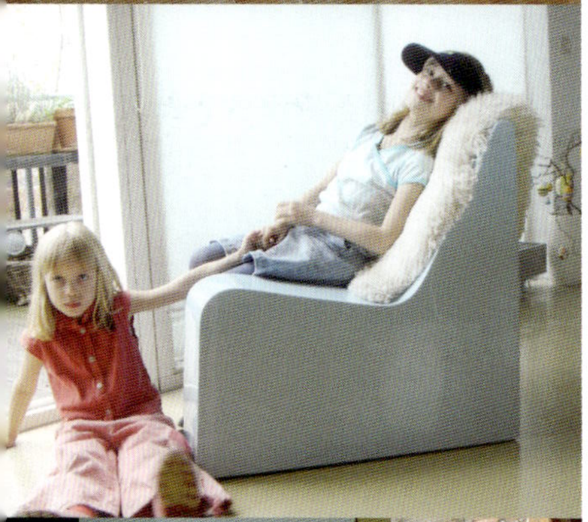

Sexy Relaxy — 2000

E&Y, Japan.
Material: polyester.
Limited edition of 50 pieces + 3 A/P

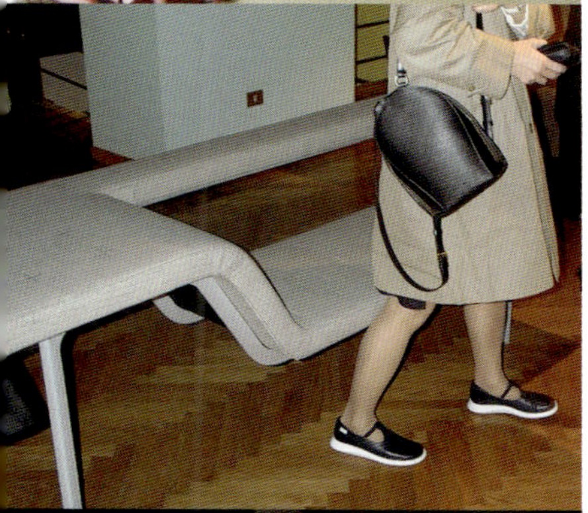

Table-bench — 2000

Sawaya&Moroni, Italy.
Materials: aluminium, upholstery.

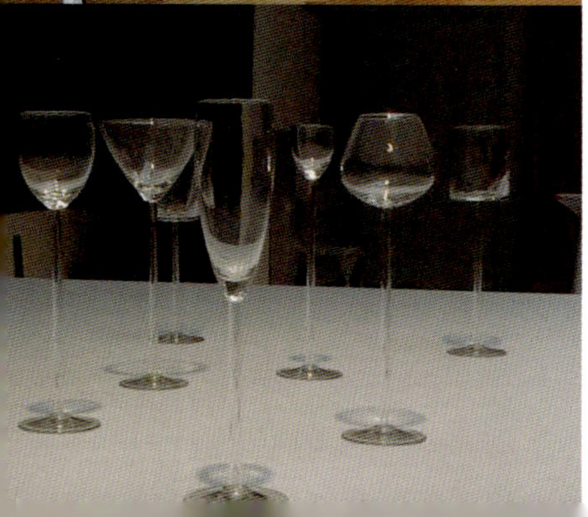

Longdrink glasses — 2000

Droog Design, the Netherlands.
Material: glass.
Ordinary, archetypal glasses on extremely long stems.

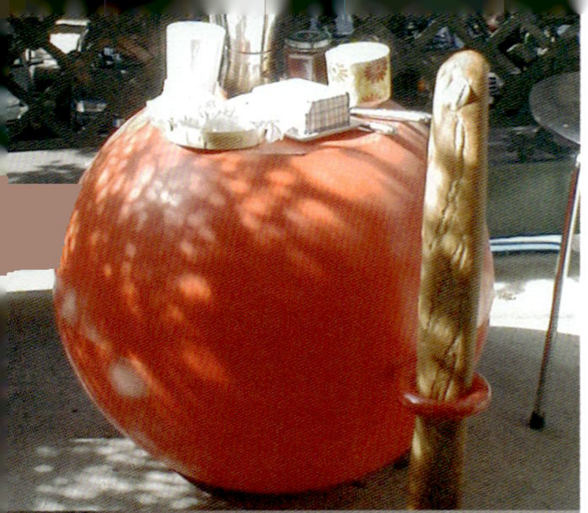

Zzzidt-object — 2000

Centraal Museum Utrecht

Material: LDPE.

Ida van Zijl, vice-director of Centraal Museum, asked me to design their garden furniture. Since the design for the garden was already finished, I didn't want to attach furniture. I looked at my own garden and saw how nice kids' toys contrasted with the green. I took the shape of a kangaroo ball and translated it into a stool/side table.

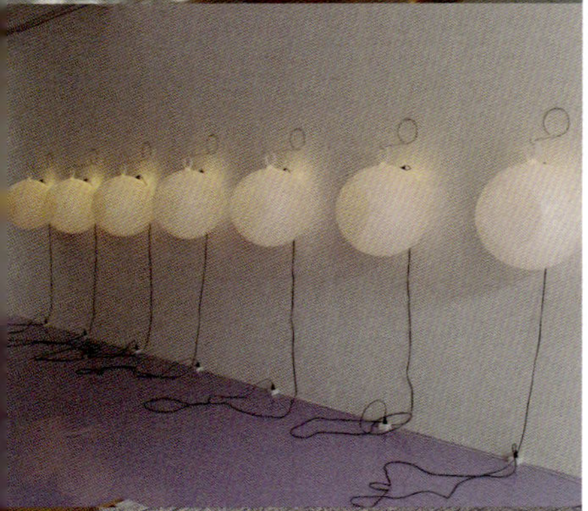

Zzzidt-light — 2000

Material: LDPE.

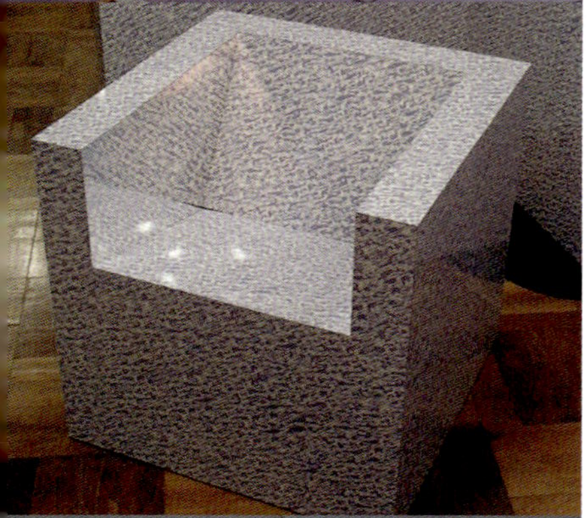

Hard-sofa — 2000

Sottsass associates/Abet Laminati, Italy.

Materials: wood, mirror, Abet Laminati.

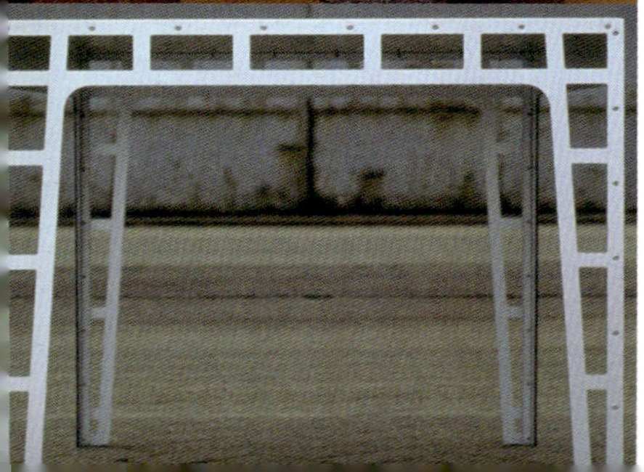

One of a kind dining table — 2000

Hidden, the Netherlands.

Material: Alucabon®.

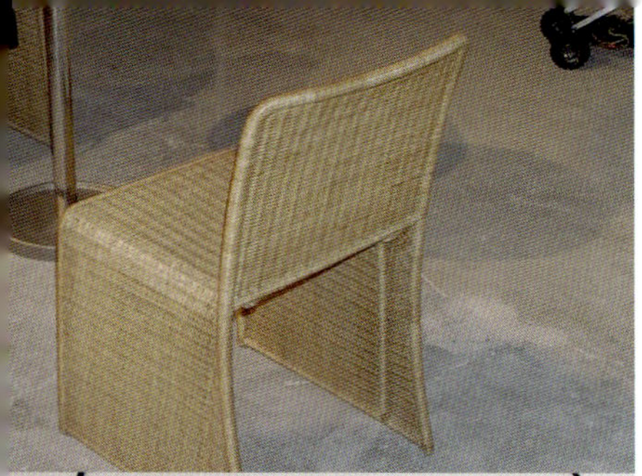

Back-up chair — 2001

IDÉE / Sputnik, Japan.
Material: rattan.

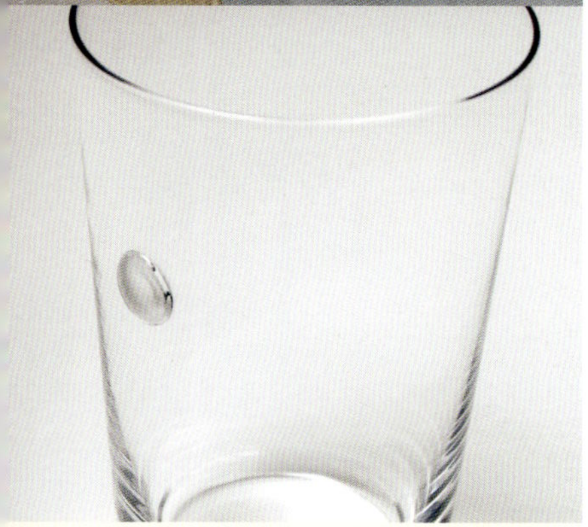

On the spot — 2001

Glass, Trico, Japan.
Material: crystal.

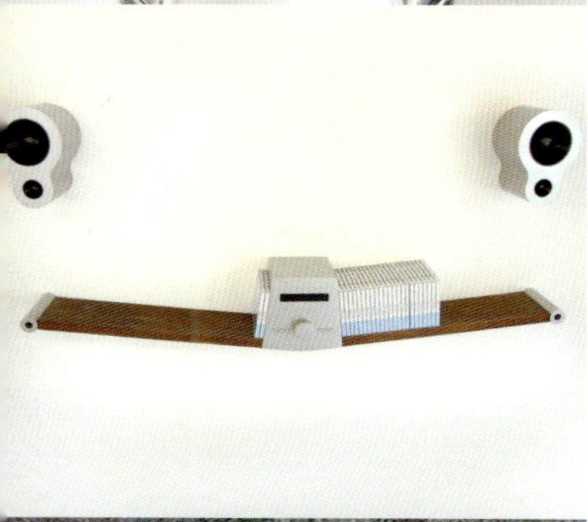

Whings — 2001

CD player + storage for Hidden, the Netherlands.
Prototype.
A wall mounted CD player with tuner and CD storage.

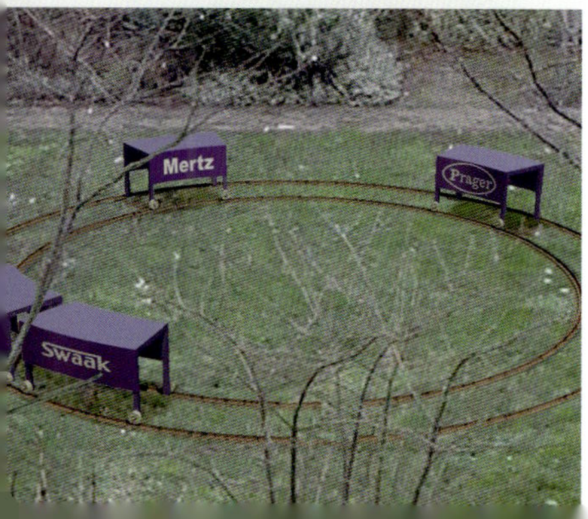

Garden bench — 2001

Dutch Design institute.
Proposal.

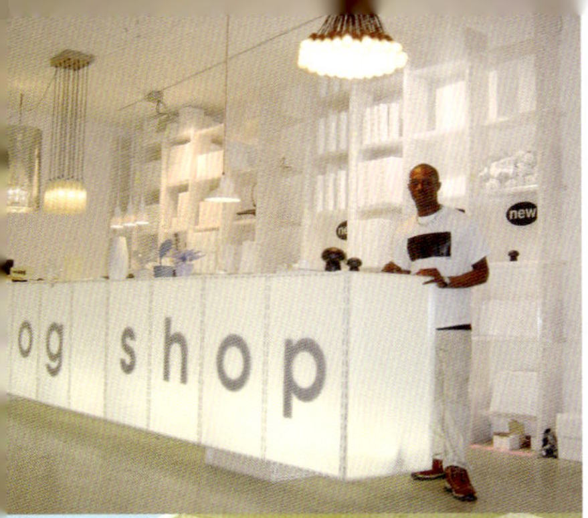

Droog shop, Droog design — 2001

Droog Design shop Milan, Droog Design, the Netherlands.
Materiasl: Lexan, tape.

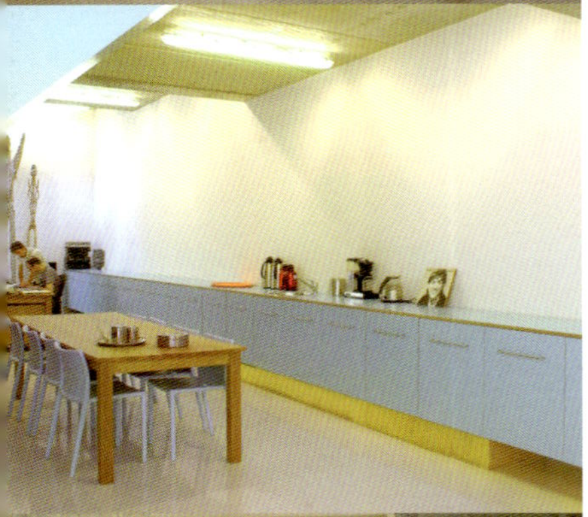

Interior studio Thonik — 2001

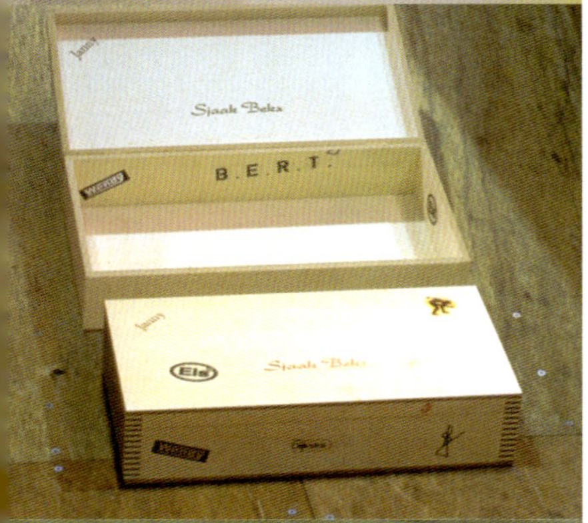

Trace box — 2001

Droog Design, the Netherlands.
Materials: walnut, maple, stickers.
Everybody who works on this box in the factory (about 20 people) leaves a trace on the outside or inside of the boxes.

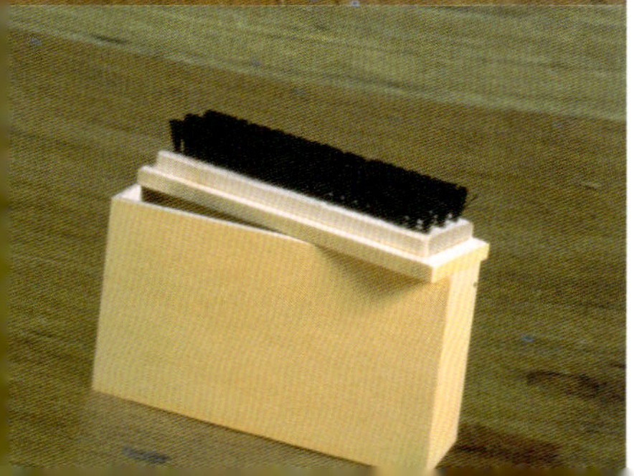

Shoe polish box — 2001

Droog Design, the Netherlands.
Materials: maple, brush.

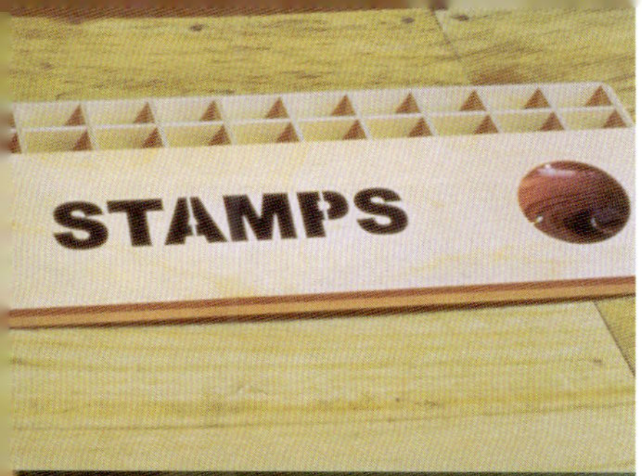

Stamps box — 2001

Droog design, the Netherlands.
Materials: maple, magnifying glass.

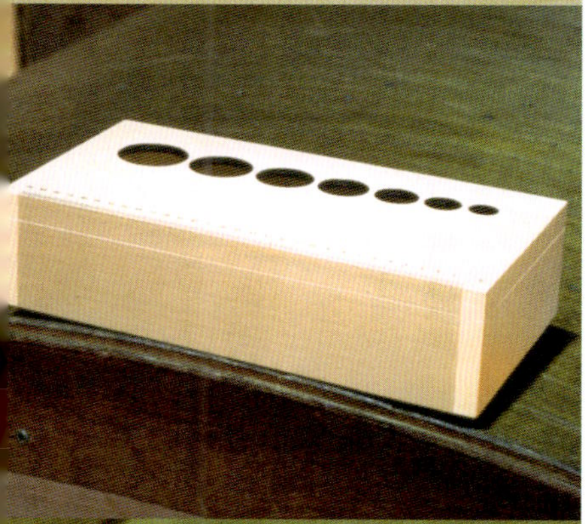

Drawing box — 2001

Droog design, the Netherlands.
Material: maple.

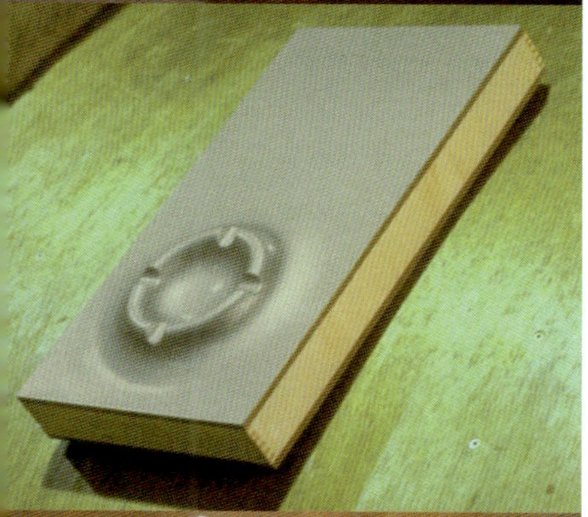

Cigar box — 2001

Droog Design, the Netherlands.
Materials: maple, stainless steel.

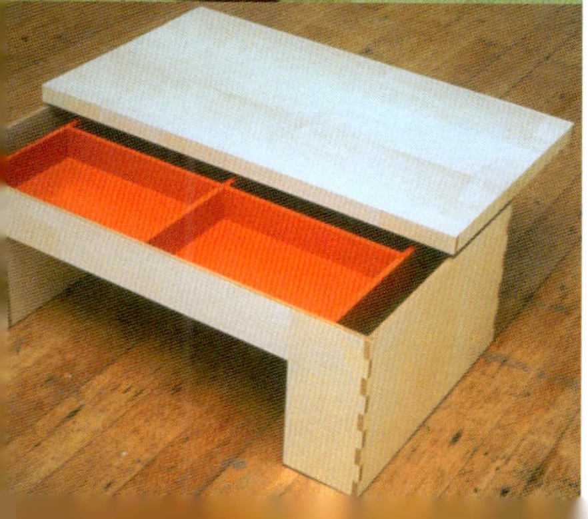

Table box — 2001

Droog Design, the Netherlands.
Material: maple.

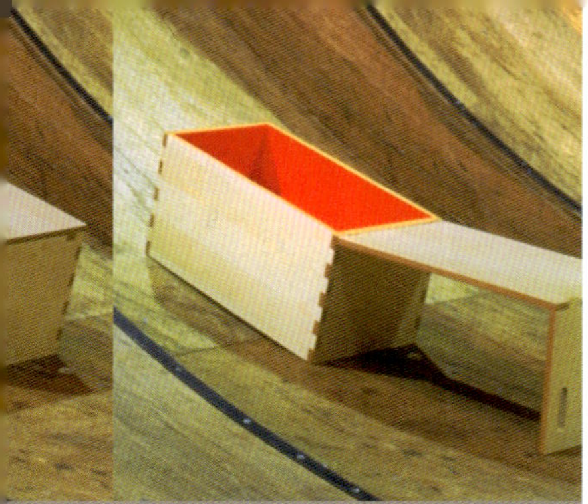

Kids box — 2001

Droog Design, the Netherlands.
Material: maple.

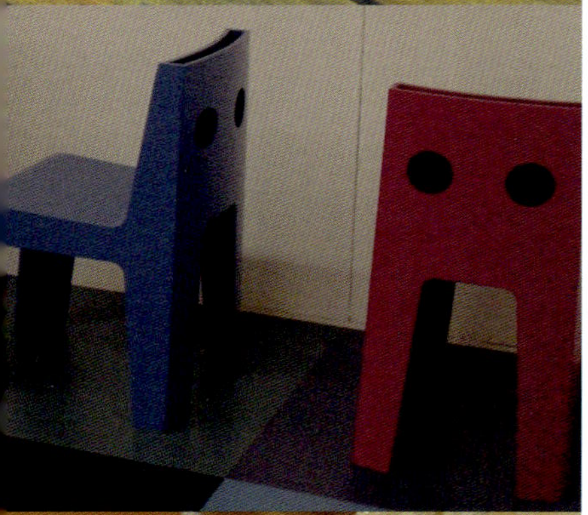

Abel children's chair — 2001

Live & Play, the Netherlands.
Material: beech plywood.

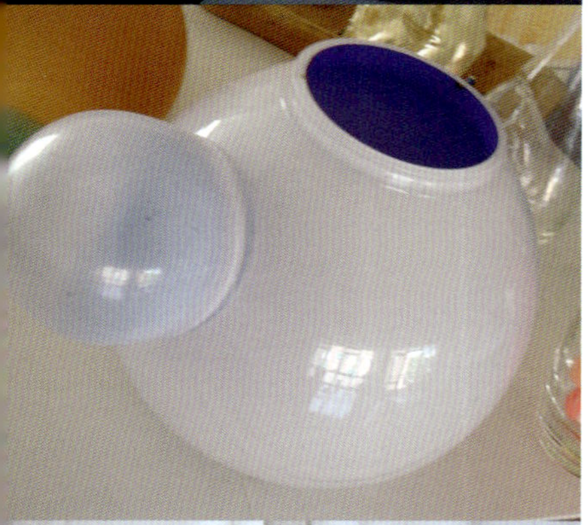

What Happened? — 2001

Serie of vases and a bowl for Covo.
Material: murano glass.
Emmanuel Babled, the art director for this project, asked me to do something untraditional with Murano blown glass. Not the virtuosity of the master glassblower but how to treat the glass was the goal. Emmanuel warned me that, if I wanted something too strange, the master would refuse. Lucky me – he was in a good mood that day, so he 'attacked, dropped and mutated' the glass for me.

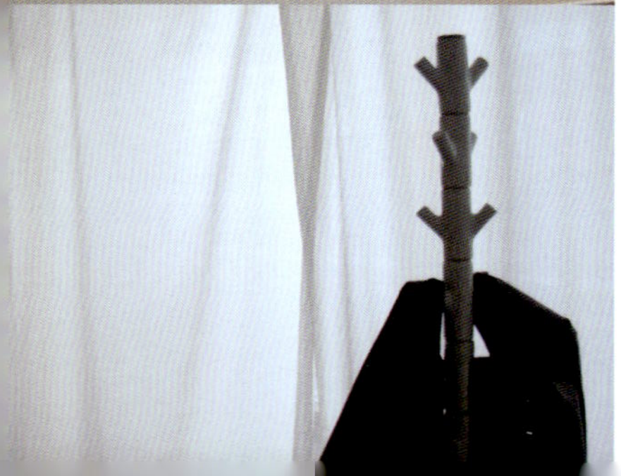

Bone — 2001

Planet, the Netherlands.
Materials: ceramic, steel.

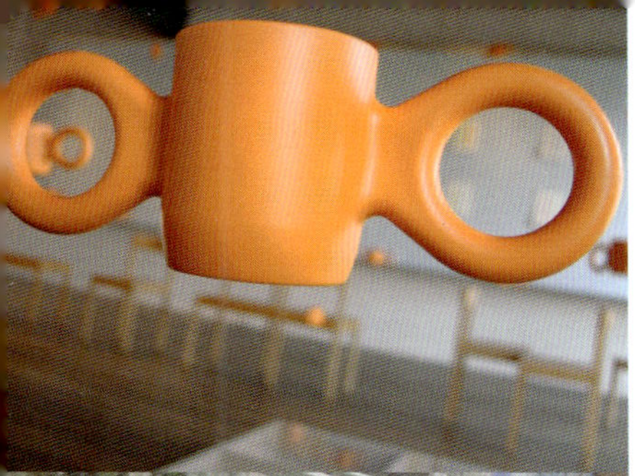

Dombo mug — 2001

RTL+ Germany. Material: PPC.
About the Dombo mug Itsuro Shibata, contributing writer for Axis magazine, wrote: His works never make you frown, rather they make you smile. Dombo, for instance, which I heard that he initially designed for his kids, is certainly a great achievement of ergonomics (based on the way kids hold cups with both hands). But the essential matter on this design is that it looks simply cute and lovely. Richard Hutten, smiling like a Zen master, shows us the possibilities of how design can expand our perception in good and simple ways. And he keeps us comfortably surprised.

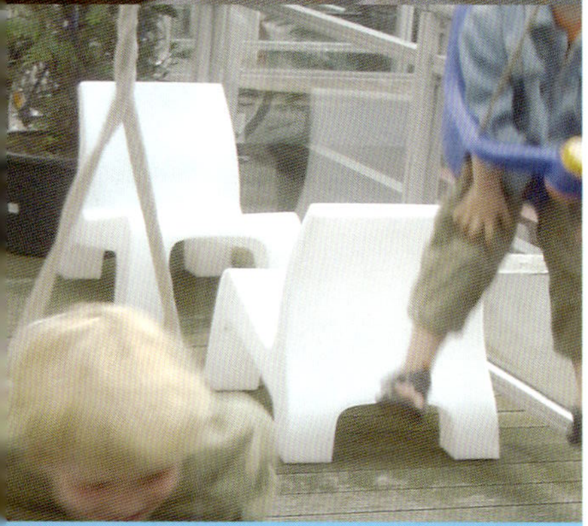

Rhino chair — 2001

Material: LDPE.

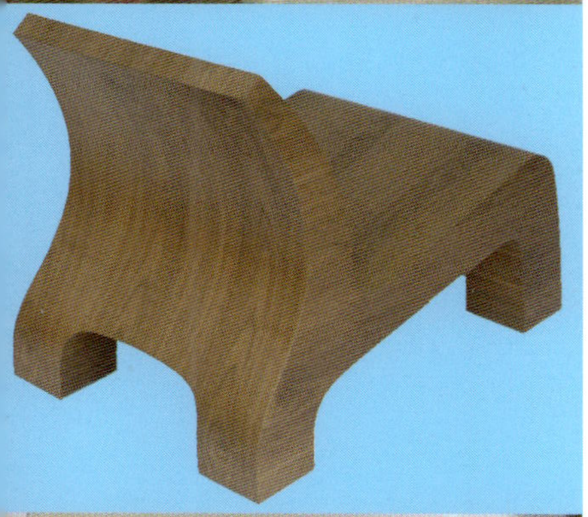

Rhino chair — 2001

Rhino chair.
Material: walnut.
Autoproduction. Limited edition of 3 pieces.

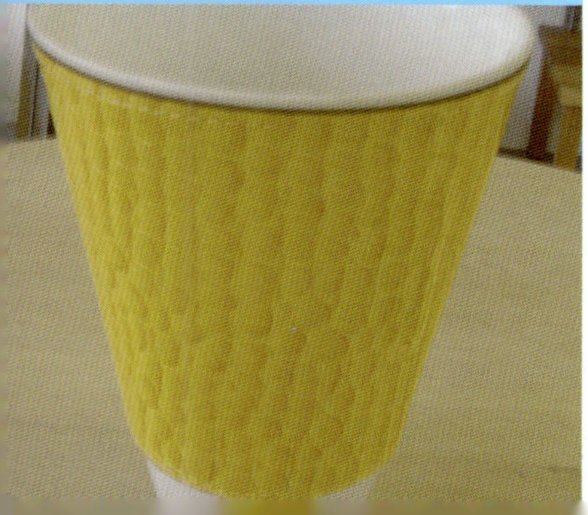

Dutch vase — 2001

Sawaya & Moroni.
Materials: porcelain, leather.

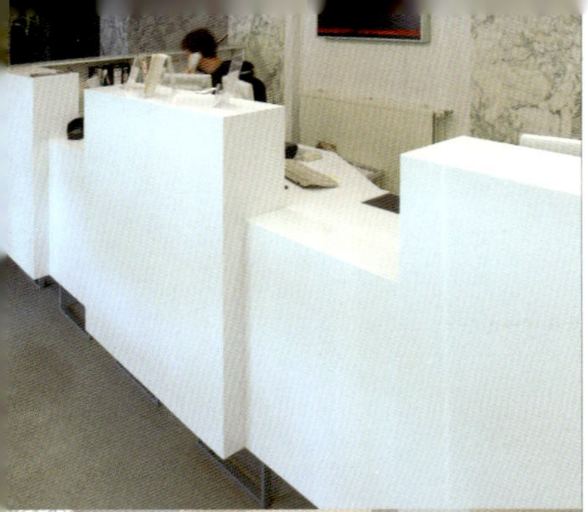

Entrance Filmmuseum — 2002

Amsterdam.

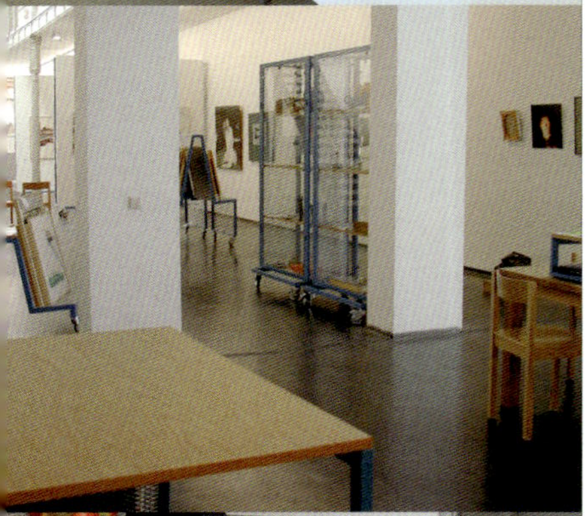

Interior CBK art centre Dordrecht — 2002

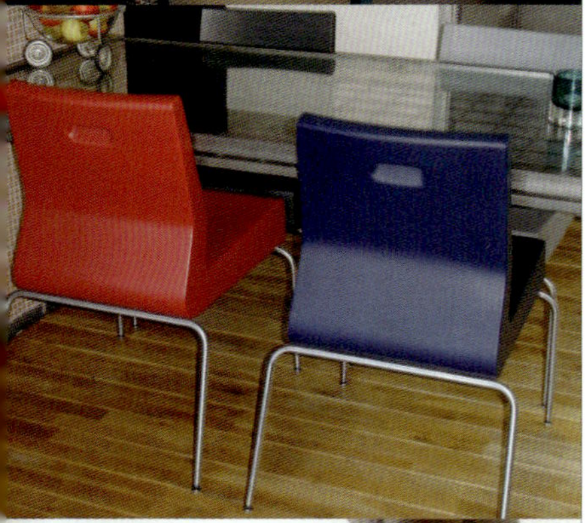

HH chair — 2002

Harvink.
Material: LDPE.

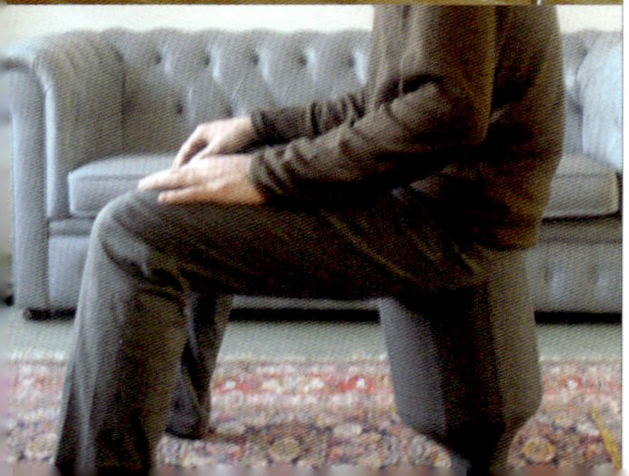

Stoolpants — 2002

ECAL, Swiss.

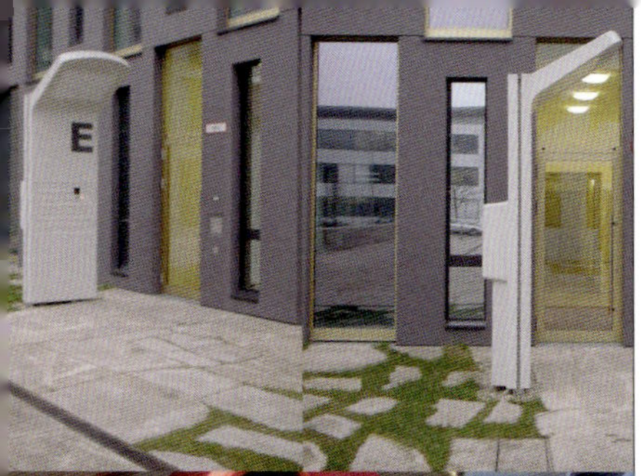

Luifel MVRDV — 2002

Building, Munich.

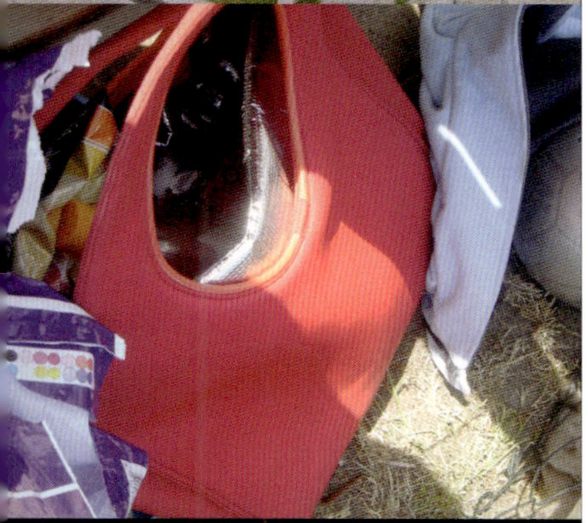

Below zero bag — 2003

Unilever/ Project X.

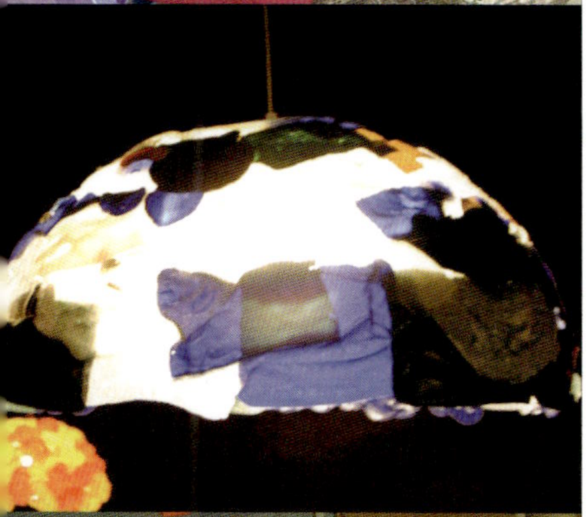

Meltingpot lamp — 2003

Autoproduction.
Limited edition of 50 pieces.

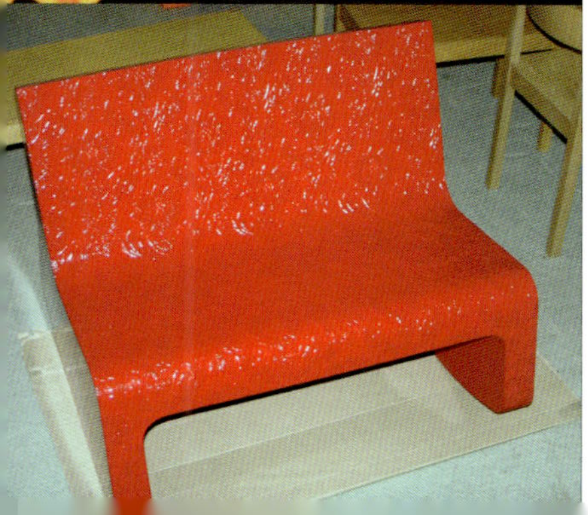

2Loveseat — 2000

Autoproduction.
Limited edition of 3 pieces.
Materilas: wood, resin.

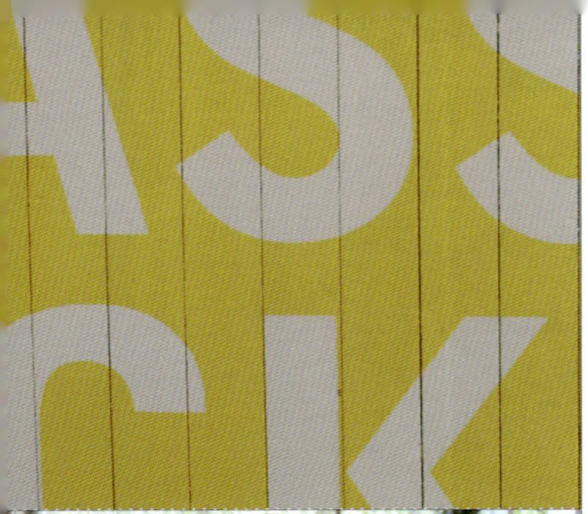

Entrance hall NAI museum of architecture — 2003 Rotterdam, the Netherlands.

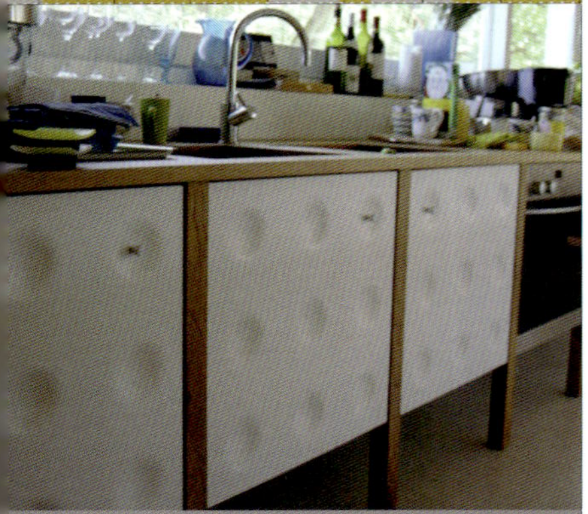

Kitchen for fam. Schuitema — 2003 Wassenaar, the Netherlands.

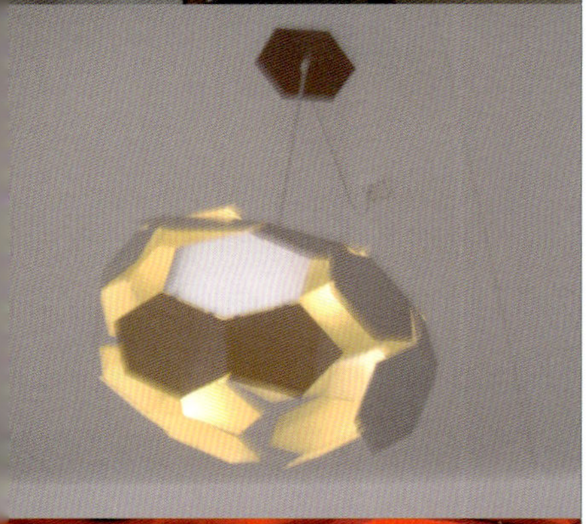

Thunderball, light, prototype — 2003 Material: stainless steel.

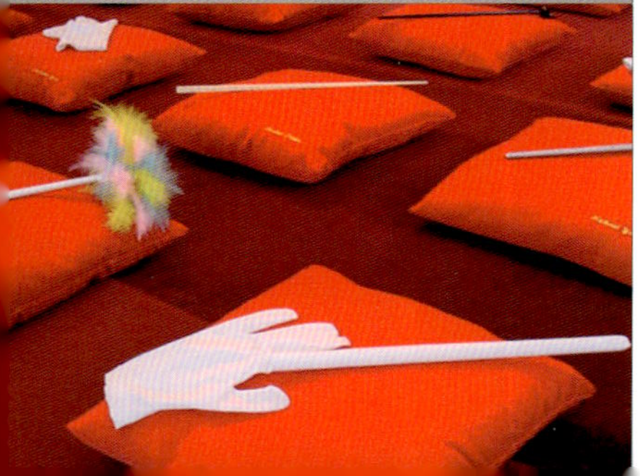

Baton — 2003 ECAI, Swiss.

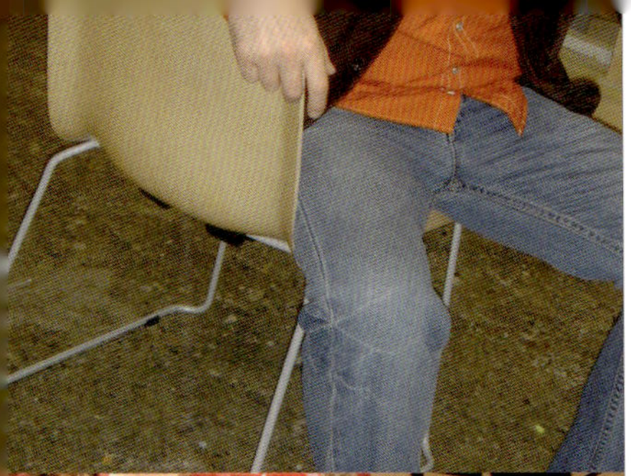

This chair — 2004

Lensvelt, the Netherlands.
Materials: 3D veneer, steel.

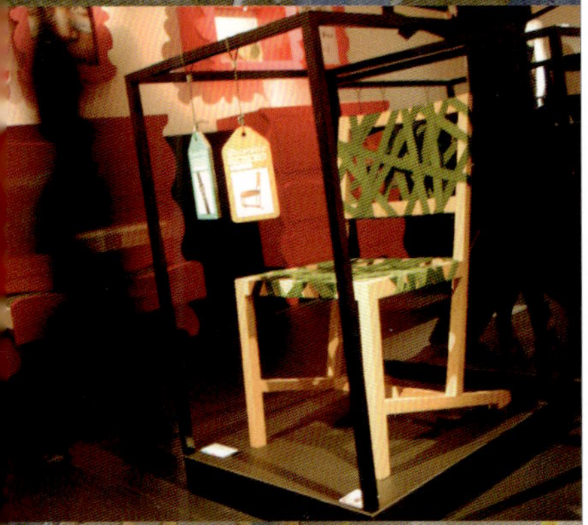

Berlage chair — 2004

Gemeentemuseum The Hague.

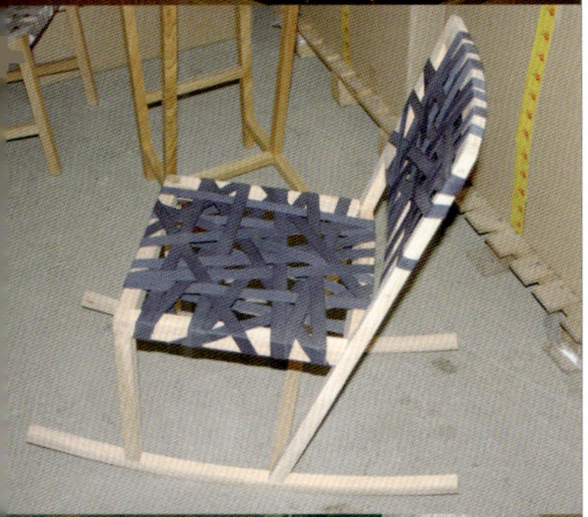

Berlage rocking chair — 2004

Gemeentemuseum The Hague.

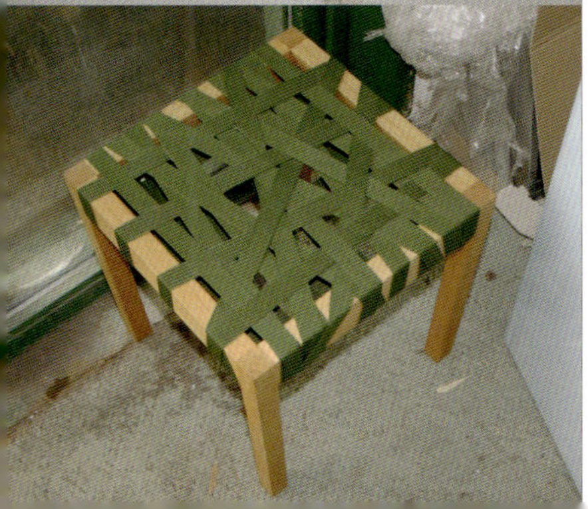

Berlage stool — 2004

Gemeentemuseum The Hague.

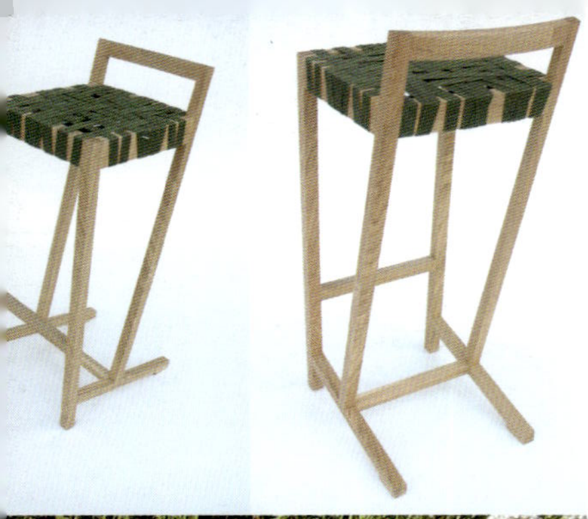

Berlage bar stool — 2004

Gemeentemuseum The Hague.

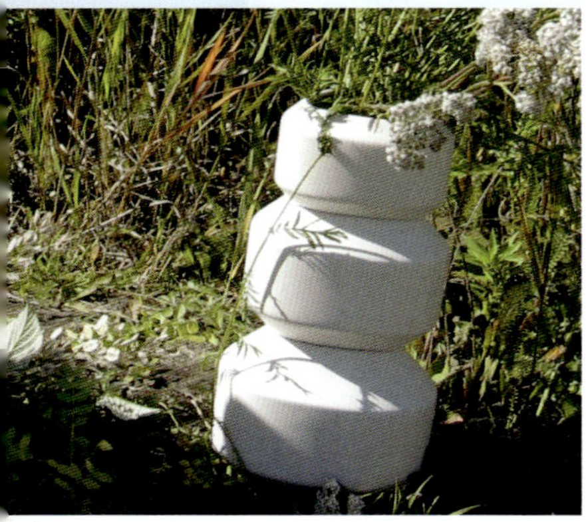

Rocket vase — 2004

Object collection, Canada.

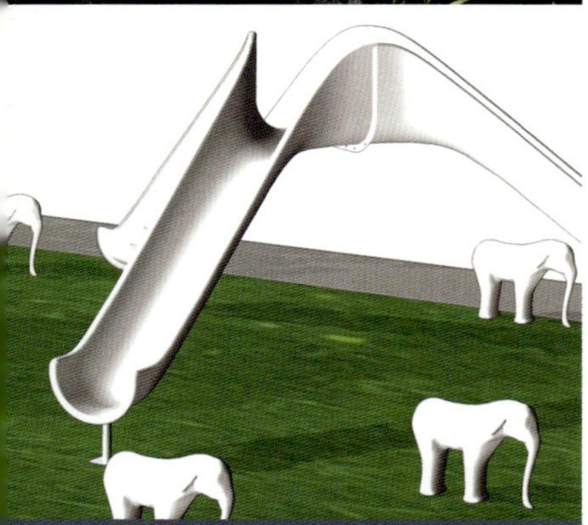

Buurt Ne9en — 2004-2006

'interior' for Kristal - Het oosten in MVRDV building.

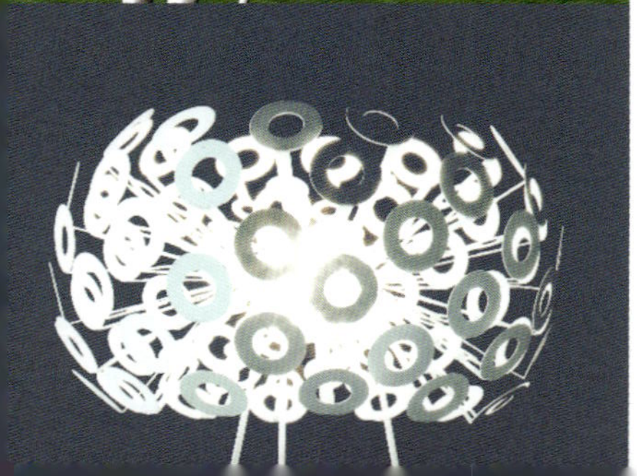

Dandelion — 2004

MOOOI
Material: steel.

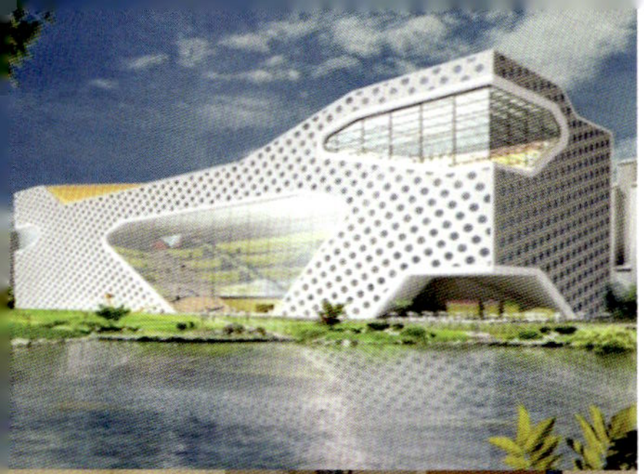

Rhea shopping mall — 2004

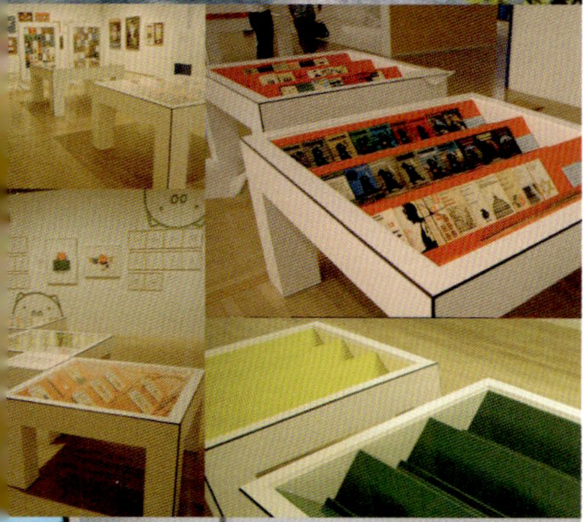

Exhibition design Dick Bruna — 2004

Suntory Museum Osaka, Japan.

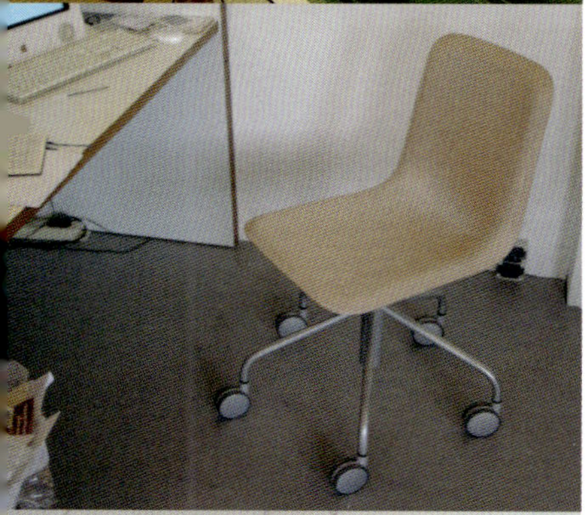

This office chair — 2005

Lensvelt office furniture.
Materials: steel, 3D veneer.

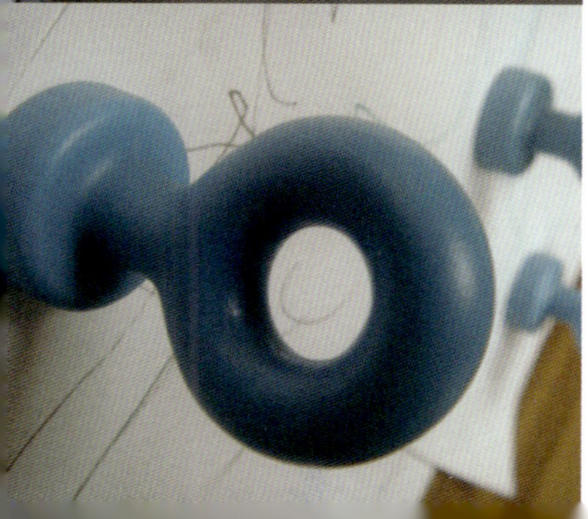

O-magnet — 2005

Kikkerland, New York.
Material: PP.

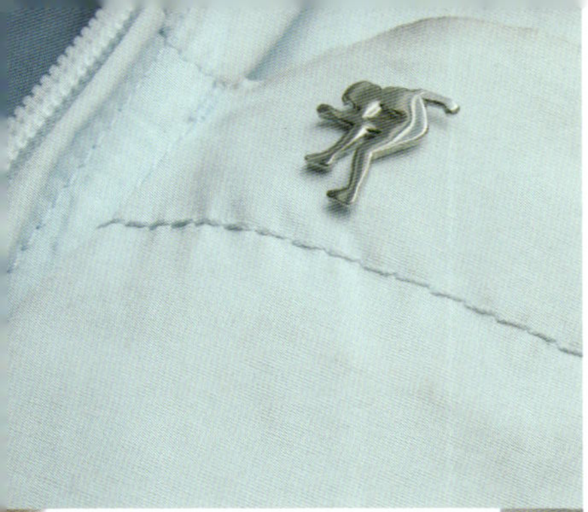

Woonbeurspin — 2005

Award, HPR woonbeurs Amsterdam.
Material: nickel.

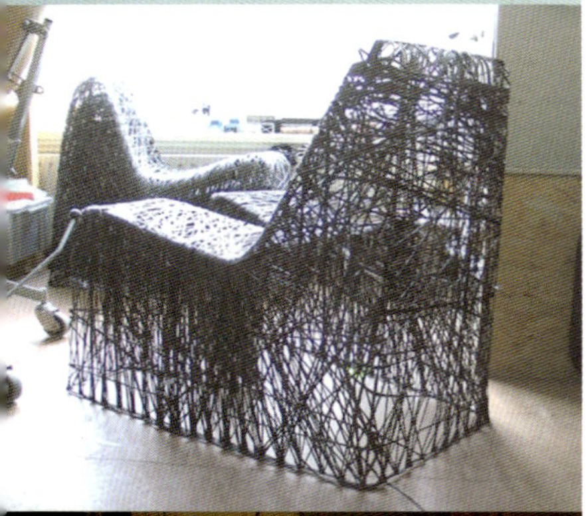

Sexy Relaxy random carbon copy — 2005

One-off, autoproduction together with Bertjan Pot.

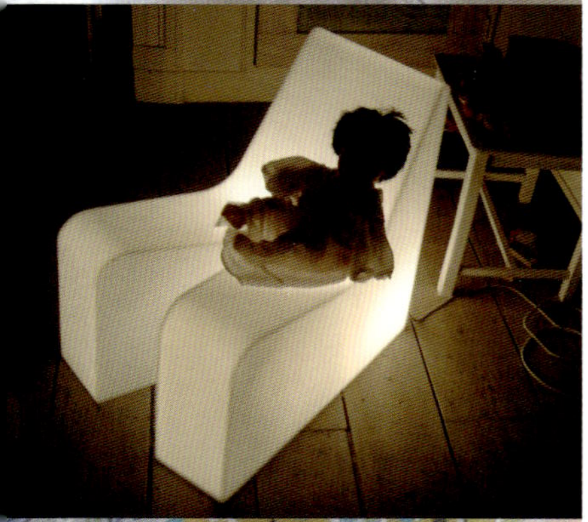

Sexy Relaxy rotation

Moulded LDPE with light.

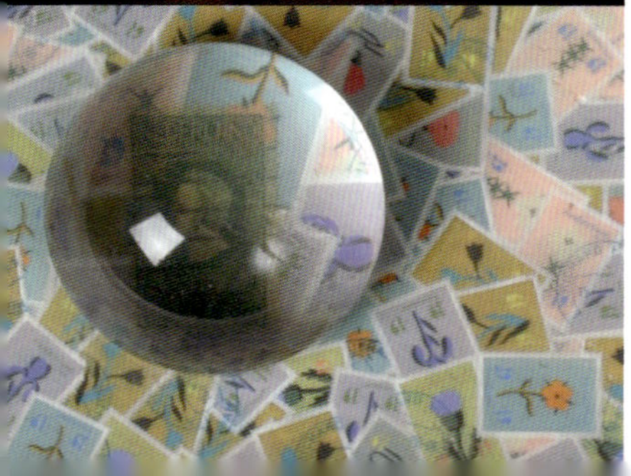

Stamp-plate

TNT Royal Dutch post.
Material: stamps.

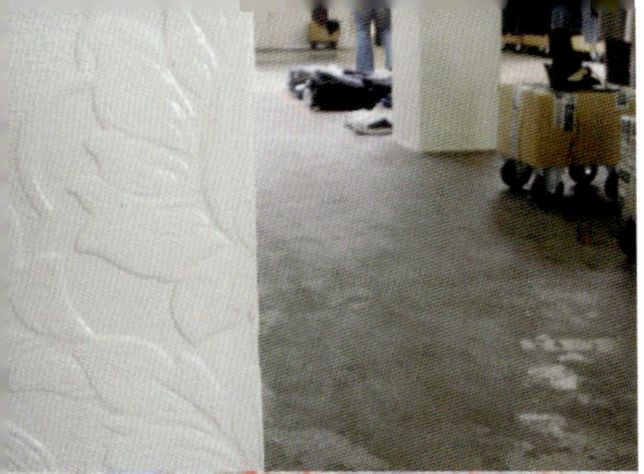

Interior fashion shop Beljon — 2005

By Margreeth Olsthoorn.

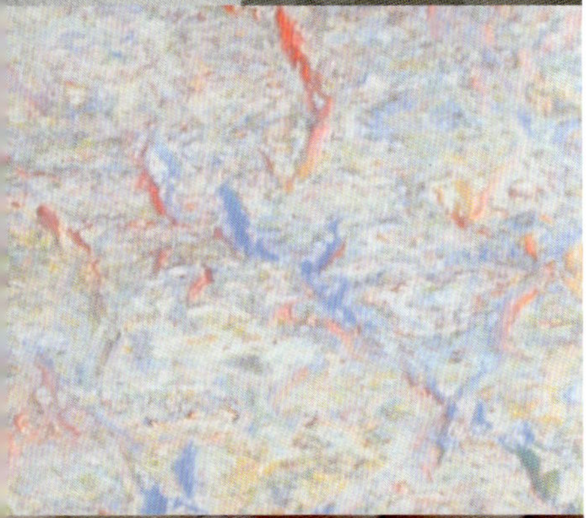

Marmoleum floor — 2006

Forbo, the Netherlands.

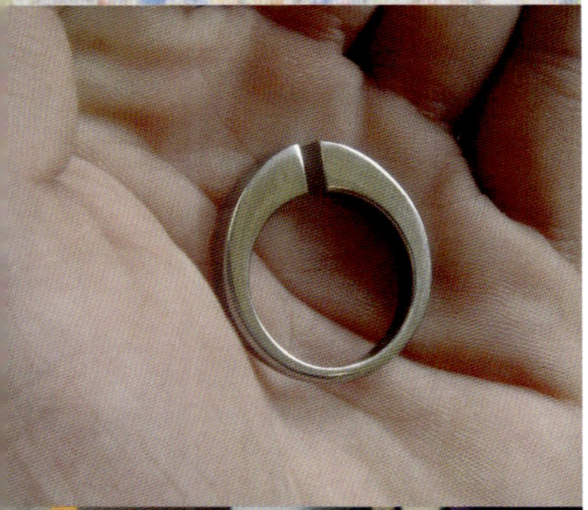

Yes Ring — 2006

Felissimo USA, Japan.
Material: stainless steel.

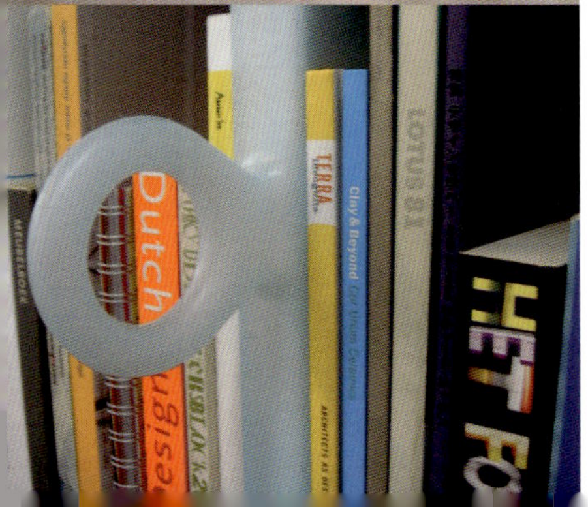

Book cover works in use — 2006

Stichting Kunstboek, Belgium.

Selection of books and catalogues

- *Catalogue Interieur '92*, Stichting Interieur, 1992
- *Catalogue Interieur '94*, Stichting Interieur, 1994
- *Progetti e territori 1994*, Veronafiere
- *The international Design Yearbook 1994*, Laurence King Publishers, 1994
- *Minimalismo – Etica delle forme nel design*, Luppeti, 1996
- *Meubelboek*, Stichting Sofa, 1996
- *De Tafel – The Table*, Stichting Interieur, 1996
- *The international Design Yearbook 1996*, Laurence King Publishers, 1996
- *Droog design 1991-1996*, Centraal Museum Utrecht, 1997
- *Re(f)use*, Cultural Connections, 1997
- *Spirit of the Nineties*, NAI Publishers, 1997
- *Interieur '98*, Stichting Interieur, 1998
- *Aperto Vetro*, Electa, 2000
- *Droog & Dutch Design*, Centraal Museum, 2000
- *200 in 2000*, kunsthal uitgeverij Waanders, 2000
- *The international Design Yearbook 2000*, Laurence King Publishers, 2000
- *Repères 2002*, Salon du Meuble de Paris, 2001
- *Un móvil en la patera, deseñando el siglo XXI*, EACC, 2001
- *Kunststof(f)*, Luc Delrue, 2002
- *Less+More*, 010 Publishers, 2002
- *Richard Hutten: taking form, making form*, 010 Publishers 2002
- *Spoon*, Phaidon Press, 2002
- *From bakelite to composite*, Stichting Kunstboek, 2002
- *Clay & beyond*, Het kruithuis, 2002
- *The Eco-design handbook*, Thames & Hudson, 2002
- *The international Design Yearbook 2002*, Laurence King Publishers, 2002
- *In Vorm*, Museum Het Valkenhof, 2003
- *Dutch Design van de 20ste eeuw*, Thoth, 2003
- *Reality Machines*, NAI publishers, 2003
- *No design, no style*, Hangaram Design Art Museum, Seoul Arts Center, 2003
- *Designers on Design*, Conran Octopus, 2004
- *Yearbook Dutch Design 03/04*, Vivid Episode Publishers, 2004
- *Furniture Collection Stedelijk Museum Amsterdam*, NAi Publishers, 2004
- *Simply Droog*, Droog Design, 2004
- *False Flat*, Phaidon press, 2004
- *The Design Encyclopedia*, MoMA NY Laurence King, 2004
- *Minimalist Design*, Birkhäuser, 2004
- *Au doigt et à la baguette*, ECAL, 2005
- Nest – *design for the interior*, NAi Publishers 2005
- *Absolute Architecture*, Michel Blanckaert, 2006
- *Under Cover*, 010 Publishers, 2006
- *Richard Hutten: Works in use*, Stichting Kunstboek, 2006

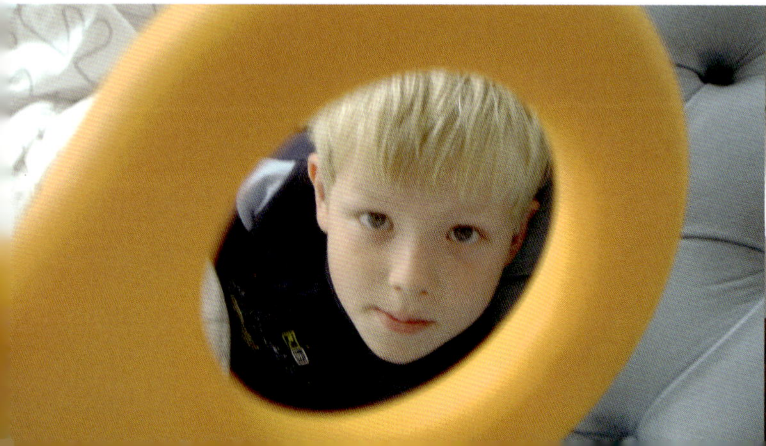

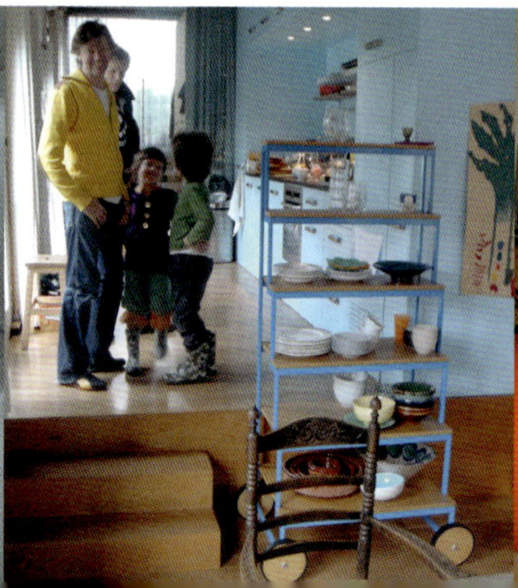

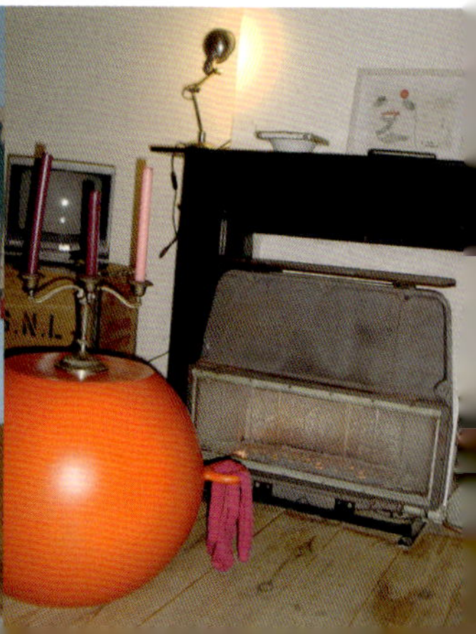

Selection of works in public collections

- **Table-chair** Stedelijk Museum Amsterdam, Vitra Museum Weil am Rhein, Centraal Museum Utrecht, FNAC Paris, Neue Sammlung Munich.
- **Table-upon-table barstool** Centraal Museum Utrecht
- **No sign of design table** Stedelijk Museum Amsterdam,
- **No sign of design chair** Stedelijk Museum Amsterdam
- **Bronto children chair** Stedelijk Museum Amsterdam, Centraal Museum Utrecht, Design Museum London, Museum für Gestaltung, Frankfurt, Design Center Seoul.
- **Bronto constructionless chair** Stedelijk Museum Amsterdam, Centraal Museum Utrecht
- **Fruittray** Medusa Museum of design, Reykjavik, MoMA San Francisco
- **=dish** MoMA San Francisco
- **Moneyplate** Centraal Museum Utrecht, MomA San Francisco, Kunst Industri Museet Copenhagen
- **One of a kind chair** Design Museum Lisbon
- **One of a kind table** Design Museum Lisbon
- **The Cross** Stedelijk Museum Amsterdam, Centraal Museum Utrecht, Parco Museum, Tokyo.
- **Sexy Relaxy** Stedelijk Museum Amsterdam, Design Museum London. FNAC Paris, Design Center Seoul.
- **Split-level** Centraal Museum Utrecht
- **S(h)it on it** Centraal Museum Utrecht
- **Zzzidt-object** Museum für Konkrete Kunst, Iglostadt, Centraal Museum Utrecht, FNAC Paris
- **Drawing box** Centraal Museum Utrecht
- **KZN birthday calendar** Design Museum London
- **Thing 6** FNAC Paris
- **Berlage Chair** Gemeente Museum The Hague, Design Center Seoul.
- **Domoor** (Dombo) mug Stedelijk Museum Amsterdam, Design Center Seoul.
- **Cup plate** Stedelijk Museum 's Hertogenbosch
- **Rhino chair** Design Center Seoul.
- **2loveseat** Stedelijk Museum Gouda
- **Centraal Museum chair** Medusa Museum of design Reykjavik, Centraal Museum Utrecht

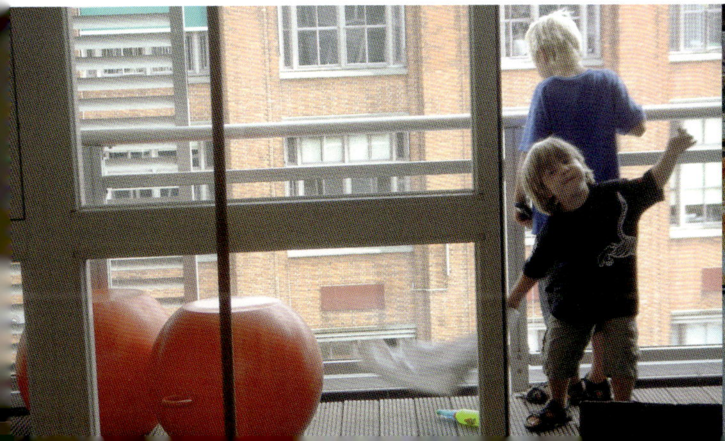

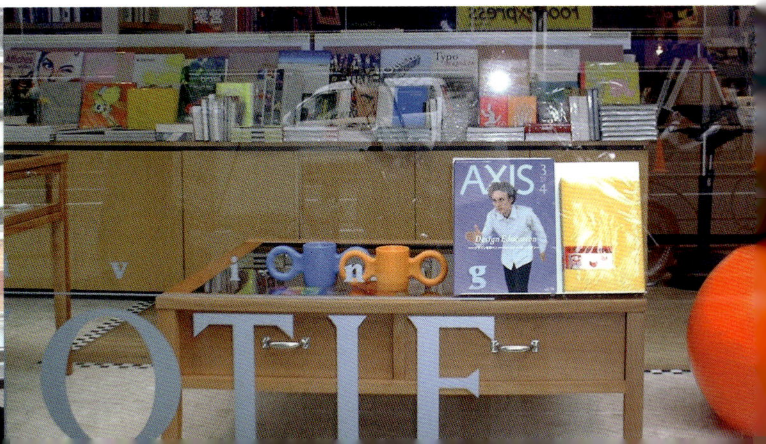

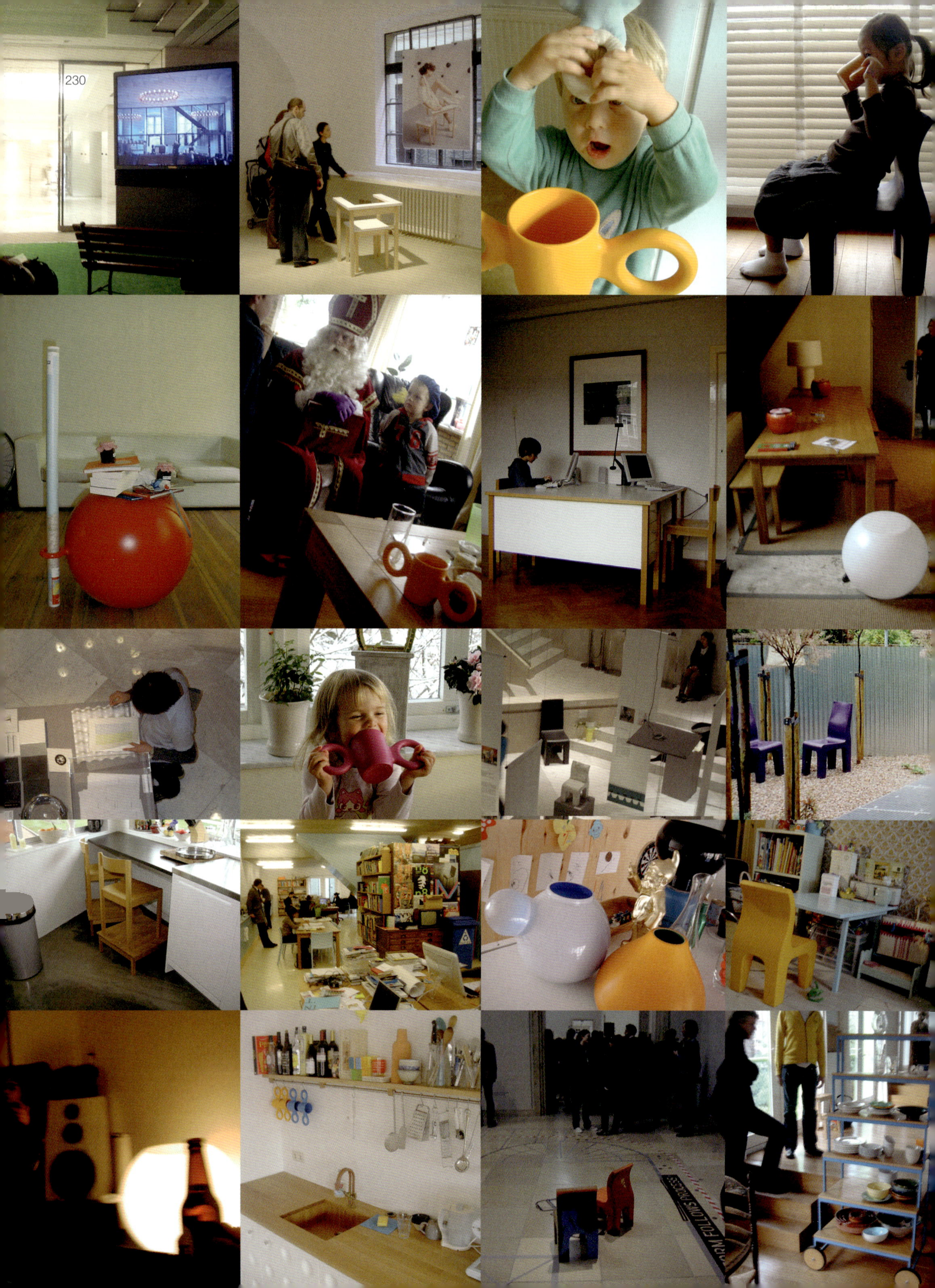

A vrai dire, je connaissais par bribes le travail de ce designer, par le biais d'expositions collectives, vues lors du salon du meuble de Milan, comme celles de Dutch Design, mais surtout à travers Droog design avec lequel il est impliqué, dès sa création, en 1993. Et dont il reste, aujourd'hui encore, l'une des figures emblématiques. Le célèbre collectif néerlandais lui a en effet permis à ses débuts – une fois diplômé de la Design Academy d'Eindhoven où il a poursuivi ses études de design industriel de 1986 à 1991 – de se faire connaître à l'échelle internationale. Mais, par ailleurs, il en a aussi assis la réputation comme quelques autres designers de sa génération (des amis aujourd'hui) tels Marcel Wanders, Jurgen Bey ou Hella Jongerius.
Pour être tout à fait honnête, cette rencontre à Tokyo avec Richard Hutten a été un peu plus déterminante pour moi. Elle m'a permis d'apprendre à mieux connaître l'esprit créatif des Hollandais, moi qui jusque-là avait exclusivement les yeux rivés sur le design italien. Il ne s'agit pas bien sûr de mettre ici en avant un style spécifiquement hollandais, dans la mesure où le design, devenu langage international, ne peut se résumer à une identité territoriale. Mais, il est indéniable que la culture contemporaine locale, dans sa façon d'intégrer à la fois la modernité et le vernaculaire, imprègne fortement les nouvelles générations de designers, conscients plus que jamais des effets pervers de la globalisation. Elle renforce une identité, la singularise. Richard Hutten ne fait pas exception à la règle.
L'invention hollandaise, dont il est un nouveau héraut, n'est pas forcément dans la révolution, la «tabula rasa» moderniste, mais dans les détournements, les récupérations, la citation d'objets, de matières ou pratiques déjà existantes, dans cette envie de proposer de nouvelles poétiques pour le quotidien. L'innovation résulte d'une réinterprétation personnelle, dans un détournement singulier, des points de vue préexistants. Culturellement, on dénote ainsi une attitude commune chez beaucoup de créateurs néerlandais actuels qui s'opère par affinités électives et qui, à l'instar de Richard Hutten, consiste à poser un autre regard sur les choses ordinaires. A en révéler une autre forme de beauté.

Le regard d'Hutten questionne avec intelligence l'usage qui devient le fil conducteur de cet ouvrage. «In use» marque un point de vue. Contextuel, il cherche à désacraliser l'objet du design, à le sortir des pages lisses et glacées des magazines à la mode et des espaces confinés du musée, pour le replacer dans un contexte réel, plus intime, plus divers, plus cru… en somme dans la «vraie vie». Volontairement «indiscret», le designer en quête de réponses aux desseins qu'il a nourris pour ses projets, cherche ici à tisser des liens, à provoquer un dialogue entre l'utilisateur et l'objet acquis, à mettre en exergue la circulation des objets qu'il a créés, soulignant, par là même, au-delà d'une expression artistique ou industrielle (la sienne), son envie d'inscrire le design dans une réalité quotidienne. Une forme d'humanisme.

Brigitte Fitoussi – Mars 2006

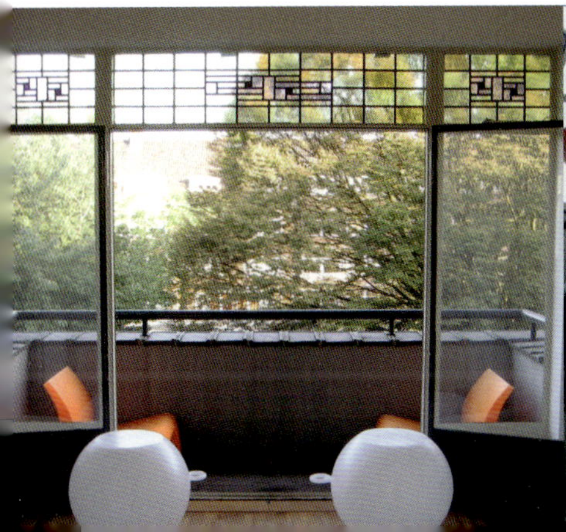

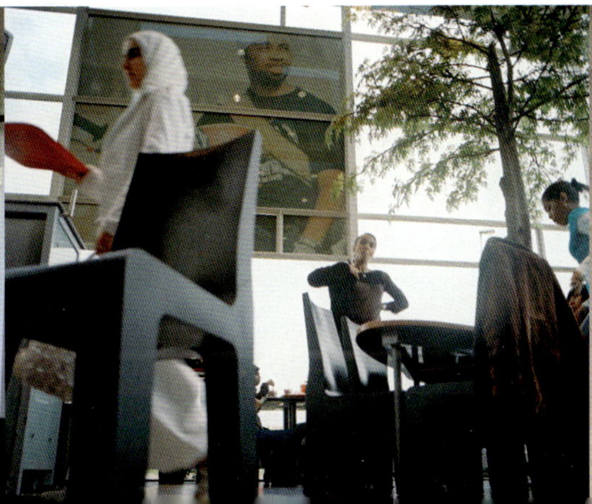

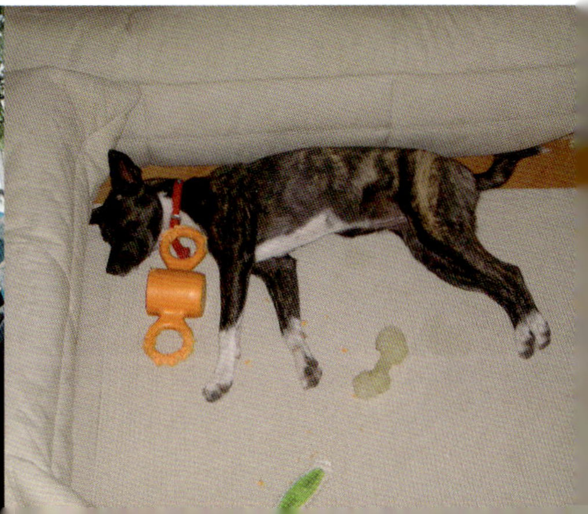

ENTRETIEN avec Richard Hutten

D'où vient cette idée de faire un livre dont le fil conducteur traite de la circulation de vos objets, et plus généralement de l'usage?

Les objets que je crée sont faits pour être utilisés. L'essence même du designer est de faire des produits qui puissent jouer un rôle dans la vie de tous les jours. Depuis que je travaille, certaines personnes m'envoient de temps en temps des photos d'objets en situation que j'ai conçus et qu'ils ont achetés. C'est toujours très agréable de voir que mes objets interfèrent dans leur quotidien, qu'ils en sont même si contents qu'ils prennent le temps de les photographier et de m'en envoyer spontanément des clichés. C'est une jolie façon de me faire partager un peu de leur intimité. C'est aussi le plus beau compliment qu'on puisse me faire. Mon travail prend en effet tout son sens à travers l'usager. Je veux créer des objets qui donnent la possibilité de tisser des liens affectifs, qui deviennent porteurs de souvenirs... La plupart des photos qui m'ont été envoyées (pour ce livre) sont du mug Dombo. Ce qui semble logique, dans la mesure où, parmi tous mes produits, c'est le seul qui depuis sa création a été vendu à plus de 100.000 exemplaires. En extrapolant, ne serait-il pas amusant, que les enfants qui grandissent actuellement avec ce mug, se disent dans 20 ans: «Utilisiez-vous un mug Dombo quand vous étiez petit?» Ce serait pour eux une façon immédiate de se remémorer cette époque. Le sens du mot «usage» est très vaste. Mes projets de design ne sont pas exclusivement fonctionnels ou seulement pensés pour le quotidien, leur utilisation est beaucoup plus étendue. L'enjeu est autant culturel que commercial. Le versant économique de l'usage prend en compte le processus complet du design: le développement, la production et la distribution. Mon but est aussi de faire attention à chaque individu impliqué dans ce processus, pour être sûr qu'il ne se contente pas seulement de faire son travail mais qu'il en tire aussi satisfaction. L'aspect culturel est également très important pour moi. Mon travail fonctionne, comme les objets, dans des livres et des expositions. Il peut ainsi influencer l'opinion des gens sur le design, leur en donner des directions ou des clés. Pour résumer, un design capable de remplir ces trois conditions est un super design. Malheureusement, c'est rare, la plupart du temps, l'accent est seulement mis sur seulement l'un d'entre eux.

Pourquoi faire un livre si rapproché du précédent?

Quatre ans les séparent, c'est somme toute déjà un certain temps. Et depuis, j'ai réalisé de nouveaux projets. Mon premier ouvrage, de nature biographique, a été conçu comme un catalogue. Il s'est fait, au fond, malgré moi. Quand l'éditeur de ce livre m'a sollicité, j'ai d'abord refusé. J'avais 33 ans et le sentiment que ma carrière ne faisait que commencer. Je pensais que c'était prématuré. Finalement, il s'est fait, mais j'ai impression d'en avoir été seulement le sujet. Bien sûr, une fois que le livre était publié, j'étais vraiment heureux: il était beau. Mais le nouvel ouvrage me permet de tenir un rôle plus prépondérant, moins passif. J'y montre volontairement mon travail dans un contexte réel, avec de vraies gens. Pas juste de belles et lisses images prêtes à être publiées dans Wallpaper. Je trouve en effet très intéressant de démontrer qu'un produit peut fonctionner dans différentes situations. Le contexte lui donne un sens. Bien sûr, ça n'empêche pas que ce contexte soit aussi un magazine ou une exposition dans un musée. Ce n'est pas négligeable, mais je préfère me confronter à la vie réelle. Prenons l'exemple de la chaise que j'ai dessinée pour le Centraal Museum à Utrecht (Pays-Bas). Elle a d'abord été créée pour le restaurant de ce musée. Puis, quelque temps après, elle a été utilisée pour la cérémonie de mariage de nos futurs roi et reine, le prince et la princesse Willem-Alexander et Maxima. Des millions de personnes dans le monde ont ainsi pu la voir à la télévision. Actuellement, elle est même utilisée dans un fast-food (Kentucky Fried Chicken restaurant), près de mon studio. C'est tout ce que j'aime, voir un de mes projets vivre aussi bien dans un lieu dit de «haute culture» comme le musée, que dans un restaurant de masse, dit de «basse culture». Je n'ai d'ailleurs aucune préférence pour l'un ou pour l'autre de ces lieux, je

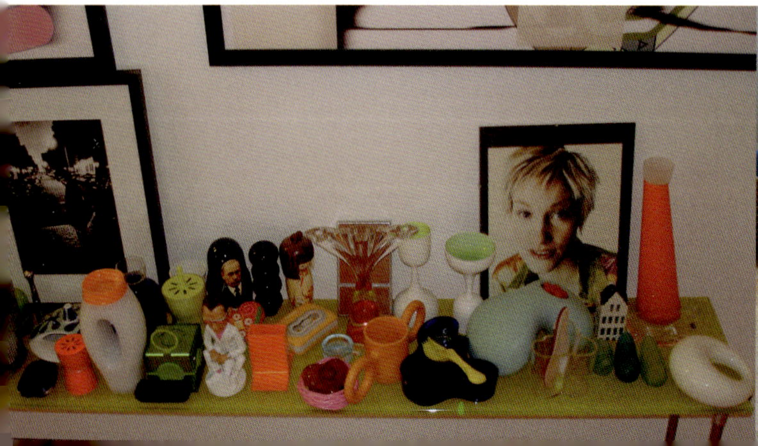

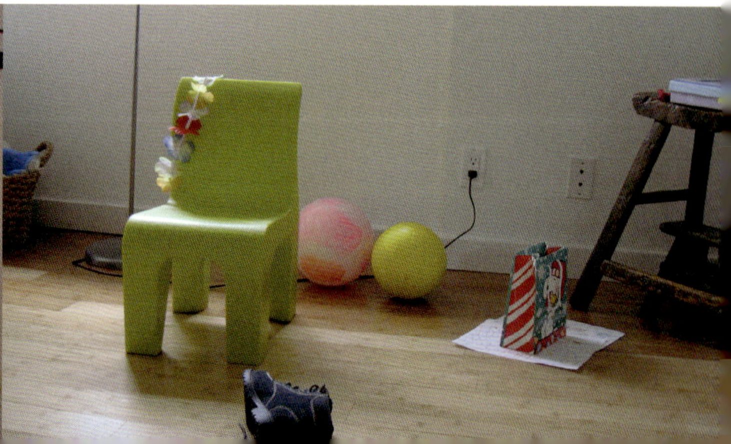

chance d'avoir une bonne équipe qui comprend ma façon de penser, qui souvent pense dans le même sens. (Quand je suis à l'autre bout du monde et que j'explique mes idées seulement par téléphone ou par mail, ils sont souvent capables de faire les premiers dessins!) Le processus dans son ensemble est presque rationnel. Chaque aspect du design a sa raison, peut être expliqué (même dans le cas où nous prenons une décision à partir de raisons émotionnelles). C'est le type de rationalisme qu'on peut voir dans la nature. Chaque aspect, par exemple, d'un papillon, a une fonction, une raison. Le résultat est toujours extraordinairement beau. Et, il existe une immense variété de papillons. C'est l'approche rationnelle qui conduit à différents types de papillons! Pourrait-on imaginer, avoir uniquement besoin d'un seul type de papillon? Ce qui est magnifique dans la nature, et là s'établit le parallèle avec le design, c'est que toutes ces variations fonctionnent très bien les unes à côté des autres. Manifestement, il existe dans la nature ce besoin de diversité, que l'on retrouve aussi chez l'Homme.

Revenons à ce dialogue avec le client…

Après avoir terminé la première phase du design, nous la présentons au client. C'est à nouveau un moment très important. Le dialogue avec le client continue. Il fait appel à ma compétence de designer, mais j'ai autant besoin de son expertise et de son savoir-faire. Nous développons donc en équipe le design, de l'esquisse jusqu'au produit fini. (Ce dialogue est ce qui, à mon avis, oppose le design à l'art. L'artiste crée une œuvre dans son atelier et, une fois finie, il la montre au monde. L'art est comme un monologue. Le designer a une palette d'investigations beaucoup plus vaste, qui l'implique davantage dans des questions contemporaines telles que la communication de masse, la production de masse ou la mondialisation. Contrairement à l'artiste, il ne réagit pas seulement en fonction de cette réalité quotidienne, il a aussi la capacité de la transformer, donnant ainsi au design un impact culturel beaucoup plus fort que l'art.) Le style n'est jamais un enjeu. Je n'ai d'ailleurs pas de goût. Le goût ne va pas dans le sens de créer quelque chose de nouveau ou de jamais vu, il en bloque au contraire toute possibilité parce qu'il se fonde sur des paramètres établis. L'un de mes objectifs est de surprendre, pas de me complaire dans un certain goût ou style qui pourrait m'en empêcher. Généralement, le design est dans la séduction, or je ne cherche pas à séduire les gens à travers mes objets, je préfère les amuser. «L'aspect sexuel» du design, la séduction, n'est pas mon but principal. Je pourrais comparer cela à l'attraction entre un homme et une femme. Ce genre d'amour – parce que le design parle d'amour – joue surtout pour moi un rôle dans le background de chacun. C'est, à mon avis, plus proche du type d'amour qu'un parent peut porter à son enfant. C'est davantage de l'ordre de la joie, de la surprise, de la composition, des rêves… Pour beaucoup de designers, la beauté, dans le sens de l'esthétique, est le principal but. Ce n'est pas le genre de beauté que je recherche. J'essaye dans mes projets d'émouvoir et de toucher les gens de plusieurs façons, par exemple avec humour et surprise. Cette approche n'est bien sûr pas unique, si on pense, par exemple, au travail de Gaetano Pesce, Ingo Maurer ou d'Achille Castiglioni, mais c'est la façon de travailler qui me convient le mieux. Elle me ressemble.

Si l'on revient à la case départ, quel rôle a joué (joue?) Droog Design dans votre parcours?

Droog a très vite repéré et apprécié mon approche du design, mon esprit aussi, me donnant d'emblée une plateforme internationale. C'était pratiquement au début de mon parcours. Je donnais alors des réponses générales à des questions générales (le plus souvent mes propres questions). Comment les gens réagissent, quel rôle jouent les objets dans leur vie…? De ces réponses sont nés des projets inédits. Aujourd'hui, je donne des réponses spécifiques à des questions spécifiques, presque toujours posées par quelqu'un d'autre. Je cherche beaucoup plus à collaborer avec des compagnies qui produisent industriellement ou qui possèdent une véritable expertise. Une grande partie de mon job consiste toujours à observer. Je ressens toujours ce besoin de me raccorder à l'existant, avec le même état d'esprit qu'à mes débuts avec Droog, mais dans une approche différente. Droog continue à se focaliser sur l'aspect culturel du design. J'éprouve, au contraire, le besoin d'étendre plus largement le champ du design, notamment à l'usage qui est devenu plus important pour moi.

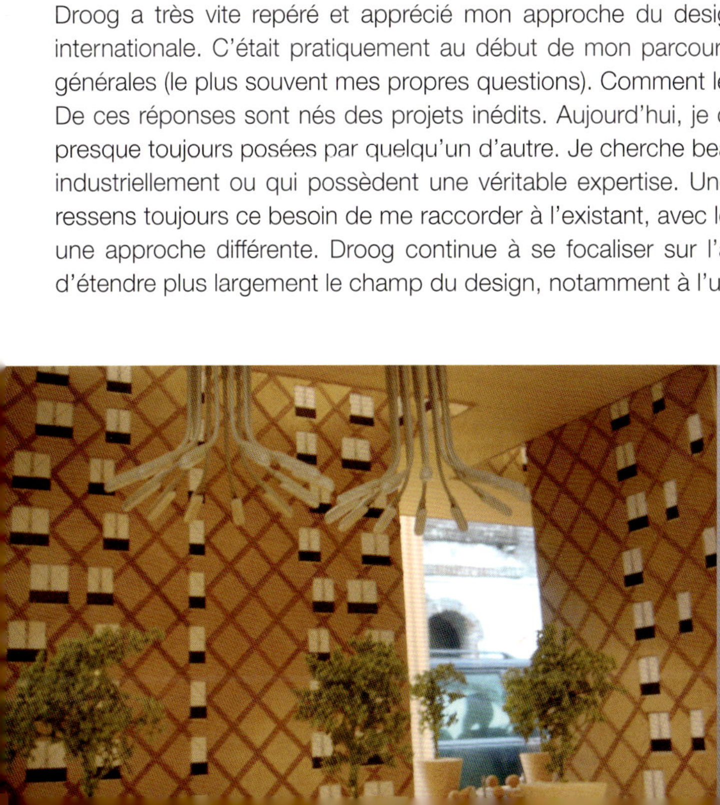

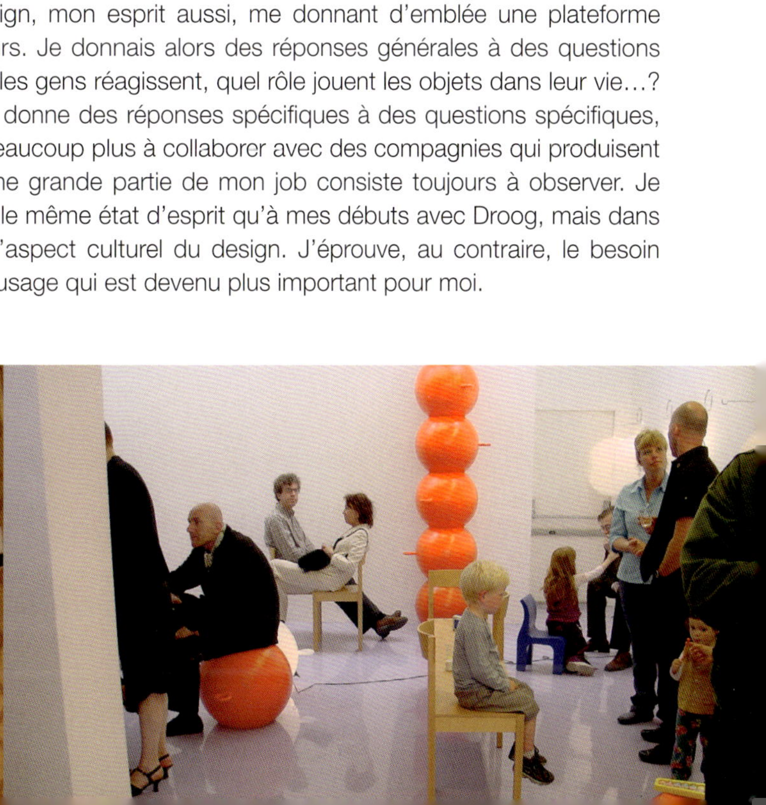

Que signifie pour vous la modernité?

La Modernité fait écho pour moi à quelque chose du passé. Beaucoup (trop) d'architectes de cette période ont été faussement interprétés. Les magnifiques bâtiments de Le Corbusier, Rietveld ou Mies van der Rohe ont donné une excuse à de nombreux architectes pour faire des bâtiments ennuyeux et insipides, des lieux sans âme. Si un projet n'a pas d'âme, quelle qu'en soit la discipline, architecture ou produit, il n'a pas de raison d'être. La modernité a été comprise comme un style et non comme une philosophie. Il est très difficile de faire un projet simple ou épuré qui ait une âme…

A première vue, vos objets semblent très simples ou nés de l'archétype, mais on s'aperçoit vite qu'ils ont toujours quelque chose en plus (de l'ironie, de l'humour, de la narration). Le décalage est-il une des clés de votre créativité?

Si j'utilise des formes existantes et/ou des archétypes, il existe plusieurs raisons à cela. Avant tout, elles permettent à l'utilisateur d'en avoir une reconnaissance immédiate, de se les approprier plus facilement. Utiliser des formes existantes, issues de l'histoire, de la nature ou de l'artisanat, me semble toujours un bon point de départ. J'aime aussi surprendre, d'où ce décalage que l'on retrouve souvent dans la conception de mes objets. Faire sourire en créant la surprise ou même en choquant. L'humour est une bonne voie pour essayer de rendre les gens heureux. Pour moi, le bonheur n'est pas superficiel, mais l'un des buts les plus importants dans la vie. «Sortir un peu du cadre», ou utiliser le décalage, est aussi un moyen de faire réfléchir. Le travail est souvent extrême, radical, dans ses points de départ et dans les limites que je me fixe, mais le résultat laisse toujours une porte ouverte à l'interprétation. Traditionnellement le design cherche à résoudre des problèmes. Moi, je ne résous pas de problèmes, je crée des possibilités. Telle une histoire dont la fin serait libre ou que le lecteur aurait à inventer…

Trois mots pour être un bon designer selon vous?

Ecouter, voir, entendre ou… penser, essayer, faire.

Vos projets dans un futur proche?

Quelques-uns seront dans ce livre avec notamment des dessins pour Christofle, Muji, l'intérieur d'un bâtiment à Amsterdam signé par les architectes MVRDV, les bureaux à l'intérieur du musée Boijmans à Rotterdam, une maison pour Ytong, un sol en linoléum pour Forbo, et une douzaine d'autres…

Qu'est-ce qui vous motive le plus dans ce métier?

La chose la plus importante pour moi est d'essayer de contenter le plus de gens possible. Plus je leur fais plaisir, plus ça me rend heureux. Je voudrais apporter une petite contribution à leurs vies quotidiennes. Ce livre est donc un hommage aux gens qui utilisent mes objets. J'aimerais fêter chaque jour de ma vie, je fais donc des outils pour la célébrer. Et au-delà de tout cela, ce métier me permet aussi de voyager très souvent, de rencontrer une foule de gens magnifiques et de beaucoup m'amuser…

Propos recueillis par Brigitte Fitoussi – Janvier-Février 2006

Brigitte Fitoussi est directrice artistique chez Christofle depuis juin 2005, responsable de toute l'image du grand orfèvre parisien. Architecte de formation (DPLG), elle partageait auparavant l'activité de journaliste, commissaire d'expositions et consultant. Chef de rubrique (design) du magazine français NUMERO, depuis 1999 (pour lequel elle continue d'écrire), elle a collaboré jusqu'ici régulièrement à de nombreuses publications françaises et étrangères. Elle est l'auteur d'une dizaine d'ouvrages consacrés au design et à l'architecture.

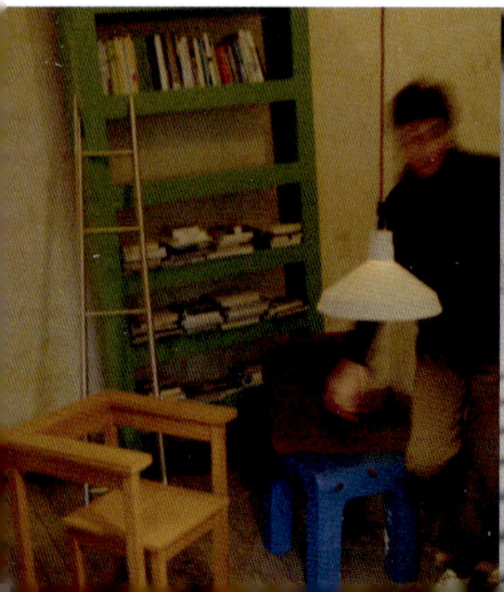

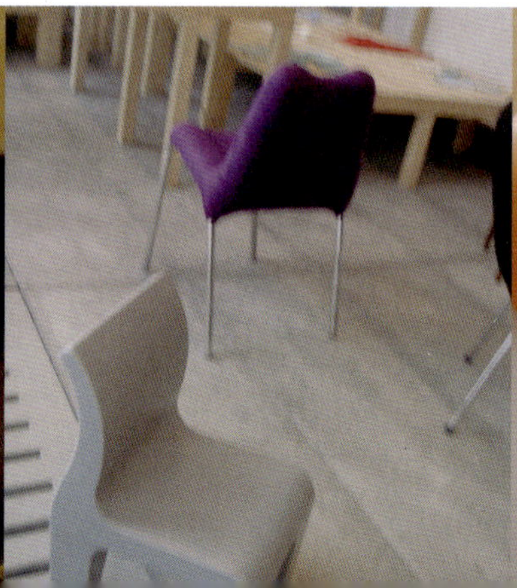

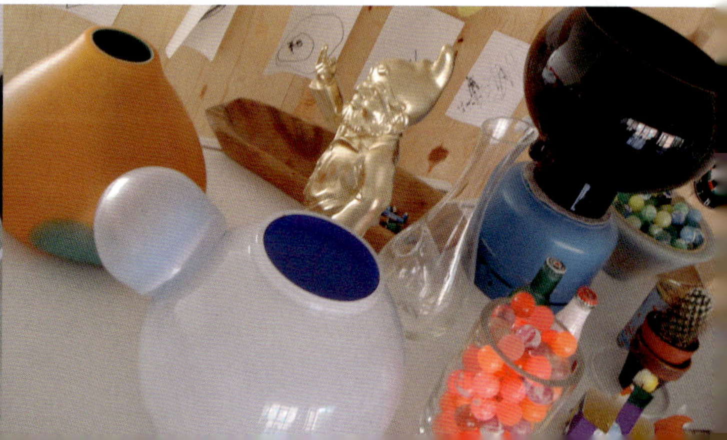

De Hutte(n) primitive à une romance de Rotterdam: pourquoi j'adore et vis avec (du mobilier) Richard Hutten

Je vis à Rotterdam, dans un environnement inspiré de Richard Hutten et avec un nombre considérable de meubles et d'objets créés par lui. Ça semble me convenir particulièrement. Rotterdam se représente elle-même comme 'Manhattan sur Meuse', mais à mon avis, le seul vrai gratte-ciel est le *Skyline Rotterdam*, l'armoire hybride de Richard Hutten créée en 1994. Grand, moderne, compartimenté et déjà en marge du firmament complexe des espaces fonctionnels du quotidien, il incarne l'engouement pour l'irrationnel, dont New York est devenu le symbole. Toutefois, on y retrouve cette urbanité typique de Rotterdam, dans laquelle la simplicité réduite et le caractère anonyme donnent à l'objet destiné au rangement (après tout, la ville s'est articulée autour du transbordement) une notion d'élégance dénuée d'absurdité.

C'est également typique du travail de Richard Hutten. Ces quinze dernières années, il a ré-assemblé les blocs de construction de base pour leur donner de nouvelles formes qui créent un nouveau type d'espace. Parfois, ces espaces peuvent être un léger enfoncement sur la surface d'une table pour faire office de plat à fruits (*Table-bowl*, 1995), ou à d'autres moments, ils donnent un tout nouveau sens, même parfois religieux, sur la place publique, comme le *Cross Bench*, en 1994. Ils ont toujours la pureté de la géométrie orthogonale et une complexité née du désir de l'artiste de remanier les rituels sociaux simples par la réorganisation des espaces où ils se déroulent.

Cela me semble être le noyau du travail de Richard Hutten. La série de chaises hautes qu'il a créées en 1996 pour ses jeunes enfants en est l'aboutissement parfait: la recombinaison de formes familières pour créer une place digne au petit enfant qui peut ainsi participer au repas de famille prend des airs d'innocence quasi lyrique. Ici, le mobilier retourne à ses racines, tout comme il sert de racine à au moins une forme de relation sociale au centre de notre société.

Bien entendu, il serait erroné de réduire Richard Hutten à ses travaux les plus anciens et plus connus. Un autre aspect de son travail, au moins, concerne l'exploration des possibilités de nouveaux matériaux, et en particulier le plastique. Quand bien même il en résulte les formes ondoyantes de sa chaise *Bronto* en 1997, et les suivantes, ou le *Fruit Tray* (1995), on peut encore déceler le mouvement vers une simplicité et une qualité génériques bien qu'elles soient à présent transposées dans des formes moins proches du romantisme inspiré par des éléments anciens de l'environnement domestique qui nourrissent le lyrisme des meubles en bois. Les deux se conjuguent dans le «Skippy Ball» réinventé par Richard Hutten en un système de siège extérieur (*Zzidt Ball*, 2000), du moins pour ceux d'entre nous qui ont grandi avec ce jouet.

Dans ses œuvres plus récentes, Richard Hutten semble se détacher du rapport aux formes familières, et donc aussi de leurs mémoires sociale (historique) ou personnelle. Et encore, il y a toujours des échos à quelque chose, même dans le chandelier abstrait fait de découpes tournant autour d'une lumière centrale qui rappellent leurs cousins en cristal, ou dans le siège qu'il vient de produire pour Lensvelt et qui réduit consciencieusement les expériences de Charles et Ray Eames et Arne Jacobsen à une base de travail sur un élément de siège élémentaire, sinon générique. Richard Hutten ne peut échapper au passé, ni au sien, ni à celui de la société, et ne le souhaite pas. Au contraire, il ne cesse d'interroger ces mémoires pour atteindre leur essence utile, ludique, charmante et par-dessus tout libre, de sorte qu'il peut ancrer profondément les choses dans notre conscient tout en laissant la porte ouverte à de nouvelles interprétations.

Aaron Betsky

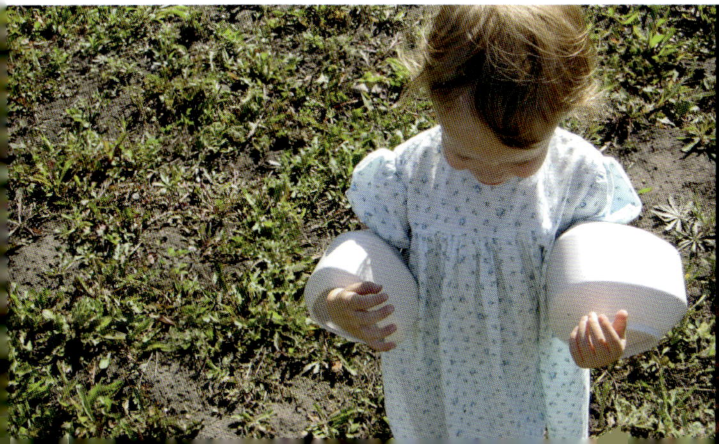

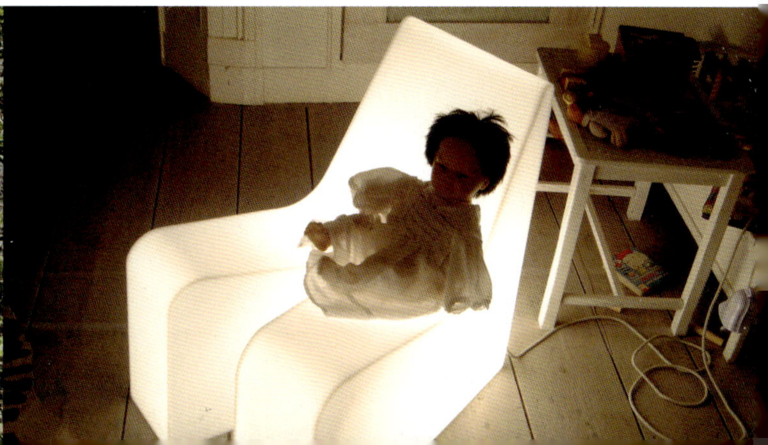

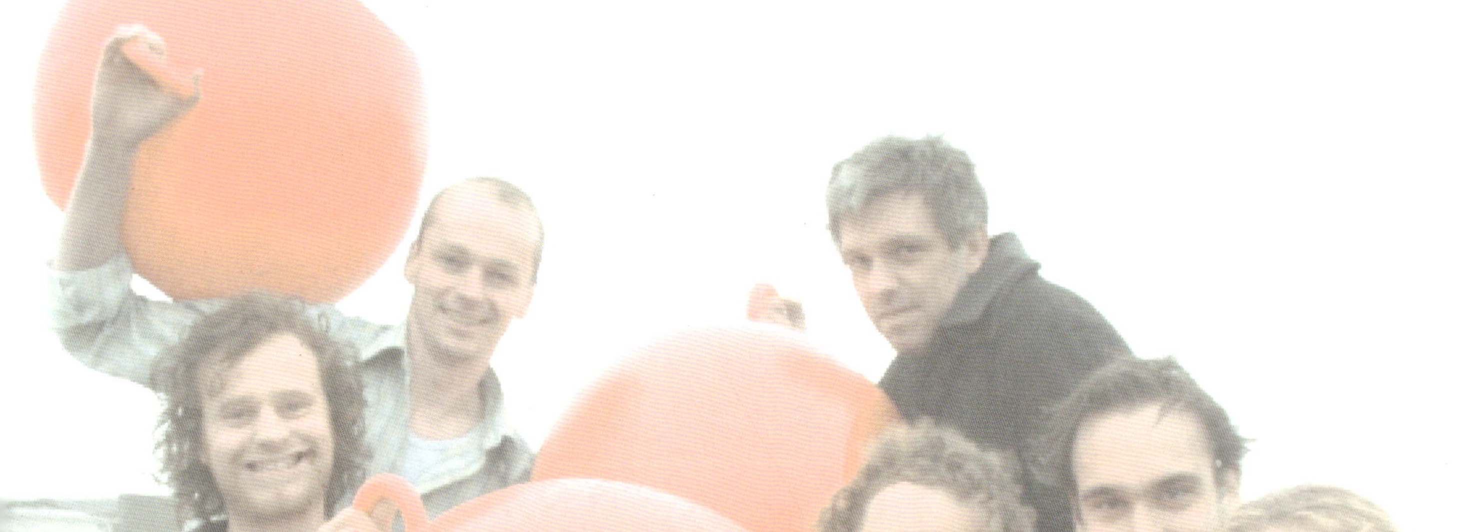

For various reasons special thanks go to:

Aad Krol, Aaron Betsky, Abel Hutten, Adalsteinn Ingólfson, Adam Ewens, Adrian Rovero, Albert van de Ark, Albert Ravestein, Alberto Perazza, Alex Wiltshire, Alexis Georgacopoulos, Alexis Tricoire, Alice Rawsthorn, Allessandro Mendini, Ann Maes, Antoine Achten, Ans Hutten, Adrie Vlasblom, Arnout Visser, Barbara Mackert, Barry Nicholson, Bart Vos, Bas Hesselink, Benjamin Budde, Bettina Fahnenbruck, Boris Braakhuis, Boris Hutten, Brenno Visser, Brigitte Fitoussi, Brooke Stoddard, Casper Vissers, Christophe Marchant, Christophe Pillet, Claudio Bartolozzi, Cok de Rooy, Daan van Trigt, Daniel Englender, Danielle van Steenbergen, David Stokes, Dick Dankers, Dick Heydra, Dick van Hoff, Dirk van Berkel, Dirk Limburg, Don Kouwenhoven, Ed Annink, Ed van Hinte, Edsko Schuitema, Elisa Astori, Emmanuel Babled, Enrico Astori, Ettore Sottsass, Felix Janssens, Francisco Polo, Gareth Williams, Geoffrey Lilge, Gerard Hutten, Gijs Bakker, Gilda Bojardi, Gisel Banel, Henrik Most, Helen van Ruiten, Herman de Jongh, Hilde Bouchez, Hitoshi Saeki, Humberto Campana, Huib Muilwijk, Ida van Zijl, Ingeborg de Roode, Ingrid de Koster, Itsuro Shibata, Jaak Van Damme, James Irvine, Jacob van Rijs, Jan Krapels, Jan van der Lande, Jeffrey Bernett, Jennifer Hudson, Jenny Shears, Jerszy Semour, Jeroen van Beuzekom, Jeroen Visser, Jeroen Wesselink, John Geresi, Joke van de Heide, Jón Snorri Sigurdsson, Josée de Pauw, Judith Seng, Judith Zoetelief, Jurgen Bey, Hannah Macmurray, Hans Lensvelt, Harm Vink, Harry Vink, Haruko Smith, Ingeborg de Roode, Irma Goedemondt, Karel Puype, Karim Rashid, Karl Lagerfeld, Karlijne Brand, Kazuaki Komura, Kazuhiko Yazaki, Kees van Graven, Keiko Sano, Koos Flinterman, Laura Housely, Leo Krol, Leon van Gerwen, Leon Ramakers, Lidewij Edelkoort, Liesbeth in 't Hout, Lotta Jonson, Louise-Anne Comeau, Marc Craane, Marcel Wanders, Marcel van de Meys, Marco Romanelli, Margareta Persson, Margreeth Olsthoorn, Marc Dubois, Mark Dytham, Markus Herchet, Martin Schomisch, Masamichi Katayama, Masanobu Tsukada, Max Borka, Max Fraser, Mayumi Matsubara, Melissa Feldman, Michael Young, Miki Amai, Mischa van de Wekke, Moniek Bucquoye, Murray Moss, Nicole Hoekstra, Nikki Gonnissen, Niko Kneepkens, Paola Antonelli, Paolo Moroni, Paola Navone, Peer de Bruyn, Peggy Brandon, Pierre Keller, Piet Knook, Pieter Lomans, Pieter Molenaar, Ranti Tjan, Raymond Molendijk, Rene Koster, Renny Ramakers, Rens Groenendijk, Reyer Kras, Reyn van der Lught, Ria van Huffel, Rieme Gleijm, Rien Hagen, Roelof Mulder, Rob Gijsbers, Ron Arad, Roni Kabessa, Saskia Hovers, Shoji Ito, Sindy Miggiels, Sjarel Ex, Stefano Fragapane, Stefano Giovannoni, Stuart Hata, Stuart Oh, Suki Sakurai, Susanne Oxenaar, Sylvia Delissen, Taco Stoppels, Terance Conran, Teruo Kurasaki, Thea Sterken, Thijs Bakker, Timo Salli, Thomas van Widdershoven, Tobias Verhoeven, Tord boontjes, Toshimichi Fukusumi Takeichiro Morinaga, Tomoki Kusakabe, Vanni Pasca, Veerle Seys, Vincent de Rijk, Willem de Roon, William Sawaya, Wilma Wisse, Wim van Krimpen, Winy Maas, Yanagase Kotoho, Yoichi Nakamuta and you.

Special thanks go to the users who submitted pictures for this book:

Tsuyoshi Matsuzawa, Yumiko Shimizu, Kosuke Okamoto, Aya Yamada, Wilfried Lentz, Vincent de Boer, Vera Bajema, Tjeert Kootstra, Tejo Remy, Erik Smidt, Roelof Mulder, Michael Young, Humberto Campana, Karim Rashid, Ed van Hinte, David Greer, Jeffrey Bernett, Renate Daniels, Ria van Huffel, Nicolas Bovesse, Marieke te Wierik, Tien Tran, Maya Epping, Martijn van Marion, Mark Moskovitz, Hans de Gooijer, Brigit Borho, Brigitte Fitoussi, Jacob van Rijs, Aad Krol, Maartje Steenkamp, Leonie Bos, Leon van Tilburg, Saskia van de Wiel, Katrien Claes, Karlijne Brand, Margreeth Olsthoorn, Juliette Ruggieri, John Geresi, Jerszy Seymour, Jason Vanderhill, Jan David Hanrath, Huib Muilwijk, Helen van Ruiten, Heike Scheller, Judith Seng, Hans Lemmerman, Gijs Haak, Geoffrey Lilge, Isabel Aranguren, Carl Lapeirre, Felix Janssens, Falco Webbink, Ellen Hoogeboom, Lowieke van Daalen, Dorien Bruijns, Martin Schömisch, Daan Braspenning, Christophe Marchant, Colin Huizing, Cees van Emst, Arnout Visser, Bas Hesselink, Boudewijn Binkhorst, Antoine Achten, Annuska Visscher, Mischa van der Wekke, Raymond Molendijk, Ann Maes, Alfred van Elk, Alexis Tricoire, Adrian Rovero, Albrecht Ruiter, Hitoshi Saeki and others.